Continuum Mechanics

Springer
*Berlin*
*Heidelberg*
*New York*
*Barcelona*
*Hong Kong*
*London*
*Milan*
*Paris*
*Tokyo*

## Advanced Texts in Physics

This program of advanced texts covers a broad spectrum of topics which are of current and emerging interest in physics. Each book provides a comprehensive and yet accessible introduction to a field at the forefront of modern research. As such, these texts are intended for senior undergraduate and graduate students at the MS and PhD level; however, research scientists seeking an introduction to particular areas of physics will also benefit from the titles in this collection.

I-Shih Liu

# Continuum Mechanics

With 28 Figures and Numerous Exercises

Springer

Professor I-Shih Liu
Universidade Federal do Rio de Janeiro
Instituto de Matematica
C.P. 68530
21945-970 Rio de Janeiro, Brasil

Liu, I-Shih, 1943-
Continuum mechanics / I-Shih Liu.
p. cm. -- (Advanced texts in physics, ISSN 1439-2674)
Includes bibliographical references.

1. Continuum mechanics. I. Title. II. Series.

QA808.2 .L58 2002
531--dc21

2002017040

ISBN 978-3-642-07702-9



Springer-Verlag Berlin Heidelberg New York
a member of BertelsmannSpringer Science+Business Media GmbH

http://www.springer.de


Printed in Germany



Cover design: *design & production* GmbH, Heidelberg

Printed on acid-free paper

To my teacher
Ingo Müller

# Preface

In this book the basic principles of continuum mechanics and thermodynamics are treated in the tradition of the rational framework established in the 1960s, typically in the fundamental memoir "The Non-Linear Field Theories of Mechanics" by Truesdell and Noll. The theoretical aspect of constitutive theories for materials in general has been carefully developed in mathematical clarity – from general kinematics, balance equations, material objectivity, and isotropic representations to the framework of rational thermodynamics based on the entropy principle. However, I make no claim that the subjects are covered completely, nor does this book cover solutions and examples that can usually be found in textbooks of fluid mechanics and linear elasticity. However, some of the interesting examples of finite deformations in elastic materials, such as biaxial stretching of an elastic membrane and inflation of a rubber balloon, are discussed.

In the last two chapters of the book, some recent developments in thermodynamic theories are considered. Specifically, they emphasize the use of Lagrange multipliers, which enables the exploitation of the entropy principle in a systematic manner for constitutive equations, and introduce some basic notions of extended thermodynamics. Although extended thermodynamics is closely related to the kinetic theory of ideal gases, very limited knowledge of kinetic theory is needed.

Earlier versions of this book have been used over the years, in the Institute of Mathematics at the Federal University of Rio de Janeiro as well as in the Institute of Applied Mechanics at the National Taiwan University, in an introductory course on continuum mechanics at the graduate level, and at the advanced undergraduate level with a simplified version. The readers are not required to have a good knowledge of either solid mechanics or fluid mechanics, but, of course, some prior acquaintance with them would be helpful.

An appendix is written at the end to provide a review of basic notions in linear algebra and tensor analysis as mathematical preliminaries for the subjects, and occasionally cross-references to it (e.g. (A.32)) are used in the text. The reader who already has a reasonable mathematical knowledge may refer to it for reference and notations. However, in introductory courses I have often put the appendix before the first chapter because most of the students may not be familiar with the notations and some basic notions. No effort has been made to compile an extensive bibliography on related works in continuum mechanics. Only those cited in the book are listed.

Examples and exercises are given to supplement the understanding of the material and sometimes to provide further insights into the subjects. Usually my students are asked to do most of the exercises to accompany the progress of learning. Their feedback on the errors and the difficulties has resulted in considerable improvement of the manuscript. Their participation is greatly appreciated.

The endeavor of writing this book depended on many ideas and work in the scientific literature. To many of the relevant researchers, acquaintances or not, are due my grateful acknowledgements for their contributions. My special acknowledgement is due to Prof. Müller for his friendship and inspirations on many of my scientific trajectories. Finally, I would like to thank my family, especially my wife Lu Ping, for their understanding and patience during many long hours of preparing the manuscript over the years.

Rio de Janeiro,
March 2002

I-Shih Liu

# Contents

# 1. Kinematics

## 1.1 Configuration and Deformation

In continuum mechanics, we are interested in material bodies that can undergo motions and deformations. The stage for such phenomena to occur is, of course, the four-dimensional space–time. In this book, we shall restrict our attention to the *Newtonian space–time* of classical mechanics. The Newtonian space–time $\mathcal{W}$ can be regarded as a product space of a three-dimensional Euclidean space $\mathcal{E}$ and the one-dimensional space of real numbers $\mathbb{R}$ through a one-to-one mapping

$$\phi : \mathcal{W} \to \mathcal{E} \times \mathbb{R}.$$

Such a mapping is called a *frame of reference.* To set the stage for any discussion hereafter a frame of reference is always chosen, either explicitly or implicitly. We will usually think of a frame of reference as an "observer". Different observers will measure space–time events in different ways. The relation between two different frames of reference will be discussed later.

In order to describe the presence of a body $\mathcal{B}$ in space, mathematically we shall identify it with a region in a three-dimensional Euclidean space $\mathcal{E}$ relative to a frame of reference. We call a one-to-one mapping from $\mathcal{B}$ into $\mathcal{E}$ a *configuration* of $\mathcal{B}$. Such an identification endows the (physical) body with the mathematical structure of a Euclidean space.

It is more convenient to choose a particular configuration of $\mathcal{B}$, say $\kappa$, as a reference,

$$\kappa : \mathcal{B} \to \mathcal{E}, \qquad \kappa(X) = \boldsymbol{X}. \tag{1.1}$$

We call $\kappa$ a *reference configuration* of $\mathcal{B}$. The coordinates of $\boldsymbol{X}$, $(X^\alpha, \alpha = 1,2,3)$ are called the *referential coordinates*, or more commonly the *material coordinates*, since the point $\boldsymbol{X}$ in the reference configuration is often identified with the material point $X$ of the body. The body $\mathcal{B}$ in the configuration $\kappa$ will be denoted by $\mathcal{B}_\kappa$.

Let $\kappa$ be a reference configuration and $\chi$ be an arbitrary configuration of $\mathcal{B}$. Then the mapping

$$\chi_\kappa = \chi \circ \kappa^{-1} : \mathcal{B}_\kappa \to \mathcal{B}_\chi, \qquad \boldsymbol{x} = \chi_\kappa(\boldsymbol{X}) = \chi(\kappa^{-1}(\boldsymbol{X})), \tag{1.2}$$

is called the *deformation* of $\mathcal{B}$ from $\kappa$ to $\chi$ (Fig. 1.1). In terms of coordinate systems $(x^i, i = 1,2,3)$ and $(X^\alpha, \alpha = 1,2,3)$ in the deformed and the reference configurations, respectively, the deformation $\chi_\kappa$ can be expressed as

$$x^i = \chi^i(X^\alpha), \tag{1.3}$$

where $\chi^i$ are called the *deformation functions.*

The *deformation gradient* of $\chi$ relative to $\kappa$, denoted by $F_\kappa$ is defined by

$$F_\kappa = \nabla_{\boldsymbol{X}} \chi_\kappa. \tag{1.4}$$

When the reference configuration $\kappa$ is chosen and understood in the context, $F_\kappa$ will be denoted simply by $F$. Since the mapping $\chi_\kappa$ in (1.2) is one-to-one and onto, $F$ is non-singular. Therefore, the determinant of $F$ must be different from zero,

$$J = \det F \neq 0. \tag{1.5}$$

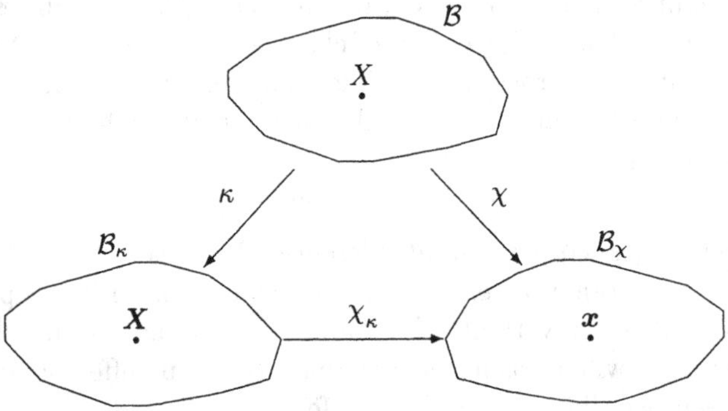


**Fig. 1.1.** Reference configuration

Relative to the coordinate systems $(X^\alpha)$ and $(x^i)$ in the reference and the deformed configurations, respectively, $F$ can be expressed in the following component form,

$$F = F^i{}_\alpha \boldsymbol{e}_i(\boldsymbol{x}) \otimes \boldsymbol{e}^\alpha(\boldsymbol{X}), \qquad F^i{}_\alpha = \frac{\partial \chi^i}{\partial X^\alpha}. \tag{1.6}$$

The deformation gradient $F$ in this expression is a two-point tensor, because it is expressed in terms of two different bases for a contravariant component at $\boldsymbol{x}$ in the deformed configuration and a covariant component at $\boldsymbol{X}$ in the reference configuration (see Example A.2.7 on p. 274). This component form is particularly simple, because it only involves partial derivatives, even if the deformation function (1.3) is not given in the Cartesian coordinate systems.

By definition, the deformation gradient $F$ is a linear transformation on the translation space $V$ of $\mathcal{E}$ (see Sect. A.2.1), $F : V \to V$, such that

$$\chi_\kappa(\boldsymbol{X}) - \chi_\kappa(\boldsymbol{X}_0) = F(\boldsymbol{X}_0)(\boldsymbol{X} - \boldsymbol{X}_0) + o(2), \tag{1.7}$$

where $o(2)$ represents the higher-order terms in $\epsilon = |\boldsymbol{X} - \boldsymbol{X}_0|$, i.e.,

$$\lim_{\epsilon \to 0} \frac{o(2)}{\epsilon} = 0.$$

In other words, the value of $F(\boldsymbol{X}_0)$ determines $\chi_\kappa(\boldsymbol{X})$ to within an error of the second-order near $\boldsymbol{X}_0$ relative to the value of $\chi_\kappa(\boldsymbol{X}_0)$.

Let $d\boldsymbol{X} = \boldsymbol{X} - \boldsymbol{X}_0$, a vector in $V$, be regarded as a small (infinitesimal) material line element in the reference configuration. Since $F$ is a linear transformation on $V$, the image of the line element in the deformed state $F(X_0)d\boldsymbol{X}$, denoted by the vector $d\boldsymbol{x} \in V$, is the first-order approximation of $\chi_\kappa(\boldsymbol{X}) - \chi_\kappa(\boldsymbol{X}_0)$ at $X_0$ from the relation (1.7). We write

$$d\boldsymbol{x} = F\, d\boldsymbol{X}. \tag{1.8}$$

Similar relations for material surface elements and volume elements in the deformed state are derived in the following example.

**Example 1.1.1** Let $da_\kappa$ and $\boldsymbol{n}_\kappa$ be a small material surface element and its unit normal in the reference configuration and $da$ and $\boldsymbol{n}$ be the corresponding ones in the deformed configuration. Also, let $dv_\kappa$ and $dv$ be small material volume elements in the reference and the deformed configurations, respectively. Then we have

$$\boldsymbol{n}\, da = JF^{-T}\boldsymbol{n}_\kappa da_\kappa, \qquad dv = |J|\, dv_\kappa. \tag{1.9}$$

*Proof.* Let the small rectangular surface element $da_\kappa$ be formed from the two line elements $d\boldsymbol{X}_1$ and $d\boldsymbol{X}_2$. Then we have from (1.8)

$$\begin{aligned} \boldsymbol{n}_\kappa &= \frac{d\boldsymbol{X}_1 \times d\boldsymbol{X}_2}{|d\boldsymbol{X}_1 \times d\boldsymbol{X}_2|}, & \boldsymbol{n} &= \frac{Fd\boldsymbol{X}_1 \times Fd\boldsymbol{X}_2}{|Fd\boldsymbol{X}_1 \times Fd\boldsymbol{X}_2|}, \\ da_\kappa &= |d\boldsymbol{X}_1 \times d\boldsymbol{X}_2|, & da &= |Fd\boldsymbol{X}_1 \times Fd\boldsymbol{X}_2|. \end{aligned}$$

Therefore, for any vector $\boldsymbol{v}$, we have from (A.32)

$$\begin{aligned} \boldsymbol{v} \cdot \boldsymbol{n}\, da &= \boldsymbol{v} \cdot Fd\boldsymbol{X}_1 \times Fd\boldsymbol{X}_2 = F(F^{-1}\boldsymbol{v}) \cdot Fd\boldsymbol{X}_1 \times Fd\boldsymbol{X}_2 \\ &= J\, F^{-1}\boldsymbol{v} \cdot d\boldsymbol{X}_1 \times d\boldsymbol{X}_2 \\ &= J\, F^{-1}\boldsymbol{v} \cdot \boldsymbol{n}_\kappa da_\kappa = \boldsymbol{v} \cdot JF^{-T}\boldsymbol{n}_\kappa da_\kappa. \end{aligned}$$

Similarly, let the small volume element $dv_\kappa$ be formed from three line elements $d\boldsymbol{X}_1$, $d\boldsymbol{X}_2$, and $d\boldsymbol{X}_3$. Then we have

$$\begin{aligned} dv &= |d\boldsymbol{x}_1 \cdot d\boldsymbol{x}_2 \times d\boldsymbol{x}_3| = |Fd\boldsymbol{X}_1 \cdot Fd\boldsymbol{X}_2 \times Fd\boldsymbol{X}_3| \\ &= |\det F|\, |d\boldsymbol{X}_1 \cdot d\boldsymbol{X}_2 \times d\boldsymbol{X}_3| = |\det F|\, dv_\kappa. \end{aligned}$$

Note that if $\det F = 1$, the deformation is volume preserving. □

### 1.1.1 Change of Reference Configuration

Let $\hat{\kappa}$ be another reference configuration of $\mathcal{B}$. The map

$$\lambda = \hat{\kappa} \circ \kappa^{-1} : \mathcal{E} \to \mathcal{E}, \qquad \hat{\boldsymbol{X}} = \lambda(\boldsymbol{X}) = \hat{\kappa}(\kappa^{-1}(\boldsymbol{X})),$$

is called a *change of reference configuration* from $\kappa$ to $\hat{\kappa}$, and the deformations from the two different reference configurations $\kappa$ and $\hat{\kappa}$ to $\chi$ are related by (Fig. 1.2)

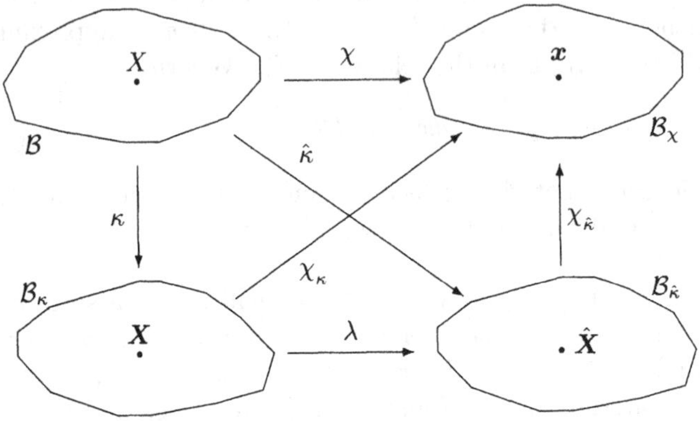


**Fig. 1.2.** Change of reference configuration

$$\chi_\kappa = \chi_{\hat{\kappa}} \circ \lambda, \qquad \chi_\kappa(\boldsymbol{X}) = \chi_{\hat{\kappa}}(\lambda(\boldsymbol{X})) = \chi_{\hat{\kappa}}(\hat{\boldsymbol{X}}),$$

which implies that

$$\nabla_{\boldsymbol{X}} \chi_\kappa = (\nabla_{\hat{\boldsymbol{X}}} \chi_{\hat{\kappa}})(\nabla_{\boldsymbol{X}} \lambda).$$

In other words, we have

$$F_\kappa = F_{\hat{\kappa}}\, P, \qquad P = \nabla_{\boldsymbol{X}}(\hat{\kappa} \circ \kappa^{-1}). \tag{1.10}$$

These are the relations between the deformation gradients with respect to two different reference configurations.

## 1.2 Strain and Rotation

The deformation gradient is a measure of local deformation of the body. In this section we shall introduce other measures of deformation that have more suggestive physical meanings, such as change of shape and orientation.

Since the deformation gradient $F$ is non-singular, by the polar decomposition theorem (see Sect. A.1.8), there exist two positive definite symmetric tensors $U$ and $V$, and an orthogonal tensor $R$, uniquely determined by $F$, such that

$$F = RU = VR \tag{1.11}$$

holds.

To interpret this decomposition in physical terms, we observe that a positive definite symmetric tensor represents a state of pure stretches along three mutually orthogonal axes (spectral theorem, Sect. A.1.8) and an orthogonal tensor represents a rotation. Therefore, (1.11) assures that any local deformation is a combination of a pure stretch and a rotation: first stretch $U$ then rotation $R$ or first rotation $R$ then stretch $V$.

We call $R$ the *rotation tensor*, while $U$ and $V$ are called the *right* and the *left stretch tensors*, respectively. The right and left, of course, refer to the position they appear at the "right" and "left" in the respective expressions of the decomposition (1.11). Both stretch tensors measure the local strain, the change of shape, while the tensor $R$ describes the local rotation, the change of orientation, experienced by material elements of the body.

Clearly we have

$$\begin{aligned} U^2 = F^T F, \qquad V^2 = F F^T, \\ \det U = \det V = |\det F|. \end{aligned} \tag{1.12}$$

Let the eigenvalues and the eigenvectors of $U$ be $v_i$ and $\boldsymbol{e}_i$, $i = 1, 2, 3$,

$$U\boldsymbol{e}_i = v_i \boldsymbol{e}_i \qquad \text{(no sum)}.$$

Then, since $V = RUR^T$, we have

$$V(R\boldsymbol{e}_i) = RUR^T(R\boldsymbol{e}_i) = RU(\boldsymbol{e}_i) = v_i(R\boldsymbol{e}_i).$$

In other words, $V$ and $U$ have the same eigenvalues and their eigenvectors differ only by the rotation $R$. The eigenvalues $v_i$ are called the *principal stretches*, and the corresponding mutually orthogonal eigenvectors are called the *principal directions.*

Given a deformation gradient $F$, the stretch tensors $U$ and $V$ are determined from the relation $(1.12)_1$ by taking the "square root" (see p. 259) of the positive definite symmetric tensors $F^T F$ and $F F^T$, respectively. In practical calculations, it is more convenient to introduce the *right* and the *left Cauchy–Green strain tensors* defined by

$$C = U^2 = F^T F, \qquad B = V^2 = F F^T, \tag{1.13}$$

respectively, as alternatives to the strain measures $U$ and $V$. In components, they are given by

$$C_{\alpha\beta} = g_{ij} F^i{}_\alpha F^j{}_\beta, \qquad B^{ij} = g^{\alpha\beta} F^i{}_\alpha F^j{}_\beta,$$

where $g_{ij}$ and $g^{\alpha\beta}$ are the covariant and the contravariant components of the corresponding metric tensor of the coordinate systems in the current and the reference configurations. In matrix form, they can be written as

$$[C_{\alpha\beta}] = [F^i{}_\alpha]^T[g_{ij}][F^j{}_\beta], \qquad [B^{ij}] = [F^i{}_\alpha][g^{\alpha\beta}][F^j{}_\beta]^T.$$

**Example 1.2.1** Consider a deformation $\boldsymbol{x} = \chi_\kappa(\boldsymbol{X})$ given in Cartesian coordinates, in both the reference and the deformed configurations, by

$$\begin{aligned} x &= X + \kappa Y, \\ y &= Y, \\ z &= Z. \end{aligned} \tag{1.14}$$

This deformation is called a *simple shear* and $\kappa > 0$ is called the amount of shear (see Fig. 1.3). We have

$$[F_{i\alpha}] = \begin{bmatrix} 1 & \kappa & 0 \\ 0 & 1 & 0 \\ 0 & 0 & 1 \end{bmatrix} \tag{1.15}$$

and since $\det F = 1$, simple shear is volume preserving. We can easily obtain

$$[C_{\alpha\beta}] = \begin{bmatrix} 1 & \kappa & 0 \\ \kappa & 1+\kappa^2 & 0 \\ 0 & 0 & 1 \end{bmatrix}.$$

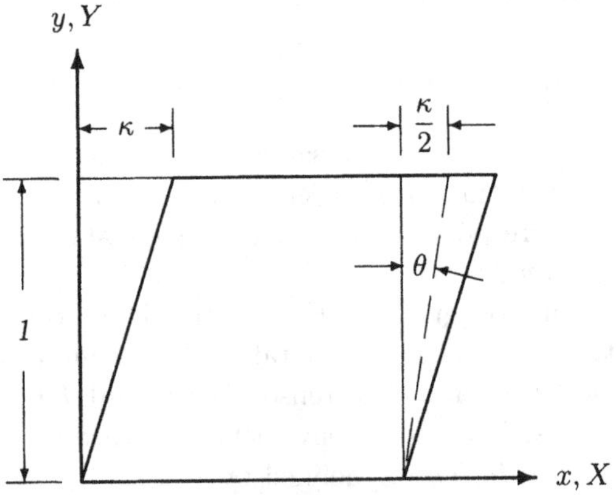


**Fig. 1.3.** Simple shear

From the square root of $C$, we obtain

$$[U_{\alpha\beta}] = \begin{bmatrix} \frac{2}{\sqrt{4+\kappa^2}} & \frac{\kappa}{\sqrt{4+\kappa^2}} & 0 \\ \frac{\kappa}{\sqrt{4+\kappa^2}} & \frac{2+\kappa^2}{\sqrt{4+\kappa^2}} & 0 \\ 0 & 0 & 1 \end{bmatrix},$$

and the principal stretches,

$$v_1,\, v_2 = \pm\tfrac{1}{2}\kappa + \tfrac{1}{2}\sqrt{4+\kappa^2}, \qquad v_3 = 1,$$

with the corresponding principal directions,

$$\boldsymbol{e}_1,\, \boldsymbol{e}_2 = (-\tfrac{1}{2}\kappa \pm \tfrac{1}{2}\sqrt{4+\kappa^2})\boldsymbol{e}_x + \boldsymbol{e}_y, \qquad \boldsymbol{e}_3 = \boldsymbol{e}_z.$$

Note that in the principal direction $\boldsymbol{e}_1$ (with the positive sign), the principal stretch, $v_1 > 1$, is an extension, while in the direction $\boldsymbol{e}_2$ (with the negative sign), the stretch, $v_2 < 1$, is a contraction. Similarly, we have

$$[B_{ij}] = \begin{bmatrix} 1+\kappa^2 & \kappa & 0 \\ \kappa & 1 & 0 \\ 0 & 0 & 1 \end{bmatrix}, \quad [V_{ij}] = \begin{bmatrix} \frac{2+\kappa^2}{\sqrt{4+\kappa^2}} & \frac{\kappa}{\sqrt{4+\kappa^2}} & 0 \\ \frac{\kappa}{\sqrt{4+\kappa^2}} & \frac{2}{\sqrt{4+\kappa^2}} & 0 \\ 0 & 0 & 1 \end{bmatrix}.$$

The rotation tensor can be calculated from $R = FU^{-1}$,

$$[R_{i\alpha}] = \begin{bmatrix} \frac{2}{\sqrt{4+\kappa^2}} & \frac{\kappa}{\sqrt{4+\kappa^2}} & 0 \\ \frac{-\kappa}{\sqrt{4+\kappa^2}} & \frac{2}{\sqrt{4+\kappa^2}} & 0 \\ 0 & 0 & 1 \end{bmatrix}.$$

Note that if we denote $\theta = \tan^{-1}(\kappa/2)$, then $R$ becomes

$$[R_{i\alpha}] = \begin{bmatrix} \cos\theta & \sin\theta & 0 \\ -\sin\theta & \cos\theta & 0 \\ 0 & 0 & 1 \end{bmatrix},$$

which is a clockwise rotation about the $z$–axis by the angle $\theta$. □

**Exercise 1.2.1** Consider a torsion and extension of a cylinder given by the deformation function in cylindrical coordinates $(R, \Theta, Z)$ and $(r, \theta, z)$ in the reference and the current configurations, respectively,

$$r = \sqrt{a}R, \quad \theta = \Theta + \tau Z, \quad z = \frac{1}{a} Z. \tag{1.16}$$

Determine the matrices $[F^i{}_\alpha]$, $[B^{ij}]$, $[C_{\alpha\beta}]$ and the matrices of their corresponding physical components. Moreover, verify that the deformation is volume preserving. Note that the *pure torsion* (for $a = 1$) is locally a simple shear of amount $\tau r$.

## 1.3 Linear Strain Tensors

The strain tensors in the previous section are introduced for finite deformations in general. In the classical linear theory, only small deformations are considered. For the passage from the general theory to the linear theory, we shall first give some geometrical meanings of the Cauchy–Green strain tensors, $C$ and $B$.

Consider two infinitesimal material line segments $d\boldsymbol{X}_1$ and $d\boldsymbol{X}_2$ in the reference configuration and their corresponding ones $d\boldsymbol{x}_1$ and $d\boldsymbol{x}_2$ in the current configuration. From (1.8), we have

$$d\boldsymbol{x}_1 \cdot d\boldsymbol{x}_2 = F d\boldsymbol{X}_1 \cdot F d\boldsymbol{X}_2 = (F^T F) d\boldsymbol{X}_1 \cdot d\boldsymbol{X}_2 = C\, d\boldsymbol{X}_1 \cdot d\boldsymbol{X}_2.$$

Therefore we can consider the following quantity of change between the reference and current configuration for length and angle,

$$d\boldsymbol{x}_1 \cdot d\boldsymbol{x}_2 - d\boldsymbol{X}_1 \cdot d\boldsymbol{X}_2 = 2E\, d\boldsymbol{X}_1 \cdot d\boldsymbol{X}_2, \tag{1.17}$$

where

$$E = \frac{1}{2}(C - 1) \tag{1.18}$$

is called the *Green–St. Venant strain tensor*, or the finite strain tensor in the reference configuration.

Similarly, since $d\boldsymbol{X}_1 = F^{-1} d\boldsymbol{x}_1$, we also have

$$d\boldsymbol{X}_1 \cdot d\boldsymbol{X}_2 = d\boldsymbol{x}_1 \cdot (F^{-T} F^{-1}) d\boldsymbol{x}_2 = B^{-1} d\boldsymbol{x}_1 \cdot d\boldsymbol{x}_2,$$

and hence

$$d\boldsymbol{x}_1 \cdot d\boldsymbol{x}_2 - d\boldsymbol{X}_1 \cdot d\boldsymbol{X}_2 = 2e\, d\boldsymbol{x}_1 \cdot d\boldsymbol{x}_2,$$

where

$$e = \frac{1}{2}(1 - B^{-1}) \tag{1.19}$$

is called the *Almansi–Hamel strain tensor*, or the finite strain tensor in the current configuration (see Sect. 31 of [72]).

Both the strain tensors $E$ and $e$ vanish when there is no deformation, i.e., $F = \mathit{1}$. For small deformations, these strain tensors are therefore small quantities.

In order to consider small deformations, we introduce the *displacement vector* from the reference configuration $\kappa$ (see Fig. 1.4),

$$\boldsymbol{u}(\boldsymbol{X}) = \chi_\kappa(\boldsymbol{X}) - \boldsymbol{X},$$

and its referential gradient,

$$H(\boldsymbol{X}) = \nabla_{\boldsymbol{X}}\, \boldsymbol{u}(\boldsymbol{X}).$$

On the other hand, we can also write the displacement vector as

$$\boldsymbol{u}(\boldsymbol{x}) = \boldsymbol{x} - \chi_\kappa^{-1}(\boldsymbol{x}),$$

and its spatial gradient as

$$h(\boldsymbol{x}) = \nabla_{\boldsymbol{x}}\, \boldsymbol{u}(\boldsymbol{x}).$$

Obviously, we have

$$F = \mathit{1} + H, \qquad F^{-1} = \mathit{1} - h,$$

and hence in terms of the displacement gradients, we obtain

$$\begin{aligned} E &= \frac{1}{2}(H + H^T + H^T H), \\ e &= \frac{1}{2}(h + h^T - h^T h). \end{aligned} \tag{1.20}$$

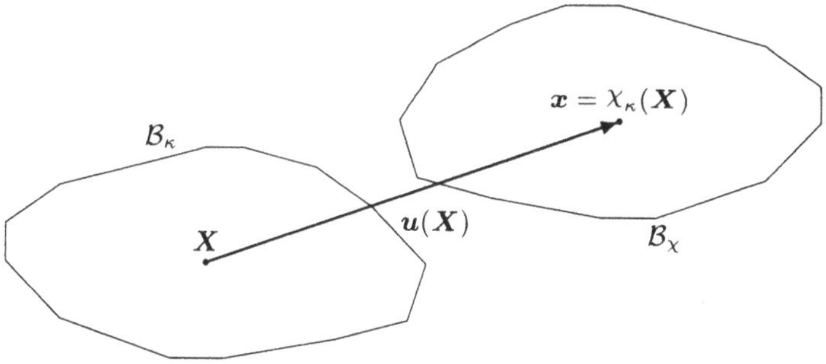


**Fig. 1.4.** Displacement vector

For small deformations, let us write

$$F = \mathit{1} + F_\varepsilon,$$

where $F_\varepsilon$ is a small quantity, $|F_\varepsilon| \ll 1$. Hence

$$H = F_\varepsilon,$$

and since $H^T H$ is then a second-order quantity, we can write from $(1.20)_1$

$$E = \widetilde{E} + o(2),$$

where

$$\widetilde{E} = \frac{1}{2}(H + H^T) \tag{1.21}$$

is called the *infinitesimal strain tensor.* This linear strain tensor is introduced by Cauchy in the classical theory of elasticity.

For small displacement gradients, the right stretch tensor $U$ and the rotation tensor $R$ can be approximated by

$$\begin{aligned} U &= \sqrt{F^T F} = 1 + \tfrac{1}{2}(H + H^T) + o(2) = 1 + \widetilde{E} + o(2), \\ R &= FU^{-1} = 1 + \tfrac{1}{2}(H - H^T) + o(2) = 1 + \widetilde{R} + o(2), \end{aligned} \tag{1.22}$$

where

$$\widetilde{R} = \frac{1}{2}(H - H^T) \tag{1.23}$$

is called the *infinitesimal rotation tensor.* Note that infinitesimal strain and rotation are the symmetric and skew-symmetric parts of the displacement gradient.

In view of (1.17) we can give geometrical meanings to the components of the infinitesimal strain tensor $\widetilde{E}$ relative to a Cartesian coordinate system. First, let $d\boldsymbol{X}_1 = d\boldsymbol{X}_2 = s_o\boldsymbol{e}_1$ be a small material line segment in the direction of the unit base vector $\boldsymbol{e}_1$ and $s$ be the deformed length. Then we have

$$s^2 - s_o^2 = 2s_o^2\,(\widetilde{E}\boldsymbol{e}_1 \cdot \boldsymbol{e}_1),$$

which implies that

$$\widetilde{E}_{11} = \frac{s^2 - s_o^2}{2s_o^2} = \frac{(s - s_o)(s + s_o)}{2s_o^2} \simeq \frac{s - s_o}{s_o}.$$

In other words, $\widetilde{E}_{11}$ is the change of length per unit original length of a small line segment in the $\boldsymbol{e}_1$-direction. The other diagonal components, $\widetilde{E}_{22}$ and $\widetilde{E}_{33}$ have similar interpretations as elongation per unit length in their respective directions.

Now, let $d\boldsymbol{X}_1 = s_o\boldsymbol{e}_1$ and $d\boldsymbol{X}_2 = s_o\boldsymbol{e}_2$ and denote the angle between the two line segments after deformation by $\theta$. Then we have

$$s_o^2\,|F\boldsymbol{e}_1||F\boldsymbol{e}_2|\cos\theta - s_o^2\cos\frac{\pi}{2} = 2s_o^2\,(\widetilde{E}\boldsymbol{e}_1 \cdot \boldsymbol{e}_2),$$

from which, if we write $\gamma = \pi/2 - \theta$, the change from its original right angle, then

$$\frac{\sin\gamma}{2} = \frac{\widetilde{E}_{12}}{|F\boldsymbol{e}_1||F\boldsymbol{e}_2|}.$$

Since $|\widetilde{E}_{12}| \ll 1$ and $|F\boldsymbol{e}_i| \simeq 1$, it follows that $\sin\gamma \simeq \gamma$ and we conclude that

$$\widetilde{E}_{12} \simeq \frac{\gamma}{2}.$$

Therefore, the component $\widetilde{E}_{12}$ is equal to one-half the change of angle between the two line segments originally along the $\boldsymbol{e}_1$- and $\boldsymbol{e}_2$-directions. Other off-diagonal components, $\widetilde{E}_{23}$ and $\widetilde{E}_{13}$ have similar interpretations as the change of angles.

Moreover, since $\det F = \det(1 + H) \simeq 1 + \operatorname{tr} H$ for small deformations, by $(1.9)_2$ we have, for a small material volume,

$$\widetilde{E}_{ii} = \operatorname{tr} H \simeq \frac{dv - dv_\kappa}{dv_\kappa}.$$

Thus the sum of the diagonal components of $\widetilde{E}_{ij}$ measures the infinitesimal change of volume per unit original volume. We call the trace of $\widetilde{E}$ the *dilatation.*

Furthermore, we can also interpret the components of the infinitesimal rotation tensor $\widetilde{R}$ as "average" local rotations. Let $d\boldsymbol{X} = s_o(\cos\theta\,\boldsymbol{e}_1 + \sin\theta\,\boldsymbol{e}_2)$ be a small line segment in the $x_1$–$x_2$ plane making an angle $\theta$ with the $x_1$ axis. We are interested in the change of angle after the deformation in the plane. After deformation, its image $d\boldsymbol{x} = Fd\boldsymbol{X}$ can be written as

$$d\boldsymbol{x} = d\boldsymbol{X} + \widetilde{E}d\boldsymbol{X} + \widetilde{R}d\boldsymbol{X}.$$

The segment $d\boldsymbol{x}$, in general, is not a vector in the $x_1$–$x_2$ plane. Let its projection vector onto the plane be denoted by $d\boldsymbol{x}_h$ and its length by $s$. If we denote the angle of rotation from $d\boldsymbol{X}$ to $d\boldsymbol{x}_h$ by $\omega$, then we have the following relation:

$$d\boldsymbol{X} \times d\boldsymbol{x} \cdot \boldsymbol{e}_3 = d\boldsymbol{X} \times d\boldsymbol{x}_h \cdot \boldsymbol{e}_3 = s_o s \,\sin\omega.$$

Therefore, we obtain

$$\begin{aligned} s_o s\,\sin\omega &= s_o^2\left(\cos^2\theta\,\widetilde{E}_{21} - \sin^2\theta\,\widetilde{E}_{12} - \sin\theta\cos\theta(\widetilde{E}_{11} - \widetilde{E}_{22})\right.\\ &\qquad \left.+ \cos^2\theta\,\widetilde{R}_{21} - \sin^2\theta\,\widetilde{R}_{12}\right)\\ &= s_o^2\left(\widetilde{R}_{21} + \cos 2\theta \widetilde{E}_{12} - \tfrac{1}{2}\sin 2\theta(\widetilde{E}_{11} - \widetilde{E}_{22})\right). \end{aligned}$$

Since $s \simeq s_o$ and the angle $\omega$ is small, we obtain

$$\omega \simeq \widetilde{R}_{21} + \cos 2\theta \widetilde{E}_{12} - \tfrac{1}{2}\sin 2\theta(\widetilde{E}_{11} - \widetilde{E}_{22}).$$

Note that the rotation angle $\omega$ in the $x_1$–$x_2$ plane depends on the orientation of the segment $d\boldsymbol{X}$. However, if we define the average rotation in the plane by

$$<\omega> = \frac{1}{2\pi}\int_0^{2\pi} \omega(\theta)\,d\theta,$$

then after integration of the above expression, it follows that

$$<\omega> \simeq \widetilde{R}_{21}.$$

Therefore, even though the rotation of an individual line segment depends on $\widetilde{R}$, as well as $\widetilde{E}$, the average local rotation at a material point in the $x_1$–$x_2$ plane is equal to the component $\widetilde{R}_{21}$. The other two non-vanishing components $\widetilde{R}_{13}$ and $\widetilde{R}_{32}$ have similar interpretations.

**Remark.** A similar treatment for small deformations, $F = 1 + F_\varepsilon$, can also be based on the spatial displacement gradient $h$. Indeed, since $F^{-1} = 1 - F_\varepsilon + o(2)$, we have $h = F_\varepsilon + o(2)$ or

$$h = H + o(2).$$

In other words, the two displacement gradients

$$\frac{\partial u_i}{\partial X_j} \quad \text{and} \quad \frac{\partial u_i}{\partial x_j}$$

have approximately the same value for small deformations. Therefore, since in the classical linear theory, the nonlinear terms are insignificant, it is usually not necessary to introduce the reference configuration in the linear theory. The classical infinitesimal strain and rotation, in the Cartesian coordinate system, are often defined as

$$\begin{aligned} \widetilde{E}_{ij} &= \frac{1}{2}\left(\frac{\partial u_i}{\partial x_j} + \frac{\partial u_j}{\partial x_i}\right), \\ \widetilde{R}_{ij} &= \frac{1}{2}\left(\frac{\partial u_i}{\partial x_j} - \frac{\partial u_j}{\partial x_i}\right), \end{aligned} \tag{1.24}$$

in the current configuration. □

**Exercise 1.3.1** Consider the simple shear deformation given in (1.14).

1) Determine the displacement vector $u$.
2) Assuming $\kappa \ll 1$, determine the infinitesimal strain and the infinitesimal rotation tensors for simple shear of the classical theory.
3) Compare with the previous results for finite shear strains.

**Exercise 1.3.2** Consider the pure torsion ($a = 1$) given in (1.16).

1) Determine the finite Green–St. Venant strain tensor.
2) Determine the infinitesimal strain and the infinitesimal rotation tensors, assuming a small displacement gradient.
3) Compare the results between finite and small strains.

**Exercise 1.3.3** Show that the infinitesimal strain tensor $\widetilde{E}_{ij}$ satisfies the so-called *compatibility condition*,

$$\varepsilon^{pik}\varepsilon^{qjl}\widetilde{E}_{ij,kl} = 0,$$

where the permutation symbols have been used (see p. 250). In other words, since the linear strain tensor is the symmetric part of the displacement gradient, its components are not independent functions. Their second partial derivatives must satisfy the compatibility condition for the existence of the displacement function. In Cartesian coordinates, they read

$$\begin{aligned}
&\widetilde{E}_{11,22} + \widetilde{E}_{22,11} - 2\,\widetilde{E}_{12,12} = 0,\\
&\widetilde{E}_{22,33} + \widetilde{E}_{33,22} - 2\,\widetilde{E}_{23,23} = 0,\\
&\widetilde{E}_{33,11} + \widetilde{E}_{11,33} - 2\,\widetilde{E}_{31,31} = 0,\\
&\widetilde{E}_{12,23} + \widetilde{E}_{23,12} - \widetilde{E}_{22,31} - \widetilde{E}_{31,22} = 0,\\
&\widetilde{E}_{23,31} + \widetilde{E}_{31,23} - \widetilde{E}_{33,12} - \widetilde{E}_{12,33} = 0,\\
&\widetilde{E}_{31,12} + \widetilde{E}_{12,31} - \widetilde{E}_{11,23} - \widetilde{E}_{23,11} = 0.
\end{aligned}$$

## 1.4 Motion

A *motion* of $\mathcal{B}$ can be regarded as a continuous sequence of configurations in time, i.e., we call $\chi = \{\chi_t,\, t \in I\!R \,|\, \chi_t : \mathcal{B} \to \mathcal{E}\}$ a *motion* of $\mathcal{B}$. The body in the configuration at time $t$ will be denoted by $\mathcal{B}_t$.

The motion $\chi$ of $\mathcal{B}$ can be expressed as a map,

$$\chi : \mathcal{B} \times I\!R \to \mathcal{E}, \qquad \boldsymbol{x} = \chi(X, t) = \chi_t(X),$$

and given a reference configuration $\kappa$, by (1.2) it can also be expressed as

$$\chi_\kappa : \mathcal{B}_\kappa \times I\!R \to \mathcal{E}, \qquad \boldsymbol{x} = \chi_\kappa(\boldsymbol{X}, t) = \chi_t(\kappa^{-1}(\boldsymbol{X})), \tag{1.25}$$

so that

$$\chi_\kappa(\,\cdot\,, t) = \chi_t \circ \kappa^{-1} : \mathcal{B}_\kappa \to \mathcal{B}_t$$

is a deformation of the body $B$ from $\kappa$ to $\chi_t$. Therefore, a motion can also be regarded as a continuous sequence of deformation from a reference configuration in time.

For a fixed material point $\boldsymbol{X}$,

$$\chi_\kappa(\boldsymbol{X}, \,\cdot\,) : I\!R \to \mathcal{E}$$

is a curve called the *path* (or *trajectory*) of the material point $X$. The *velocity* $\boldsymbol{v}$ and the *acceleration* $\boldsymbol{a}$ are defined as the first and the second time derivative of the position as it moves along the path of the material point $\boldsymbol{X}$.

$$\begin{aligned} \boldsymbol{v} &: \mathcal{B}_\kappa \times I\!R \to V & \boldsymbol{v} &= \frac{\partial \chi_\kappa(\boldsymbol{X},t)}{\partial t}, \\ \boldsymbol{a} &: \mathcal{B}_\kappa \times I\!R \to V & \boldsymbol{a} &= \frac{\partial^2 \chi_\kappa(\boldsymbol{X},t)}{\partial t^2}, \end{aligned} \tag{1.26}$$

where $V$ is the translation space of $\mathcal{E}$. The velocity and the acceleration are vector quantities. Here, of course, we have assumed that $\chi_\kappa(\boldsymbol{X},t)$ is twice differentiable with respect to $t$. For simplicity, hereafter we shall assume that all functions are smooth enough for the conditions needed in the context, without their regularities explicitly specified.

### 1.4.1 Material and Spatial Descriptions

A material body is endowed with some physical properties whose values may change along with the deformation of the body in a motion. A quantity defined on a motion can be described in essentially two different ways: either by the evolution of its value along the path of a material point or by the change of its value at a fixed location in the current configuration of the body. The former is called a material description and the latter a spatial description. We shall make them more precise below.

For a given motion $\chi$ and a fixed reference configuration $\kappa$, consider a quantity, with its value in some space $W$, defined on the motion of $\mathcal{B}$ by a function

$$f : \mathcal{B}_\kappa \times I\!R \to W. \tag{1.27}$$

Then it can also be defined on the current configuration at any time $t$,

$$\tilde{f}(\,\cdot\,,t) : \mathcal{B}_t \to W,$$

by

$$\tilde{f}(\boldsymbol{x},t) = \tilde{f}(\chi_\kappa(\boldsymbol{X},t),t) = f(\boldsymbol{X},t). \tag{1.28}$$

As a custom in continuum mechanics, one usually denotes the functions $f$ and $\tilde{f}$ by the same symbol, since they have the same value at the corresponding point, and write, by an abuse of notations,

$$f = f(\boldsymbol{X},t) = f(\boldsymbol{x},t),$$

and call the former the *material* description (or *referential* description to emphasize the presence of a reference of configuration) and the latter the *spatial* description of the function $f$. Sometimes the material description is referred to as the *Lagrangian* description and the spatial description as the *Eulerian* description.

Possible confusions may arise in this abuse of notations, especially when differentiations are involved. Of course, one way of avoiding this is to write out explicitly the variables concerned, for example, $\partial_t f(\boldsymbol{X}, t)$ to mean the time derivative with $\boldsymbol{X}$ held fixed in the material description and $\partial_t f(\boldsymbol{x}, t)$ with $\boldsymbol{x}$ held fixed in the spatial description. In continuum mechanics, however, it is usually preferable to avoid such confusions by using different notations for differentiation in these situations.

In the material description, the time derivative is denoted by a dot or by $d/dt$ while the differential operators such as gradient, divergence and curl are denoted by Grad, Div and Curl, respectively, beginning with capital letters:

$$\dot{f} = \frac{df}{dt} = \frac{\partial f(\boldsymbol{X}, t)}{\partial t}, \qquad \operatorname{Grad} f = \nabla_{\boldsymbol{X}} f(\boldsymbol{X}, t), \text{ etc.}$$

In the spatial description, the time derivative is the usual partial derivative $\partial/\partial t$ and the differential operators are indicated by lower-case letters, grad, div and curl:

$$\frac{\partial f}{\partial t} = \frac{\partial f(\boldsymbol{x}, t)}{\partial t}, \qquad \operatorname{grad} f = \nabla_{\boldsymbol{x}} f(\boldsymbol{x}, t), \text{ etc.}$$

The relations between these notations can easily be obtained. Indeed, let $\psi$ be a scalar field and $\boldsymbol{u}$ be a vector field. we have

$$\dot{\psi} = \frac{\partial \psi}{\partial t} + (\operatorname{grad} \psi) \cdot \boldsymbol{v}, \qquad \dot{\boldsymbol{u}} = \frac{\partial \boldsymbol{u}}{\partial t} + (\operatorname{grad} \boldsymbol{u})\boldsymbol{v}, \tag{1.29}$$

and

$$\operatorname{Grad} \psi = F^T \operatorname{grad} \psi, \qquad \operatorname{Grad} \boldsymbol{u} = (\operatorname{grad} \boldsymbol{u})F. \tag{1.30}$$

In particular, taking the velocity $\boldsymbol{v}$ for $\boldsymbol{u}$ in the last relation we have the following formula for the spatial velocity gradient:

$$\operatorname{grad} \boldsymbol{v} = \dot{F}F^{-1}, \tag{1.31}$$

since $\operatorname{Grad} \boldsymbol{v} = \operatorname{Grad} \dot{\boldsymbol{x}} = \dot{F}$.

We call $\dot{f}$ the *material time derivative* of $f$, which is the time derivative of $f$ following the path of the material point. Therefore, by the definition (1.26), we can write the velocity $\boldsymbol{v}$ and the acceleration $\boldsymbol{a}$ as

$$\boldsymbol{v} = \dot{\boldsymbol{x}}, \qquad \boldsymbol{a} = \ddot{\boldsymbol{x}},$$

and hence by $(1.29)_2$,

$$\boldsymbol{a} = \dot{\boldsymbol{v}} = \frac{\partial \boldsymbol{v}}{\partial t} + (\operatorname{grad} \boldsymbol{v})\boldsymbol{v}, \tag{1.32}$$

the acceleration is the material time derivative of the velocity.

**Example 1.4.1** We have the following identities:

$$\begin{aligned} &\operatorname{Grad} J = J \operatorname{div}(F^T), \\ &\operatorname{div}(J^{-1}F^T) = 0. \end{aligned} \tag{1.33}$$

*Proof.* Regarding $J$ as a function of $F$ and employing the definition of gradient (A.45), we have, for an arbitrary vector $\boldsymbol{a}$,

$$\begin{aligned} \boldsymbol{a} \cdot \operatorname{Grad} J &= (\operatorname{Grad} J)[\boldsymbol{a}] = J(F(\boldsymbol{X}+\boldsymbol{a})) - J(F(\boldsymbol{X})) + o(|\boldsymbol{a}|) \\ &= (\partial_F J)[F(\boldsymbol{X}+\boldsymbol{a}) - F(\boldsymbol{X})] = (\partial_F J)[\operatorname{Grad}(\chi_\kappa(\boldsymbol{X}+\boldsymbol{a}) - \chi_\kappa(\boldsymbol{X}))] \\ &= (\partial_F J)[\operatorname{Grad}(\operatorname{Grad}\chi_\kappa[\boldsymbol{a}])] = (\partial_F J)[\operatorname{Grad}(F\boldsymbol{a})] \\ &= J\,(F^{-T})[\operatorname{Grad}(F\boldsymbol{a})] = J \operatorname{tr}(F^{-1}\operatorname{Grad}(F\boldsymbol{a})) \\ &= J \operatorname{tr}(\operatorname{grad}(F\boldsymbol{a})) = J\,\boldsymbol{a}\cdot\operatorname{div}(F^T), \end{aligned}$$

which proves the first relation. In the above derivation, we have made successive use of the relations, $F = \operatorname{Grad}\chi_\kappa$, (A.48), (1.30) and (A.75). The error terms, like the one shown in the first line, are omitted for simplicity.

To prove the second relation of (1.33), we obtain from (A.76)

$$\begin{aligned} \boldsymbol{a}\cdot\operatorname{div}(J^{-1}F^T) &= \operatorname{div}(J^{-1}F\boldsymbol{a}) \\ &= F\boldsymbol{a}\cdot\operatorname{grad} J^{-1} + J^{-1}\operatorname{div}(F\boldsymbol{a}) \\ &= -J^{-2}\boldsymbol{a}\cdot\operatorname{Grad} J + J^{-1}\boldsymbol{a}\cdot\operatorname{div}(F^T), \end{aligned}$$

which is equal to zero by the first identity. □

**Exercise 1.4.1** Verify the following identities similar to (1.33):

$$\begin{aligned} &\operatorname{Div}(JF^{-T}) = 0, \\ &\operatorname{grad} J = -J \operatorname{Div}(F^{-T}). \end{aligned} \tag{1.34}$$

**Exercise 1.4.2** Derive the physical components of velocity and acceleration in (use Exercise A.2.11)

1) cylindrical coordinate system: $\boldsymbol{x} = (r(X,t), \theta(X,t), z(X,t))$

$$\boldsymbol{v} = \dot r\,\boldsymbol{e}_{\langle r\rangle} + r\dot\theta\,\boldsymbol{e}_{\langle\theta\rangle} + \dot z\,\boldsymbol{e}_{\langle z\rangle},$$

$$\boldsymbol{a} = (\ddot r - r\dot\theta^2)\,\boldsymbol{e}_{\langle r\rangle} + (r\ddot\theta + 2\dot r\dot\theta)\,\boldsymbol{e}_{\langle\theta\rangle} + \ddot z\,\boldsymbol{e}_{\langle z\rangle};$$

2) spherical coordinate system: $\boldsymbol{x} = (r(X,t), \theta(X,t), \phi(X,t))$

$$\boldsymbol{v} = \dot r\,\boldsymbol{e}_{\langle r\rangle} + r\dot\theta\,\boldsymbol{e}_{\langle\theta\rangle} + r\dot\phi\sin\theta\,\boldsymbol{e}_{\langle\phi\rangle},$$

$$\begin{aligned} \boldsymbol{a} = {} & (\ddot r - r\dot\theta^2 - r\dot\phi^2\sin^2\theta)\,\boldsymbol{e}_{\langle r\rangle} + (r\ddot\theta + 2\dot r\dot\theta - r\dot\phi^2\sin\theta\cos\theta)\,\boldsymbol{e}_{\langle\theta\rangle} \\ & + (r\ddot\phi\sin\theta + 2\dot r\dot\phi\sin\theta + 2r\dot\theta\dot\phi\cos\theta)\,\boldsymbol{e}_{\langle\phi\rangle}. \end{aligned}$$

**Exercise 1.4.3** Given a motion

$$\chi_\kappa(\boldsymbol{X},t) = \boldsymbol{x}_\circ(t) + Q(t)(\boldsymbol{X} - \boldsymbol{X}_\circ), \tag{1.35}$$

where $Q(t)$ is a time-dependent orthogonal tensor.

1) Show that the velocity and the acceleration are given by

$$\boldsymbol{v} = \dot{\boldsymbol{x}}_\circ + \boldsymbol{\omega} \times (\boldsymbol{x} - \boldsymbol{x}_\circ),$$
$$\boldsymbol{a} = \ddot{\boldsymbol{x}}_\circ + \dot{\boldsymbol{\omega}} \times (\boldsymbol{x} - \boldsymbol{x}_\circ) + \boldsymbol{\omega} \times (\boldsymbol{\omega} \times (\boldsymbol{x} - \boldsymbol{x}_\circ)),$$

where the angular velocity $\boldsymbol{\omega}$ is defined as the axial vector of the skew-symmetric tensor $Q\dot{Q}^T$ (see (A.29), $\boldsymbol{\omega} = \langle Q\dot{Q}^T \rangle$).

2) Show that, for this motion, the shape (length and angle) of any material element does not change. It is called a rigid motion.

**Exercise 1.4.4** Given an acceleration field in the spatial description,

$$\boldsymbol{a}(\boldsymbol{x},t) = k^2 x \boldsymbol{e}_x + k^2 y \boldsymbol{e}_y, \tag{1.36}$$

where $\{\boldsymbol{e}_i\}$ is the natural basis of the Cartesian coordinate system.

1) Determine the deformation $\boldsymbol{x} = \chi_\kappa(\boldsymbol{X},t)$ of this motion, with the following initial conditions:

$$x(X,Y,0) = X, \qquad \dot{x}(X,Y,0) = kX,$$
$$y(X,Y,0) = Y, \qquad \dot{y}(X,Y,0) = -kY.$$

2) Find the velocity field in the material and the spatial descriptions.
3) Sketch the path of this motion.

## 1.5 Relative Deformation

In practice, for a given motion, the reference configuration $\kappa$ is often chosen as the configuration at some instant $t = t_0$, $\kappa = \chi(\cdot, t_0)$. But such a choice is unnecessary, in general. The configuration $\kappa$ can be chosen independently of any motion.

We may even choose the (current) configuration $\chi_t$ as the reference configuration so that past and future deformations can be described relative to it. Such a choice is sometimes desirable, especially for fluids, and it is worthwhile also to introduce it here.

We denote the position of the material point $X$ at time $\tau$ by $\boldsymbol{\xi}$,

$$\boldsymbol{\xi} = \chi(X,\tau) = \chi_\tau(X). \tag{1.37}$$

Then, we can write

$$\boldsymbol{\xi} = \chi_{(t)}(\boldsymbol{x},\tau) = \chi_\tau(\chi_t^{-1}(\boldsymbol{x})), \tag{1.38}$$

where $\chi_{(t)}(\cdot,\tau) = \chi_\tau \circ \chi_t^{-1} : \mathcal{B}_t \to \mathcal{B}_\tau$ is the deformation at time $\tau$ relative to the configuration at time $t$, or simply called the *relative deformation* (Fig. 1.5). The *relative deformation gradient* $F_t$ is defined by

$$F_t(\boldsymbol{x},\tau) = \nabla_{\boldsymbol{x}}\chi_{(t)}(\boldsymbol{x},\tau), \tag{1.39}$$

that is, the deformation gradient at time $\tau$ with respect to the configuration at time $t$. Of course, if $\tau = t$,

$$F_t(\boldsymbol{x},t) = \mathit{1}. \tag{1.40}$$

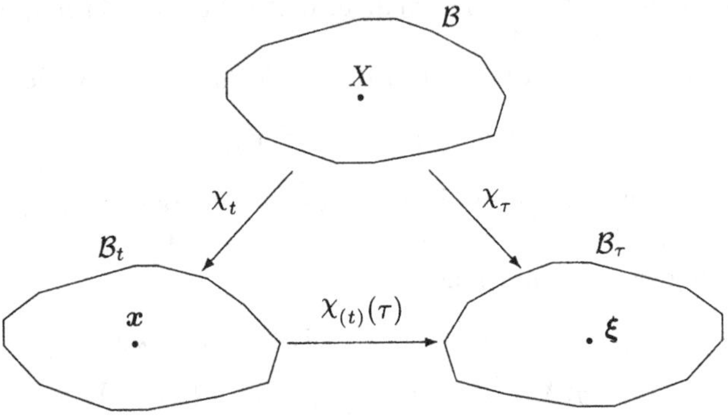


**Fig. 1.5.** Relative deformation

Let $\kappa$ be a fixed reference configuration, then we can also express the relative deformation as

$$\boldsymbol{\xi} = \chi_{(t)}(\boldsymbol{x},\tau) = \chi_\kappa(\chi_\kappa^{-1}(\boldsymbol{x},t),\tau), \tag{1.41}$$

from which we can show that

$$F_\kappa(\boldsymbol{X},\tau) = F_t(\boldsymbol{x},\tau)F_\kappa(\boldsymbol{X},t). \tag{1.42}$$

This relation is best exemplified in the schematic diagram Fig. 1.6.

**Example 1.5.1** Consider a motion in the $x$–$y$ plane relative to the reference configuration $\kappa$ in the Cartesian coordinate system,

$$\boldsymbol{x} = \chi_\kappa(X,Y,t) = (Xe^t,\, Y(t+1)), \tag{1.43}$$

or, in terms of deformation functions,

$$x = Xe^t, \qquad y = Y(t+1),$$

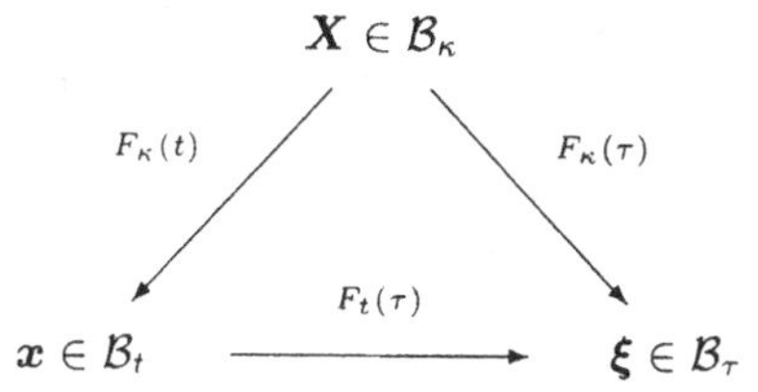


**Fig. 1.6.** Relative deformation gradient

which yield

$$X = xe^{-t}, \qquad Y = \frac{y}{t+1},$$

or

$$\boldsymbol{X} = \chi_\kappa^{-1}(x, y, t) = (xe^{-t}, \frac{y}{t+1}).$$

Therefore, from (1.41) the relative deformation is given by

$$\begin{aligned}\boldsymbol{\xi} &= \chi_\kappa(\chi_\kappa^{-1}(x, y, t), \tau) = \chi_\kappa(xe^{-t}, \frac{y}{t+1}, \tau)\\ &= (xe^{\tau-t}, \frac{\tau+1}{t+1}y).\end{aligned}$$

From (1.43) and the above function, the deformation gradients can be calculated:

$$\begin{aligned}F_\kappa(t) &= e^t \boldsymbol{e}_x \otimes \boldsymbol{e}_x + (t+1)\,\boldsymbol{e}_y \otimes \boldsymbol{e}_y,\\ F_t(\tau) &= e^{\tau-t}\boldsymbol{e}_x \otimes \boldsymbol{e}_x + \frac{\tau+1}{t+1}\,\boldsymbol{e}_y \otimes \boldsymbol{e}_y,\end{aligned}$$

where $\{\boldsymbol{e}_x, \boldsymbol{e}_y\}$ is the natural basis of the Cartesian coordinate system. The relations (1.40) and (1.42) can be checked easily.

We can also obtain the path of the material point $(X_0, Y_0)$ in this motion, by eliminating time $t$ from the deformation function,

$$\begin{cases} y = Y_0\Big(\log\dfrac{x}{X_0} + 1\Big), & \text{for} \quad X_0 \neq 0,\\ x = 0, & \text{for} \quad X_0 = 0.\end{cases}$$

Note that, for this motion, the reference configuration is the configuration of the body at the instant $t = 0$. (What is the image of a square in the reference configuration at any instant in the motion?) □

**Exercise 1.5.1** Given a velocity field

$$\boldsymbol{v}(\boldsymbol{x}, t) = u(y)\boldsymbol{e}_x. \tag{1.44}$$

1) Determine the relative deformation $\boldsymbol{\xi} = \chi_{(t)}(\boldsymbol{x}, \tau)$ of this motion.
2) Show that

$$F_t(\tau) = \mathit{1} + (\tau - t)\kappa N, \tag{1.45}$$

where $\kappa = du/dy$, and $N = \boldsymbol{e}_x \otimes \boldsymbol{e}_y$.

**Exercise 1.5.2** At fixed time $t$, the solution curves of the velocity field $\boldsymbol{v}(\boldsymbol{x},t)$ are called the *streamlines* of the motion at time $t$, i.e., if $\boldsymbol{x}=\boldsymbol{x}(s)$, $\boldsymbol{x}(0)=\boldsymbol{x}_\circ$ is the streamline passing through $\boldsymbol{x}_\circ$ at time $t$, then

$$\frac{d\boldsymbol{x}(s)}{ds}=\boldsymbol{v}(\boldsymbol{x}(s),t).$$

Determine the streamlines of the motion given in the Cartesian coordinate system by

$$\boldsymbol{x}=(Xe^{t^2},\,Ye^{t},\,Z).$$

## 1.6 Rate of Deformation

Whereas the deformation gradient measures the local deformation, the material time derivative of deformation gradient measures the rate at which such changes occur. Another measure for the rate of deformation, more commonly used in fluid mechanics, is the spatial gradient of velocity. They are related by (1.31), $\operatorname{grad}\boldsymbol{v}=\dot{F}F^{-1}$.

The material time derivative of deformation gradient, $\dot{F}$, is the rate of change of deformation relative to the reference configuration. Similarly, we can define

$$L(\boldsymbol{x},t)=\frac{\partial}{\partial\tau}F_t(\boldsymbol{x},\tau)\Big|_{\tau=t}=\dot{F}_t(\boldsymbol{x},t), \tag{1.46}$$

to be the rate of change of deformation relative to the current configuration. From (1.42), $F_t(\tau)=F(\tau)F(t)^{-1}$, by taking the derivative with respect to $\tau$, we have

$$\begin{aligned}\dot{F}_t(\tau)&=\dot{F}(\tau)F(t)^{-1}\\&=(\operatorname{grad}\boldsymbol{v}(\tau))F(\tau)F(t)^{-1}=(\operatorname{grad}\boldsymbol{v}(\tau))F_t(\tau),\end{aligned}$$

and since $F_t(t)=1$, we conclude that

$$L=\operatorname{grad}\boldsymbol{v}. \tag{1.47}$$

In other words, the velocity gradient, $\operatorname{grad}\boldsymbol{v}$, can also be interpreted as the rate of change of deformation relative to the current configuration.

Moreover, if we apply the polar decomposition to the relative deformation gradient $F_t(\boldsymbol{x},\tau)$,

$$F_t=R_tU_t=V_tR_t,$$

by holding $\boldsymbol{x}$ and $t$ fixed and taking the derivative of $F_t$ with respect to $\tau$, we obtain

$$\dot{F}_t(\tau)=R_t(\tau)\dot{U}_t(\tau)+\dot{R}_t(\tau)U_t(\tau),$$

and hence by putting $\tau = t$, we have

$$L(t) = \dot{U}_t(t) + \dot{R}_t(t). \tag{1.48}$$

If we denote

$$D(t) = \dot{U}_t(t), \qquad W(t) = \dot{R}_t(t), \tag{1.49}$$

we can show easily that

$$D^T = D, \qquad W^T = -W. \tag{1.50}$$

Therefore, the relation (1.48) is just a decomposition of the tensor $L$ into its symmetric and skew-symmetric parts, or from (1.47) we have

$$\begin{aligned} D &= \frac{1}{2}(\operatorname{grad} \boldsymbol{v} + \operatorname{grad} \boldsymbol{v}^T), \\ W &= \frac{1}{2}(\operatorname{grad} \boldsymbol{v} - \operatorname{grad} \boldsymbol{v}^T). \end{aligned} \tag{1.51}$$

In view of (1.49) the symmetric part of the velocity gradient, $D$, is called the *rate of strain tensor* or simply the *stretching tensor*, and the skew-symmetric part of the velocity gradient, $W$, is called the *rate of rotation tensor* or simply the *spin tensor*.

Since the spin tensor $W$ is skew-symmetric, it can be represented as an axial vector $\boldsymbol{w}$ (see Sect. A.1.6). The components of $\boldsymbol{w}$ are usually defined by $w^i = e^{ijk} W_{kj}$, hence it follows that

$$\boldsymbol{w} = \operatorname{curl} \boldsymbol{v}. \tag{1.52}$$

The vector $\boldsymbol{w}$ is usually called the *vorticity* vector in fluid dynamics.[1]

Other tensors of interest, in terms of the relative deformation, are the *Rivlin Ericksen tensors* by taking the time derivatives of $C_t(\tau) = F_t(\tau)^T F_t(\tau)$,

$$A_n(\boldsymbol{x}, t) = C_t^{(n)}(\boldsymbol{x}, t) = \frac{\partial^n}{\partial \tau^n} C_t(\boldsymbol{x}, \tau)\Big|_{\tau = t}, \qquad n = 1, 2, 3, \cdots. \tag{1.53}$$

In particular, for $n = 1$,

$$A_1(\boldsymbol{x}, t) = \dot{C}_t(\boldsymbol{x}, t) = \dot{F}_t(\boldsymbol{x}, t)^T + \dot{F}_t(\boldsymbol{x}, t) = L^T + L,$$

which implies that

$$A_1 = 2\, D. \tag{1.54}$$

Therefore, $A_1$ is just twice the strain rate tensor. More generally, the Rivlin Ericksen tensor of order $n$ is a measure for the strain rate of higher order.

[1] In terms of the notation (A.29), the vorticity vector $\boldsymbol{w}$ and the spin tensor $W$ are related by $\boldsymbol{w} = \langle -2W \rangle$ and not by $\boldsymbol{w} = \langle W \rangle$, as introduced in Sect. A.1.6.

**Exercise 1.6.1** Show that the spin tensor $W$ can also be defined as

$$W = \dot{R}R^T.$$

**Exercise 1.6.2** Verify that

$$\dot{\boldsymbol{v}} = \frac{\partial \boldsymbol{v}}{\partial t} + \frac{1}{2}\operatorname{grad}(\boldsymbol{v}\cdot\boldsymbol{v}) + 2W\boldsymbol{v},$$

or in terms of the vorticity vector $\boldsymbol{w}$,

$$\dot{\boldsymbol{v}} = \frac{\partial \boldsymbol{v}}{\partial t} + \frac{1}{2}\operatorname{grad}(\boldsymbol{v}\cdot\boldsymbol{v}) + \boldsymbol{w}\times\boldsymbol{v}. \tag{1.55}$$

**Exercise 1.6.3** Consider the velocity field given by (1.44) (Exercise 1.5.1), Show that

$$\begin{aligned} A_1 &= \kappa(N + N^T), \\ A_2 &= 2\,\kappa^2 N^T N, \\ A_3 &= 0. \end{aligned} \tag{1.56}$$

**Exercise 1.6.4** Show that for the rigid motion given by (1.35), the stretching tensor $D = 0$ and the angular velocity $\boldsymbol{\omega}$ is the negative axial vector of the spin tensor $W$ (see (A.29), $\boldsymbol{\omega} = -\langle W\rangle$).

## 1.7 Change of Frame and Objective Tensors

A frame of reference can be interpreted as an *observer* who observes an event in terms of position and time with a ruler and a clock. Different observers may use different rulers and clocks and come up with different results for the same event. However, if the same units of measure for their rulers and clocks are used, they should obtain the same *distance* and the same *time lapse* between any two events under observation, even though the values of their observations may still be different. We shall impose these requirements on a change of frame from one to another.

Let $\phi$ and $\phi^*$ be two frames of reference. We call

$$* = \phi^* \circ \phi^{-1} : \mathcal{E}\times I\!R \to \mathcal{E}\times I\!R$$

$$* : (\boldsymbol{x}, t) \mapsto (\boldsymbol{x}^*, t^*)$$

a *change of frame* from $\phi$ to $\phi^*$ (Fig. 1.7). For convenience, we have denoted the map $\phi^*\circ\phi^{-1}$ by the symbol $*$.

Since we shall restrict ourselves to the use of frames of reference that yield the same distance and the same time lapse between two events, a change

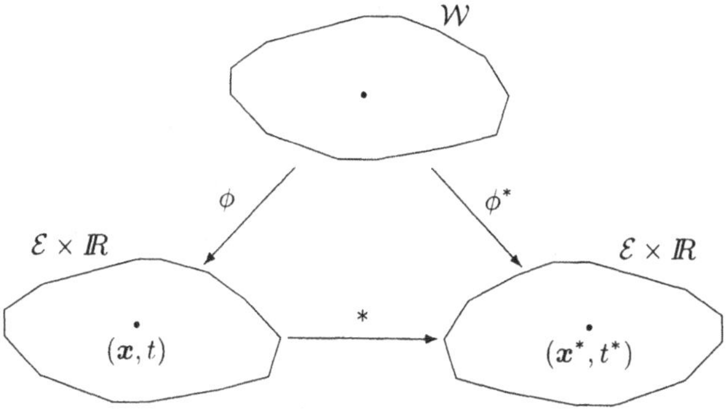


**Fig. 1.7.** Change of frame

of frame $*$, in general, must be of the following form:

$$\begin{aligned} \boldsymbol{x}^* &= Q(t)(\boldsymbol{x} - \boldsymbol{x}_\circ) + \boldsymbol{c}(t), \\ t^* &= t + a, \end{aligned} \tag{1.57}$$

for some $a \in I\!R$, $\boldsymbol{x}_\circ \in \mathcal{E}$, $\boldsymbol{c}(t) \in \mathcal{E}$, and $Q(t) \in \mathcal{O}(V)$, where $\mathcal{O}(V)$ is the orthogonal group on $V$.

A change of frame $*$ defined by (1.57) is a time-dependent rigid transformation, also referred to as a *Euclidean transformation.* It is the most general form of the change of frame. The collection of all Euclidean transformations will be called the *Euclidean class.*

A change of frame gives rise to a linear map on $V$ in the following way: Let $*$ be a change of frame from $\phi$ to $\phi^*$ defined by (1.57). For any vector $\boldsymbol{u} \in V$, there exist $\boldsymbol{x}_1, \boldsymbol{x}_2 \in \mathcal{E}$ in the frame $\phi$, such that

$$\boldsymbol{u} = \boldsymbol{x}_2 - \boldsymbol{x}_1.$$

Let $\boldsymbol{x}_1^*$ and $\boldsymbol{x}_2^*$ be the corresponding points in the frame $\phi^*$, and denote

$$\boldsymbol{u}^* = \boldsymbol{x}_2^* - \boldsymbol{x}_1^*,$$

then from (1.57)

$$\boldsymbol{u}^* = Q(t)(\boldsymbol{x}_2 - \boldsymbol{x}_1) = Q(t)\boldsymbol{u}. \tag{1.58}$$

Therefore, we can define a linear map $\boldsymbol{Q}^* : V \to V$ taking $\boldsymbol{u}$ to $\boldsymbol{u}^*$ by

$$\boldsymbol{Q}^*(\boldsymbol{u}) = \boldsymbol{u}^* = Q(t)\boldsymbol{u}.$$

That is,

$$\boldsymbol{Q}^* = Q(t). \tag{1.59}$$

More generally, let $\mathcal{S}_n$ denote the $n$-th-order tensor space of $V$, i.e., $\mathcal{S}_n = \overset{n}{\otimes} V$, $n = 1, 2, 3, \cdots$. Then a change of frame $*$ gives rise to a linear map on $\mathcal{S}_n$, which will also be denoted by $\boldsymbol{Q}^*$,

$$\boldsymbol{Q}^* : \mathcal{S}_n \to \mathcal{S}_n$$

in the following way: For any $\boldsymbol{u}_1, \cdots, \boldsymbol{u}_n \in V$, we define

$$\boldsymbol{Q}^*(\boldsymbol{u}_1 \otimes \cdots \otimes \boldsymbol{u}_n) = Q\boldsymbol{u}_1 \otimes \cdots \otimes Q\boldsymbol{u}_n, \tag{1.60}$$

and extend it linearly to the space $\mathcal{S}_n$. In particular, for $n = 1$ we have (1.59), and for $n = 2$ we have

$$\boldsymbol{Q}^*(\boldsymbol{u}_1 \otimes \boldsymbol{u}_2) = Q\boldsymbol{u}_1 \otimes Q\boldsymbol{u}_2 = Q\,(\boldsymbol{u}_1 \otimes \boldsymbol{u}_2)\,Q^T.$$

Hence for any $T \in V \otimes V$, we have

$$T^* = \boldsymbol{Q}^*(T) = QTQ^T. \tag{1.61}$$

For $M \in \mathcal{S}_n$ the condition $M^* = \boldsymbol{Q}^*(M)$ in components is given by

$$M^*_{i_1 \cdots i_n} = Q_{i_i}^{\ j_1} \cdots Q_{i_n}^{\ j_n} M_{j_1 \cdots j_n}.$$

In the physical interpretation we can say that $\boldsymbol{u}$ and $\boldsymbol{u}^*$, $T$ and $T^*$ so related, are the same vector and the same tensor as observed in two different frames $\phi$ and $\phi^*$. Moreover, a scalar independent of observers should have the same value in any frame, hence we may also define $\boldsymbol{Q}^*$ on $\mathcal{S}_n$ for $n = 0$, i.e., $\mathcal{S}_0 = I\!R$, by

$$\boldsymbol{Q}^* : I\!R \to I\!R \qquad \boldsymbol{Q}^* = 1. \tag{1.62}$$

Note that $\boldsymbol{Q}^* : \mathcal{S}_n \to \mathcal{S}_n$ depends only on the rotational part $Q(t)$ of the change of the frame $*$. Therefore we may also refer to $\boldsymbol{Q}^*$ as the *induced linear map* on $\mathcal{S}_n$ by the rotation $Q(t)$ (in the change of frame).

Let $\Phi$ denote the set of all frames. Suppose $f$ is a map

$$f : \Phi \to \mathcal{S}_n.$$

We call $f(\phi)$ the value of $f$ observed in the frame $\phi$. Let $\Sigma$ be the Euclidean class of change of frame and $\Psi$ be a subclass of $\Sigma$.

**Definition.** An observable quantity $f$ is said to be *frame-indifferent* with respect to $\Psi$, if

$$f(\phi^*) = \boldsymbol{Q}^* f(\phi) \tag{1.63}$$

for any change of frame from $\phi$ to $\phi^*$ belong to the class $\Psi$.

If $\Psi = \Sigma$, we simply say that $f$ is *frame-indifferent* or *objective* (with respect to Euclidean transformations). More specifically, a time-dependent scalar $s$, vector $\boldsymbol{u}$ or tensor $T$ is called an *objective* scalar, vector or tensor quantity, respectively, if, relative to any change of frame given by a Euclidean transformation (1.57),

$$\begin{aligned} s^*(t^*) &= s(t), \\ \boldsymbol{u}^*(t^*) &= Q(t)\,\boldsymbol{u}(t), \\ T^*(t^*) &= Q(t)\,T(t)\,Q(t)^T. \end{aligned} \tag{1.64}$$

We say that objective quantities are *invariant* under a change of observers.

**Example 1.7.1** Let $W$ be an objective skew-symmetric tensor, show that its associated axial vector $\langle W \rangle$, (see (A.29)), is not an objective vector.

Since $W$ is objective, we have

$$W^* = QWQ^T \quad \forall Q \in \mathcal{O}(V).$$

Let the associated axial vectors be

$$\boldsymbol{w} = \langle W \rangle, \qquad \boldsymbol{w}^* = \langle W^* \rangle,$$

and express $W = \frac{1}{2} W^{ij} \boldsymbol{e}_i \wedge \boldsymbol{e}_j$. Then we have, for any $\boldsymbol{u} \in V$,

$$\begin{aligned} \boldsymbol{w}^* \cdot Q\boldsymbol{u} &= \langle QWQ^T \rangle \cdot Q\boldsymbol{u} = \tfrac{1}{2} W^{ij} \langle Q(\boldsymbol{e}_i \wedge \boldsymbol{e}_j) Q^T \rangle \cdot Q\boldsymbol{u} \\ &= \tfrac{1}{2} W^{ij} \langle Q\boldsymbol{e}_i \wedge Q\boldsymbol{e}_j \rangle \cdot Q\boldsymbol{u} = \tfrac{1}{2} W^{ij}\, Q\boldsymbol{e}_i \times Q\boldsymbol{e}_j \cdot Q\boldsymbol{u} \\ &= \tfrac{1}{2} (\det Q) W^{ij} \boldsymbol{e}_i \times \boldsymbol{e}_j \cdot \boldsymbol{u} = \tfrac{1}{2} (\det Q) W^{ij} \langle \boldsymbol{e}_i \wedge \boldsymbol{e}_j \rangle \cdot \boldsymbol{u} \\ &= (\det Q) \langle W \rangle \cdot \boldsymbol{u} = (\det Q) \boldsymbol{w} \cdot \boldsymbol{u}. \end{aligned}$$

It follows that

$$\boldsymbol{w}^* = (\det Q) Q\boldsymbol{w} = \pm Q\boldsymbol{w},$$

since $Q$ is orthogonal, $\det Q = \pm 1$. Therefore, the associated axial vector $\boldsymbol{w}$ is not objective. □

### 1.7.1 Transformation Property of Motion

In mechanics, we know that motions depend on observers and, consequently, the velocity and the acceleration of a motion are not objective, in general. We shall see how they transform under a change of observers.

Let $\chi$ be a motion, and $*$ be a change of frame from $\phi$ to $\phi^*$, then

$$\boldsymbol{x} = \chi(X, t), \qquad \boldsymbol{x}^* = \chi^*(X, t^*), \qquad X \in \mathcal{B}.$$

From (1.57), we have

$$\begin{aligned} &\chi^*(X, t^*) = Q(t)(\chi(X, t) - \boldsymbol{x}_\circ) + \boldsymbol{c}(t), \\ &t^* = t + a. \end{aligned} \tag{1.65}$$

Therefore, it follows that

$$\dot{\boldsymbol{x}}^* = \frac{\partial \chi^*}{\partial t^*} = \dot{Q}(\boldsymbol{x} - \boldsymbol{x}_\circ) + Q\dot{\boldsymbol{x}} + \dot{\boldsymbol{c}}$$

or

$$\dot{\boldsymbol{x}}^* - Q\dot{\boldsymbol{x}} = \Omega(\boldsymbol{x}^* - \boldsymbol{c}) + \dot{\boldsymbol{c}}, \tag{1.66}$$

where

$$\Omega(t) = \dot{Q}(t)Q(t)^T \tag{1.67}$$

is called the *angular velocity tensor* or *spin tensor* of $\phi^*$ relative to $\phi$. The tensor $\Omega$ is skew-symmetric, i.e.,

$$\Omega^T = -\Omega. \tag{1.68}$$

Moreover, taking again the derivative of (1.66), we obtain

$$\ddot{\boldsymbol{x}}^* - Q\ddot{\boldsymbol{x}} = \ddot{\boldsymbol{c}} + 2\Omega(\dot{\boldsymbol{x}}^* - \dot{\boldsymbol{c}}) + (\dot{\Omega} - \Omega^2)(\boldsymbol{x}^* - \boldsymbol{c}). \tag{1.69}$$

The relations (1.66) and (1.69) show that the velocity and the acceleration are not objective vectors.

Note that if the rotation of the change of frame $Q(t)$ is constant in time and $\boldsymbol{c}(t)$ is linear in time, then we have $\ddot{\boldsymbol{x}}^* = Q\ddot{\boldsymbol{x}}$. Such a change of frame is called a *Galilean transformation*, which is given by

$$\begin{aligned} \boldsymbol{x}^* &= Q(\boldsymbol{x} - \boldsymbol{x}_\circ) + \boldsymbol{V}t + \boldsymbol{c}_\circ, \\ t^* &= t + a, \end{aligned} \tag{1.70}$$

for any constant $\boldsymbol{c}_\circ$, $\boldsymbol{V}$, and $Q$. The relation (1.69) shows that the acceleration is frame indifferent with respect to Galilean transformations.

On the other hand, the relation (1.66) shows that the velocity is not frame indifferent with respect to Galilean transformations. However, it also shows that the velocity is frame indifferent with respect to time-independent *rigid transformations*, defined as changes of frame given by

$$\begin{aligned} \boldsymbol{x}^* &= Q(\boldsymbol{x} - \boldsymbol{x}_\circ) + \boldsymbol{c}_\circ, \\ t^* &= t + a, \end{aligned} \tag{1.71}$$

where $\boldsymbol{c}_\circ$ and $Q$ are time independent.

### 1.7.2 Property of Some Kinematic Quantities

Let $\kappa$ be a reference configuration, and $\chi$ be a motion. Relative to a change of frame (1.57), we can write

$$\boldsymbol{X} = \kappa_\phi(X), \qquad \boldsymbol{X}^* = \kappa_{\phi^*}(X), \qquad X \in \mathcal{B}. \tag{1.72}$$

In order to see how a reference configuration may be affected by a change of frame, we choose the configuration $\kappa$ as the configuration occupied by the body at some instant $t_0$,

$$\boldsymbol{X} = \chi(X, t_0), \qquad \boldsymbol{X}^* = \chi^*(X, t_0^*),$$

where the $\chi$ and $\chi^*$ are related by (1.65). It follows immediately that

$$\boldsymbol{X}^* = Q(t_0)(\boldsymbol{X} - \boldsymbol{x}_\circ) + \boldsymbol{c}(t_0).$$

Therefore, in general, we may denote the change from $\kappa_\phi$ to $\kappa_{\phi^*}$ by the following transformation in a change of frame,

$$\boldsymbol{X}^* = \kappa_{\phi^*}(\kappa_\phi^{-1}(\boldsymbol{X})) = K(\boldsymbol{X} - \boldsymbol{x}_\circ) + \boldsymbol{c}_\circ, \tag{1.73}$$

where $K$ is a constant orthogonal tensor.

On the other hand, the motion relative to the change of frame is given by

$$\boldsymbol{x} = \chi_\kappa(\boldsymbol{X}, t), \qquad \boldsymbol{x}^* = \chi_\kappa^*(\boldsymbol{X}^*, t^*),$$

and from (1.57) we have

$$\chi_\kappa^*(\boldsymbol{X}^*, t^*) = Q(t)(\chi_\kappa(\boldsymbol{X}, t) - \boldsymbol{x}_\circ) + \boldsymbol{c}(t).$$

Therefore, we obtain for the deformation gradient,

$$F^*(\boldsymbol{X}^*, t^*) = Q(t)F(\boldsymbol{X}, t)\, K^T,$$

or simply

$$F^* = QFK^T, \tag{1.74}$$

where $K$ is a fixed orthogonal tensor due to the change of frame for the reference configuration.[2] This shows that the deformation gradient is not an objective tensor.

With polar decompositions of $F$ and $F^*$, (1.74) gives

$$R^*U^* = QRUK^T, \qquad V^*R^* = QVRK^T.$$

By the uniqueness of such decompositions, we conclude that

$$U^* = KUK^T, \quad V^* = QVQ^T, \quad R^* = QRK^T, \tag{1.75}$$

and hence

$$C^* = KCK^T, \qquad B^* = QBQ^T. \tag{1.76}$$

Therefore, we have shown that $V$ and $B$ are objective tensors, while $R$, $U$, and $C$ are not objective quantities.

If we differentiate (1.74) with respect to time, we obtain

$$\dot{F}^* = Q\dot{F}K^T + \dot{Q}FK^T.$$

[2] It is often assumed that the reference configuration is unaffected by the change of frame, in the sense that there is some instant $t_0$ such that $Q(t_0)$ is an identity tensor and hence $F^* = QF$ holds. On the other hand, if we define $\tilde{F}^* = \nabla_{\boldsymbol{X}}\chi_\kappa^*$ (compare with $F^* = \nabla_{\boldsymbol{X}^*}\chi_\kappa^*$ ) then it follows that $\tilde{F}^* = QF$.

With $\dot{F} = LF$ by (1.31), we have

$$L^*F^* = QLFK^T + \dot{Q}FK^T = QLQ^TF^* + \dot{Q}Q^TF^*,$$

and since $F^*$ is non-singular, it gives

$$L^* = QLQ^T + \dot{Q}Q^T = QLQ^T + \Omega. \tag{1.77}$$

Moreover, with $L = D + W$, it becomes

$$D^* + W^* = Q(D + W)Q^T + \Omega,$$

and by separating symmetric and skew-symmetric parts, we obtain

$$D^* = QDQ^T, \qquad W^* = QWQ^T + \Omega. \tag{1.78}$$

Therefore, while the tensors $L$ and $W$ are not objective, the rate of strain tensor $D$ is an objective quantity.

Now, let us consider an objective vector field $\boldsymbol{u}(\boldsymbol{x}, t)$. For any change of frame $*$,

$$\boldsymbol{u}^*(\boldsymbol{x}^*, t^*) = Q(t)\boldsymbol{u}(\boldsymbol{x}, t). \tag{1.79}$$

By taking the gradient with respect to $\boldsymbol{x}$, we obtain

$$(\nabla_{\boldsymbol{x}^*}\boldsymbol{u}^*)(\nabla_{\boldsymbol{x}}\chi^*) = Q\,\nabla_{\boldsymbol{x}}\boldsymbol{u},$$

and by $\nabla_{\boldsymbol{x}}\chi^* = Q(t)$ from (1.65), this implies that

$$(\operatorname{grad}\boldsymbol{u})^* = Q\,(\operatorname{grad}\boldsymbol{u})\,Q^T. \tag{1.80}$$

Therefore, $\operatorname{grad}\boldsymbol{u}$ is an objective tensor field. Similarly, we can show that if $f$ is an objective tensor field of order $n$ then $\operatorname{grad} f$ is an objective tensor field of order $n+1$.

On the other hand, if we express the relation (1.79) in the material coordinate,

$$\boldsymbol{u}^*(\boldsymbol{X}^*, t^*) = Q(t)\boldsymbol{u}(\boldsymbol{X}, t),$$

then by taking the material time derivative and the gradient with respect to $\boldsymbol{X}$, we obtain

$$\dot{\boldsymbol{u}}^* = Q\dot{\boldsymbol{u}} + \dot{Q}\boldsymbol{u}, \qquad (\operatorname{Grad}\boldsymbol{u})^* = Q\,(\operatorname{Grad}\boldsymbol{u})\,K^T, \tag{1.81}$$

where $K = \nabla_{\boldsymbol{X}}\boldsymbol{X}^*$ from (1.73). Therefore, even though $\boldsymbol{u}$ is an objective vector, its material time derivative $\dot{\boldsymbol{u}}$ is not objective, in general, and its referential gradient $\operatorname{Grad}\boldsymbol{u}$ being a second-order tensor, is not objective either.

Similarly, if $\psi$ is an objective scalar field, we have

$$\dot{\psi}^* = \dot{\psi}, \qquad (\operatorname{grad}\psi)^* = Q\ \operatorname{grad}\psi, \qquad (\operatorname{Grad}\psi)^* = K\ \operatorname{Grad}\psi. \tag{1.82}$$

In other words, $\dot{\psi}$ and $\operatorname{grad}\psi$ are objective scalar and vector fields, respectively, while $\operatorname{Grad}\psi$ is not an objective vector field.

**Example 1.7.2** For an objective vector field $\boldsymbol{u}$, if we define

$$\overset{\circ}{\boldsymbol{u}} = \dot{\boldsymbol{u}} - W\boldsymbol{u}, \tag{1.83}$$

where $W$ is the spin tensor of the motion, then $\overset{\circ}{\boldsymbol{u}}$ is objective.

Note that from (1.81) and $(1.78)_2$,

$$\begin{aligned}(\overset{\circ}{\boldsymbol{u}})^* &= \dot{\boldsymbol{u}}^* - W^*\boldsymbol{u}^* \\ &= Q\dot{\boldsymbol{u}} + \dot{Q}\boldsymbol{u} - (QWQ^T + \dot{Q}Q^T)Q\boldsymbol{u} \\ &= Q(\dot{\boldsymbol{u}} - W\boldsymbol{u}) = Q\,\overset{\circ}{\boldsymbol{u}}\,.\end{aligned}$$

This derivative is called the *corotational time derivative*, which measures the time rate of change experienced by material particles rotating along with the motion. Indeed, we can define the corotational time derivative as

$$\overset{\circ}{\boldsymbol{u}}(t) = \lim_{h\to 0}\frac{1}{h}\big(\boldsymbol{u}(t+h) - P(t+h)\boldsymbol{u}(t)\big), \tag{1.84}$$

where the linear transformation $P : V \to V$ is chosen to be the relative rotation of the motion, $P(\tau) = R_t(\tau)$ with respect to the time $t$. Therefore, $R_t(t+h)$ rotates the vector $\boldsymbol{u}(t)$ with the motion to a vector at time $(t+h)$, which can then be compared with the vector $\boldsymbol{u}(t+h)$ at the same instant. Since

$$R_t(t+h) = R_t(t) + \dot{R}_t(t)h + o(2)$$

and from (1.49)

$$R_t(t) = 1, \qquad \dot{R}_t(t) = W(t),$$

we have

$$\overset{\circ}{\boldsymbol{u}}(t) = \lim_{h\to 0}\frac{1}{h}\big(\boldsymbol{u}(t+h) - \boldsymbol{u}(t) - h\,W(t)\boldsymbol{u}(t)\big),$$

and hence (1.83) follows. □

**Exercise 1.7.1** Show that the material time derivative is independent of the frame, by verifying directly in Cartesian components that

$$\frac{\partial}{\partial t} + v_i\frac{\partial}{\partial x_i} = \frac{\partial}{\partial t^*} + v_i^*\frac{\partial}{\partial x_i^*}.$$

**Exercise 1.7.2** Let $\psi$ and $S$ be objective scalar and objective second-order tensor fields, respectively.

1) Show that their spatial gradients, $\operatorname{grad}\psi$ and $\operatorname{grad} S$, are also objective.
2) Show that $\dot{\psi}$ is objective but $\dot{S}$ is not.

**Exercise 1.7.3** Show that the corotational time derivative of an objective tensor field $S$ defined as

$$\overset{\circ}{S}(t) = \lim_{h\to 0} \frac{1}{h}\big(S(t+h) - P(t+h)S(t)P^T(t+h)\big), \tag{1.85}$$

with $P(\tau) = R_t(\tau)$, implies that

$$\overset{\circ}{S} = \dot{S} - WS + SW,$$

and show that it is objective.

**Exercise 1.7.4** Let $\boldsymbol{u}$ and $S$ be objective vector and objective second-order tensor fields, respectively. Show that the *convected time derivatives* defined by

$$\overset{\triangle}{\boldsymbol{u}} = \dot{\boldsymbol{u}} - L\boldsymbol{u},$$

$$\overset{\triangle}{S} = \dot{S} - LS - SL^T,$$

are objective. Show that the convective time derivative can be defined similar to (1.84) and (1.85) with $P(\tau) = F_t(\tau)$, the relative deformation gradient from time $t$ to $\tau$.

**Exercise 1.7.5** Show that

1) the relative right stretch tensor $U_t$ is objective, while the relative rotation tensor $R_t$ is not objective;
2) the Rivlin–Ericksen tensor $A_n$ is objective.

# 2. Balance Laws

## 2.1 General Balance Equation

The basic laws of mechanics can all be expressed, in general, in the following form,

$$\frac{d}{dt}\int_{\mathcal{P}_t} \psi \, dv = \int_{\partial\mathcal{P}_t} \Phi_\psi \boldsymbol{n} \, da + \int_{\mathcal{P}_t} \sigma_\psi \, dv, \tag{2.1}$$

for any bounded regular subregion of the body, called a part $\mathcal{P} \subset \mathcal{B}$ and the vector field $\boldsymbol{n}$, the outward unit normal to the boundary of the region $\mathcal{P}_t$ in the current configuration. The quantities $\psi$ and $\sigma_\psi$ are tensor fields of certain order $m$, and $\Phi_\psi$ is a tensor field of order $m+1$, say $m = 0$ or $m = 1$, so that $\psi$ is a scalar or vector quantity, and, respectively, $\Phi_\psi$ is a vector or second-order tensor quantity.

The relation (2.1), called the *general balance* of $\psi$ in integral form, is interpreted as asserting that the rate of increase of the quantity $\psi$ in a part $\mathcal{P}$ of a body is affected by the inflow of $\psi$ through the boundary of $\mathcal{P}$ and the growth of $\psi$ within $\mathcal{P}$. We call $\Phi_\psi$ the *flux* of $\psi$ and $\sigma_\psi$ the *supply* of $\psi$. In general, the supply $\sigma_\psi$ may contain contributions of supplies from external sources and internal productions due to the motion of the body.

It follows from the balance equation (2.1) that if the body is isolated, i.e., $\Phi_\psi = 0$ on $\partial\mathcal{B}$, and is free from supplies, $\sigma_\psi = 0$, then the total amount of the quantity $\psi$ is constant in time. In other words, the quantity $\psi$ is *conserved* and the equation is called the *conservation law* of $\psi$.

We are interested in the local forms of the integral balance (2.1) at a point in the region $\mathcal{B}_t$. The derivation of local forms rests upon certain assumptions on the smoothness of the tensor fields $\psi$, $\Phi_\psi$, and $\sigma_\psi$. Here not only regular points, where all the tensor fields are smooth, but also singular points, where they may suffer jump discontinuities, will be considered.

First of all, we need the following theorem, which is a three-dimensional version of the formula,

$$\frac{d}{dt}\int_{g(t)}^{f(t)} \psi(x,t)\, dx = \int_{g(t)}^{f(t)} \frac{\partial\psi}{\partial t}\, dx + \psi(f(t),t)\, \dot{f}(t) - \psi(g(t),t)\, \dot{g}(t),$$

in calculus for differentiation under the integral sign on a moving interval. Note that the last two terms are the product of the values of the function $\psi$ and the outward velocity at the end points.

**Theorem 2.1.1** (Transport Theorem). *Let $V(t)$ be a regular region in $\mathcal{E}$ and $u_n(\boldsymbol{x},t)$ be the outward normal speed of a surface point $\boldsymbol{x} \in \partial V(t)$. Then for any smooth tensor field $\psi(\boldsymbol{x},t)$, we have*

$$\frac{d}{dt}\int_V \psi\, dv = \int_V \frac{\partial \psi}{\partial t}\, dv + \int_{\partial V} \psi u_n\, da. \tag{2.2}$$

*Proof.* By definition,

$$\begin{aligned}\frac{d}{dt}\int_V \psi\, dv &= \lim_{h\to 0}\frac{1}{h}\Big\{\int_{V(t+h)} \psi(\boldsymbol{x},t+h)\, dv - \int_{V(t)} \psi(\boldsymbol{x},t)\, dv\Big\}\\ &= \lim_{h\to 0}\frac{1}{h}\Big\{\int_{V(t+h)} \psi(\boldsymbol{x},t+h)\, dv - \int_{V(t)} \psi(\boldsymbol{x},t+h)\, dv\Big\}\\ &\quad + \lim_{h\to 0}\frac{1}{h}\Big\{\int_{V(t)} \psi(\boldsymbol{x},t+h)\, dv - \int_{V(t)} \psi(\boldsymbol{x},t)\, dv\Big\}\\ &= \lim_{h\to 0}\frac{1}{h}\int_{V(t+h)-V(t)} \psi(\boldsymbol{x},t+h)\, dv + \int_{V(t)} \frac{\partial\psi}{\partial t}(\boldsymbol{x},t)\, dv.\end{aligned} \tag{2.3}$$

The region $V(t+h) - V(t)$ is swept out by $\partial V$ in the time interval $(t, t+h)$. In other words, a small volume element $\Delta v$, as shown in Fig. 2.1, is equal to

$$\Delta v = u_n h \Delta a.$$

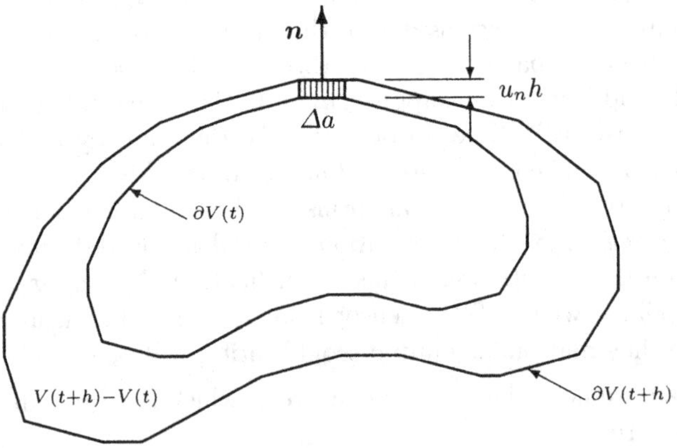


**Fig. 2.1.** Region with moving boundary

Therefore, the first term on the right-hand side of $(2.3)_3$ becomes

$$\begin{aligned}\lim_{h\to 0}\frac{1}{h}\int_{V(t+h)-V(t)}\psi(\boldsymbol{x},t+h)\,dv &= \lim_{h\to 0}\frac{1}{h}\int_{\partial V(t)}\psi(\boldsymbol{x},t+h)u_n(\boldsymbol{x},t)h\,da\\ &= \lim_{h\to 0}\int_{\partial V(t)}\psi(\boldsymbol{x},t+h)u_n(\boldsymbol{x},t)\,da\\ &= \int_{\partial V(t)}\psi(\boldsymbol{x},t)u_n(\boldsymbol{x},t)\,da,\end{aligned}$$

which proves (2.2). □

In this theorem, the surface speed $u_n(\boldsymbol{x},t)$ needs only to be defined on the boundary $\partial V$. If $V(t)$ is a material region $\mathcal{P}_t$, i.e., it always consists of the same material points of a part $\mathcal{P}\subset\mathcal{B}$, then $u_n=\dot{\boldsymbol{x}}\cdot\boldsymbol{n}$ and (2.2) becomes

$$\frac{d}{dt}\int_{\mathcal{P}_t}\psi\,dv=\int_{\mathcal{P}_t}\frac{\partial\psi}{\partial t}\,dv+\int_{\partial\mathcal{P}_t}\psi\,\dot{\boldsymbol{x}}\cdot\boldsymbol{n}\,da. \tag{2.4}$$

Now we shall extend the above transport theorem to a material region containing a surface across which $\psi$ may suffer a jump discontinuity.

An oriented smooth surface $\mathcal{S}$ in a material region $\mathcal{V}$ is called a *singular surface* relative to a field $\psi$ defined on $\mathcal{V}$, if $\psi$ is smooth in $\mathcal{V}-\mathcal{S}$ and suffers a jump discontinuity across $\mathcal{S}$. The *jump* of $\psi$ is defined as

$$[\![\psi]\!]=\psi^+-\psi^-, \tag{2.5}$$

where $\psi^+$ and $\psi^-$ are the one-side limits from the two regions of $\mathcal{V}$ separated by $\mathcal{S}$ and designated as $\mathcal{V}^+$ and $\mathcal{V}^-$, respectively.

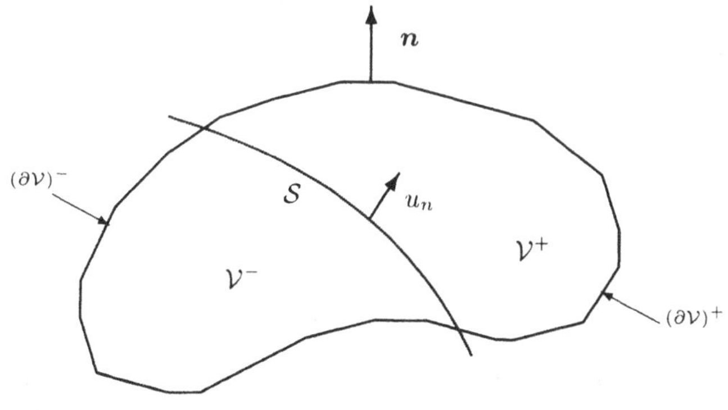


**Fig. 2.2.** Singular surface

Let $u_n$ be the normal speed of $\mathcal{S}$ with the direction pointing into $\mathcal{V}^+$ and $\boldsymbol{n}$ be the outward unit normal of $\partial\mathcal{V}$ (Fig. 2.2). Denote

$$(\partial\mathcal{V})^{\pm} = \partial\mathcal{V}^{\pm} \cap \partial\mathcal{V}.$$

Since both $(\partial\mathcal{V})^+$ and $(\partial\mathcal{V})^-$ are material surfaces, their normal surface speed is $\dot{\boldsymbol{x}} \cdot \boldsymbol{n}$. Clearly,

$$\mathcal{V} = \mathcal{V}^+ \cup \mathcal{V}^-, \quad \partial\mathcal{V} = (\partial\mathcal{V})^+ \cup (\partial\mathcal{V})^-. \tag{2.6}$$

Since $\mathcal{S}$ need not be a material surface, $\mathcal{V}^+$ and $\mathcal{V}^-$ need not be material regions in general.

Suppose that $\psi(\boldsymbol{x}, t)$ and $\dot{\boldsymbol{x}}(\boldsymbol{x}, t)$ are smooth in $\mathcal{V}^+$ and $\mathcal{V}^-$, then the transport theorem (2.2) implies that

$$\frac{d}{dt}\int_{\mathcal{V}^+} \psi\, dv = \int_{\mathcal{V}^+} \frac{\partial\psi}{\partial t}\, dv + \int_{(\partial\mathcal{V})^+} \psi\, \dot{\boldsymbol{x}} \cdot \boldsymbol{n}\, da + \int_{\mathcal{S}} \psi^+(-u_n)\, da, \tag{2.7}$$

$$\frac{d}{dt}\int_{\mathcal{V}^-} \psi\, dv = \int_{\mathcal{V}^-} \frac{\partial\psi}{\partial t}\, dv + \int_{(\partial\mathcal{V})^-} \psi\, \dot{\boldsymbol{x}} \cdot \boldsymbol{n}\, da + \int_{\mathcal{S}} \psi^- u_n\, da. \tag{2.8}$$

Adding (2.7) and (2.8), we obtain, by the use of (2.6) the following transport theorem in a material region containing a singular surface:

**Theorem 2.1.2.** *Let $\mathcal{V}(t)$ be a material region in $\mathcal{E}$ and $\mathcal{S}(t)$ be a singular surface relative to the tensor field $\psi(\boldsymbol{x}, t)$ that is smooth elsewhere. Then we have*

$$\frac{d}{dt}\int_{\mathcal{V}} \psi\, dv = \int_{\mathcal{V}} \frac{\partial\psi}{\partial t}\, dv + \int_{\partial\mathcal{V}} \psi\, \dot{\boldsymbol{x}} \cdot \boldsymbol{n}\, da - \int_{\mathcal{S}} [\![\psi]\!]\, u_n\, da, \tag{2.9}$$

*where $u_n(\boldsymbol{x}, t)$ is the normal speed of a surface point $\boldsymbol{x} \in \mathcal{S}(t)$ and $[\![\psi]\!]$ is the jump of $\psi$ across $\mathcal{S}$.*

**Example 2.1.1** Let $V(t)$ be the volume of a part $\mathcal{P} \subset \mathcal{B}$ in a motion, then

$$V(t) = \int_{\mathcal{P}_t} dv,$$

and by taking $\psi = 1$ in (2.9), it follows that

$$\frac{dV}{dt} = \int_{\partial\mathcal{P}_t} \dot{\boldsymbol{x}} \cdot \boldsymbol{n}\, da = \int_{\mathcal{P}_t} \operatorname{div} \dot{\boldsymbol{x}}\, dv. \tag{2.10}$$

Therefore, if the motion is *incompressible*, i.e., the volume of any part $\mathcal{P}$ in $\mathcal{B}$ remains constant in the motion, then the divergence of the velocity must vanish,

$$\operatorname{div} \dot{\boldsymbol{x}} = 0.$$

□

### 2.1.1 Field Equation and Jump Condition

For a material region $\mathcal{V}$ containing a singular surface $\mathcal{S}$, the equation of general balance in integral form (2.1) becomes

$$\int_{\mathcal{V}} \frac{\partial \psi}{\partial t}\, dv + \int_{\partial \mathcal{V}} \psi\, \dot{\boldsymbol{x}} \cdot \boldsymbol{n}\, da - \int_{\mathcal{S}} [\![\psi]\!]\, u_n\, da = \int_{\partial \mathcal{V}} \boldsymbol{\Phi}_{\psi} \boldsymbol{n}\, da + \int_{\mathcal{V}} \sigma_{\psi}\, dv. \quad (2.11)$$

A point $\boldsymbol{x}$ is called *regular* if there is a material region containing $\boldsymbol{x}$ in which all the tensor fields in (2.1) are smooth. And a point $\boldsymbol{x}$ is called *singular* if it is a point on a singular surface relative to $\psi$ and $\boldsymbol{\Phi}_{\psi}$.

We can obtain the local balance equation at a regular point as well as at a singular point from the above integral equation. First, we consider a small material region $\mathcal{V}$ containing $\boldsymbol{x}$, such that $\mathcal{V} \cap \mathcal{S} = \emptyset$. By the use of the divergence theorem, (2.11) becomes

$$\int_{\mathcal{V}} \left\{ \frac{\partial \psi}{\partial t} + \operatorname{div}(\psi \otimes \dot{\boldsymbol{x}} - \boldsymbol{\Phi}_{\psi}) - \sigma_{\psi} \right\} dv = 0. \quad (2.12)$$

Since the integrand is smooth and (2.12) holds for any $\mathcal{V}$, such that $\boldsymbol{x} \in \mathcal{V}$, and $\mathcal{V} \cap \mathcal{S} = \emptyset$, the integrand must vanish at $\boldsymbol{x}$ (see the proposition on p. 280). Therefore, we obtain the balance equation at a regular point, usually called the field equation.

**Field equation.** *At a regular point* $\boldsymbol{x}$, *the general balance equation* (2.11) *reduces to*

$$\frac{\partial \psi}{\partial t} + \operatorname{div}(\psi \otimes \dot{\boldsymbol{x}} - \boldsymbol{\Phi}_{\psi}) - \sigma_{\psi} = 0. \quad (2.13)$$

The quantity $\psi \otimes \dot{\boldsymbol{x}}$ in (2.13) is called the *convective flux* of $\psi$ and this notation should be understood as $\psi \dot{\boldsymbol{x}}$ when $\psi$ is a scalar quantity.

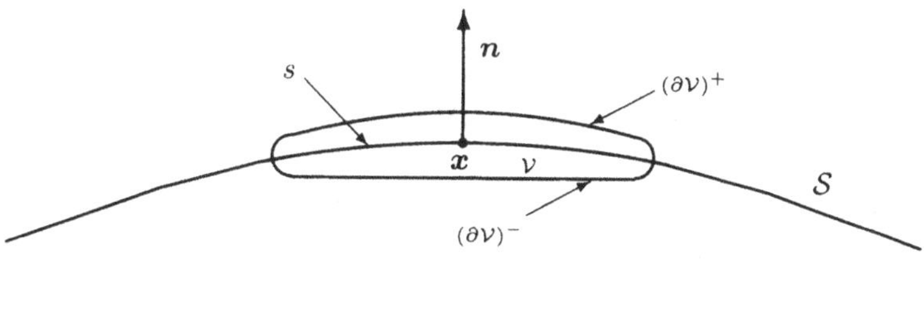


**Fig. 2.3.** At a singular point

Next, we consider a singular point $\boldsymbol{x}$, i.e., $\boldsymbol{x} \in \mathcal{S}$. Let $\mathcal{V}$ be an arbitrary material region around $\boldsymbol{x}$, and $s = \mathcal{V} \cap \mathcal{S}$ (Fig. 2.3). We shall take the limit by shrinking $(\partial\mathcal{V})^+$ and $(\partial\mathcal{V})^-$ down to $s$ in such a way that the volume of $\mathcal{V}$ tends to zero, while the area of $s$ remains unchanged. If $\partial_t\psi$ and $\sigma_\psi$ are bounded in $\mathcal{V}$ then the volume integrals vanish in the limit and (2.11) becomes

$$\int_s \{[\![\psi\,(\dot{\boldsymbol{x}}\cdot\boldsymbol{n}-u_n)]\!]-[\![\Phi_\psi]\!]\boldsymbol{n}\}\,da=0. \tag{2.14}$$

Since the integrand is smooth on $s$ and (2.14) holds for any $s$ containing $\boldsymbol{x}$, the integrand must vanish at $\boldsymbol{x}$. We obtain

**Jump condition.** *At a singular point $\boldsymbol{x}$, the general balance equation* (2.11) *reduces to*

$$[\![\psi\,(\dot{\boldsymbol{x}}\cdot\boldsymbol{n}-u_n)]\!]-[\![\Phi_\psi]\!]\boldsymbol{n}=0, \tag{2.15}$$

*if, in addition,* $\dfrac{\partial\psi}{\partial t}$ *and* $\sigma_\psi$ *are bounded in the neighborhood of* $\boldsymbol{x}$.

The jump condition (2.15) is also known as the *Rankine–Hugoniot equation.* It can also be written as

$$[\![\psi\,U]\!]+[\![\Phi_\psi]\!]\,\boldsymbol{n}=0, \tag{2.16}$$

where

$$U^\pm=u_n-\dot{\boldsymbol{x}}^\pm\cdot\boldsymbol{n} \tag{2.17}$$

are called the *local speeds of propagation* of $\mathcal{S}$ relative to the motion of the body. If $\mathcal{S}$ is a material surface then $U^\pm=0$.

### 2.1.2 Balance Equations in Material Coordinates

Sometimes, for solid bodies, it is more convenient to use the material description. The corresponding relations for the balance equation (2.13) and the jump condition (2.15) in the reference coordinates can be derived in a similar manner. We begin with the integral form (2.1), now written in the reference configuration $\kappa$,

$$\frac{d}{dt}\int_{\mathcal{P}_\kappa}\psi_\kappa\,dv_\kappa=\int_{\partial\mathcal{P}_\kappa}\Phi^\psi_\kappa\boldsymbol{n}_\kappa\,da_\kappa+\int_{\mathcal{P}_\kappa}\sigma^\psi_\kappa\,dv_\kappa. \tag{2.18}$$

In view of the relations for volume elements and surface elements (1.9), the corresponding quantities are defined as

$$\psi_\kappa=|J|\,\psi,\quad \Phi^\psi_\kappa=|J|\,\Phi_\psi F^{-T},\quad \sigma^\psi_\kappa=|J|\,\sigma_\psi. \tag{2.19}$$

The transport theorem (2.2) remains valid for $\psi_\kappa(\boldsymbol{X},t)$ in a movable region $V(t)$,

$$\frac{d}{dt}\int_V\psi_\kappa dv_\kappa=\int_V\dot{\psi}_\kappa dv_\kappa+\int_{\partial V}\psi_\kappa U_\kappa da_\kappa, \tag{2.20}$$

where $U_\kappa(\boldsymbol{X},t)$ is the outward normal speed of a surface point $\boldsymbol{X} \in \partial V(t)$. Therefore, for a singular surface $\mathcal{S}_\kappa(t)$ moving across a material region $\mathcal{V}_\kappa$ in the reference configuration, by a similar argument, we can obtain from (2.11)

$$\int_{\mathcal{V}_\kappa} \dot{\psi}_\kappa dv_\kappa - \int_{\mathcal{S}_\kappa} [\![\psi_\kappa]\!] U_\kappa da_\kappa = \int_{\partial\mathcal{V}_\kappa} \Phi^\psi_\kappa \boldsymbol{n}_\kappa da_\kappa + \int_{\mathcal{V}_\kappa} \sigma^\psi_\kappa dv_\kappa, \tag{2.21}$$

where $U_\kappa$ is the normal speed of the surface points on $\mathcal{S}_\kappa$, while the normal speed of the surface points on $\partial\mathcal{V}_\kappa$ is zero, since a material region is a fixed region in the reference configuration. From this equation, we obtain the local balance equation and the jump condition in the reference configuration,

$$\dot{\psi}_\kappa - \operatorname{Div} \Phi^\psi_\kappa - \sigma^\psi_\kappa = 0, \qquad [\![\psi_\kappa]\!] U_\kappa + [\![\Phi^\psi_\kappa]\!] \boldsymbol{n}_\kappa = 0, \tag{2.22}$$

where $U_\kappa$and $\boldsymbol{n}_\kappa$ are the normal speed and the unit normal vector of the singular surface.

**Example 2.1.2** Another way of obtaining the balance equation $(2.22)_1$ in the reference configuration is to derive from (2.13) directly. Multiplying (2.13) by $J$, with the definition (2.19), we obtain immediately the desired equation $(2.22)_1$ by the use of the following relations:

$$\dot{\psi}_\kappa = J\Big(\frac{\partial\psi}{\partial t} + \operatorname{div}(\psi \otimes \dot{\boldsymbol{x}})\Big), \qquad \operatorname{Div} \Phi^\psi_\kappa = J \operatorname{div} \Phi_\psi. \tag{2.23}$$

The first relation follows from $(1.29)_1$ and

$$\dot{J} = J\, F^{-T} \cdot \dot{F} = J \operatorname{div} \dot{\boldsymbol{x}}$$

from the use of (A.48) and (1.31). The second relation is an identity. By taking $\Phi^\psi_\kappa$ to be a vector quantity $\boldsymbol{u}$, we have

$$\begin{aligned} J \operatorname{div}(J^{-1} F \boldsymbol{u}) &= J\big(\boldsymbol{u} \cdot \operatorname{div}(J^{-1} F^T) + \operatorname{tr}(J^{-1} F \operatorname{grad} \boldsymbol{u})\big) \\ &= \operatorname{tr}(\operatorname{Grad} \boldsymbol{u}) = \operatorname{Div} \boldsymbol{u}, \end{aligned}$$

by the successive use of $(A.76)_2$, $(1.33)_2$ and $(1.30)_2$. If this result is applied to $\boldsymbol{u} = S^T \boldsymbol{v}$ for any constant vector $\boldsymbol{v}$, the identity $(2.23)_2$ is proved for a tensor quantity $\Phi^\psi_\kappa = S$.

**Example 2.1.3** *Kinematic compatibility condition*

From the deformation gradient $F = \operatorname{Grad} \chi_\kappa$ and the velocity $\dot{\boldsymbol{x}} = \partial_t \chi_\kappa$, the existence of the deformation function $\chi_\kappa(\boldsymbol{X},t)$ requires the following integrability condition:

$$\dot{F} = \operatorname{Grad} \dot{\boldsymbol{x}}. \tag{2.24}$$

If we integrate the above equation over an arbitrary part $\mathcal{P}_\kappa$ of the body in the reference configuration $\kappa$ and use the divergence theorem (see Exercise A.2.7), we obtain

$$\frac{d}{dt}\int_{\mathcal{P}_\kappa} F\, dv_\kappa = \int_{\partial\mathcal{P}_\kappa} \dot{\boldsymbol{x}} \otimes \boldsymbol{n}_\kappa\, da_\kappa,$$

which can be regarded as a special case of (2.18). If the region contains a singular surface, then at a singular point, it follows from the jump condition $(2.22)_2$ that

$$U_\kappa [\![F]\!] + [\![\dot{\boldsymbol{x}}]\!] \otimes \boldsymbol{n}_\kappa = 0. \tag{2.25}$$

This equation is usually called the kinematic compatibility condition at the singular surface. It is equivalent to the following relations:

$$[\![\dot{\boldsymbol{x}}]\!] = -U_\kappa \boldsymbol{a}, \qquad [\![F]\!] = \boldsymbol{a} \otimes \boldsymbol{n}_\kappa, \tag{2.26}$$

where $\boldsymbol{a} = [\![F\boldsymbol{n}_\kappa]\!]$ is the normal jump of $F$ at the singular surface. □

**Exercise 2.1.1** Let a moving singular surface $\mathcal{S}(t)$ in the reference configuration $\kappa$ be characterized by the condition $f(\boldsymbol{X}, t) = 0$.

1) Establish the following relations concerning the unit normals and the normal speeds in the reference and the spatial configurations as given in $(2.22)_2$ and (2.15):

$$\begin{aligned} \boldsymbol{n}_\kappa &= \frac{\operatorname{Grad} f}{|\operatorname{Grad} f|}, & U_\kappa &= -\frac{\dot{f}}{|\operatorname{Grad} f|}, \\ \boldsymbol{n} &= \frac{\operatorname{grad} f}{|\operatorname{grad} f|}, & u_n &= -\frac{\partial f/\partial t}{|\operatorname{grad} f|}. \end{aligned} \tag{2.27}$$

2) The following relations hold:

$$\boldsymbol{n}_\kappa = \frac{F^T\boldsymbol{n}}{|F^T\boldsymbol{n}|}, \qquad U_\kappa = \frac{U}{|F^T\boldsymbol{n}|}, \tag{2.28}$$

where $U$ is the local speed in the spatial configuration defined by (2.17).

## 2.2 Conservation of Mass

In continuum mechanics, we assume that for any configuration $\kappa$ of a body $\mathcal{B}$, there exists a positive integrable function

$$\rho_\kappa : \mathcal{B}_\kappa \to I\!R^+,$$

such that for any part $\mathcal{P} \subset \mathcal{B}$,

$$M(\mathcal{P}) = \int_{\mathcal{P}_\kappa} \rho_\kappa dv_\kappa. \tag{2.29}$$

We call $M(\mathcal{P})$ the *mass* of the part $\mathcal{P}$, and $M$ the *mass distribution* of $\mathcal{B}$. $\rho_\kappa$ is called the *mass density* in the configuration $\kappa$. We assume that the mass distribution, and consequently the mass density, is a frame-indifferent quantity.

By assuming the existence of a mass-density function, we are interested in bodies that can be regarded as *continuous media* only. Discrete mass distributions, such as concentrated masses, will not be considered in this book.

Since mass density depends on the configuration, let $\kappa_1$ and $\kappa_2$ be two configurations, and denote

$$\lambda = \kappa_2 \circ \kappa_1^{-1} : \mathcal{B}_{\kappa_1} \to \mathcal{B}_{\kappa_2},$$

the change of configuration from $\kappa_1$ to $\kappa_2$. We then have

$$M(\mathcal{P}) = \int_{\mathcal{P}_{\kappa_1}} \rho_{\kappa_1} dv_{\kappa 1} = \int_{\mathcal{P}_{\kappa_2}} \rho_{\kappa_2} dv_{\kappa_2},$$

which implies that

$$\int_{\mathcal{P}_{\kappa_1}} \rho_{\kappa_1} dv_{\kappa_1} = \int_{\mathcal{P}_{\kappa_1}} \rho_{\kappa_2} |\det \nabla\lambda|\, dv_{\kappa_1}.$$

Since this is valid for any part $\mathcal{P} \subset \mathcal{B}$, it follows that

$$\rho_{\kappa_1} = |\det \nabla\lambda|\, \rho_{\kappa_2}. \tag{2.30}$$

Thus, the density in one configuration determines the density in any other configuration.

Now, consider a motion $\chi$ of a body $\mathcal{B}$ and let $\rho(\boldsymbol{x}, t)$ be the mass density in the current configuration. Since the material is neither destroyed nor created in any motion, in the absence of chemical reactions, we have

**Conservation of mass.** *The total mass of any part $\mathcal{P} \subset \mathcal{B}$ does not change in any motion,*

$$\frac{d}{dt} M(\mathcal{P}) = \frac{d}{dt} \int_{\mathcal{P}_t} \rho dv = 0. \tag{2.31}$$

By comparison, it is a special case of the general balance equation (2.1) with no flux and no supply,

$$\psi = \rho, \qquad \Phi_\psi = 0, \qquad \sigma_\psi = 0,$$

and hence from (2.13) and (2.15) we obtain the local expressions for mass conservation and the jump condition at a singular point,

$$\frac{\partial \rho}{\partial t} + \operatorname{div}(\rho\, \dot{\boldsymbol{x}}) = 0, \qquad [\![\rho\,(\dot{\boldsymbol{x}} \cdot \boldsymbol{n} - u_n)]\!] = 0. \tag{2.32}$$

The equation (2.31) states that the total mass of any part is constant in time. In particular, we have

$$M(\mathcal{P}) = \int_{\mathcal{P}_\kappa} \rho_\kappa dv_\kappa = \int_{\mathcal{P}_t} \rho\, dv, \tag{2.33}$$

which implies that

$$\rho_\kappa(\boldsymbol{X}) = \rho(\boldsymbol{X}, t)|J(\boldsymbol{X}, t)|,$$

or

$$\rho = \frac{\rho_\kappa}{|\det F|}. \tag{2.34}$$

This is another form of the conservation of mass in the material description, which also follows from the general expression (2.22) and (2.19). This equation is equivalent to

$$\frac{d}{dt}(\rho J) = 0,$$

which, by the use of the relation $\partial_F J = J F^{-T}$ (see (A.48)) and (1.31), can be written as

$$\dot{\rho} + \rho \operatorname{div} \dot{\boldsymbol{x}} = 0. \tag{2.35}$$

In fluid mechanics, this equation is also referred to as the *equation of continuity.*

**Exercise 2.2.1** Show that the mass conservation (2.35) implies that for any tensor quantity $\psi$,

$$\rho \dot{\psi} = \frac{\partial \rho \psi}{\partial t} + \operatorname{div}(\rho \psi \otimes \boldsymbol{v}). \tag{2.36}$$

**Exercise 2.2.2** Let $\psi(\boldsymbol{x}, t)$ be an arbitrary function. Show that for any $\mathcal{P} \subset \mathcal{B}$,

$$\frac{d}{dt} \int_{\mathcal{P}_t} \psi \rho\, dv = \int_{\mathcal{P}_t} \dot{\psi} \rho\, dv, \tag{2.37}$$

provided that the mass is conserved. This is another version of the transport formula (2.4).

## 2.3 Laws of Dynamics

Let $\mathcal{X}$ be a motion of $\mathcal{B}$. We define the *linear momentum* $\boldsymbol{P}$ of a part $\mathcal{P} \subset \mathcal{B}$ in the motion $\mathcal{X}_t$ by

$$\boldsymbol{P}(\mathcal{P},t) = \int_{\mathcal{P}_t} \rho\, \dot{\boldsymbol{x}}\, dv, \tag{2.38}$$

which is a vector quantity, and the *angular momentum* $\boldsymbol{H}_{\boldsymbol{x}_\circ}$ of $\mathcal{P}$ with respect to $\boldsymbol{x}_\circ \in \mathcal{E}$ in the motion $\mathcal{X}_t$ by

$$\boldsymbol{H}_{\boldsymbol{x}_\circ}(\mathcal{P},t) = \int_{\mathcal{P}_t} \rho(\boldsymbol{x} - \boldsymbol{x}_\circ) \wedge \dot{\boldsymbol{x}}\, dv. \tag{2.39}$$

The angular momentum, also called the *moment of momentum*, is a skew-symmetric tensor quantity, and hence, more commonly, it is represented as an axial vector quantity. Therefore, the operator $\wedge$ will be regarded either as the exterior tensor product or as the vector product, depending on the interpretation.

In postulating the laws of dynamics, we follow the classical approach established by Newton and Euler, according to which motions are produced by the action of *forces* and *moments* upon the body. We denote

| | |
|---|---|
| $\boldsymbol{f}(\mathcal{P},t)$ | total force acting on $\mathcal{P} \subset \mathcal{B}$ at time $t$, |
| $\boldsymbol{m}_{\boldsymbol{x}_\circ}(\mathcal{P},t)$ | total moment acting on $\mathcal{P} \subset \mathcal{B}$ with respect to $\boldsymbol{x}_\circ$ at time $t$. |

The force is a vector quantity, while the moment is an axial vector or a skew-symmetric tensor quantity. The properties of forces and moments will be discussed later.

**Laws of dynamics.**

1) *There exists a frame of reference $\phi$, called an inertial frame, such that for any motion relative to which if $\boldsymbol{f}(\mathcal{P},t) = 0$ then $\dot{\boldsymbol{P}}(\mathcal{P},t) = 0$ for any $\mathcal{P} \subset \mathcal{B}$.*
2) *For any motion relative to an inertial frame,*

$$\dot{\boldsymbol{P}}(\mathcal{P},t) = \boldsymbol{f}(\mathcal{P},t), \tag{2.40}$$

$$\dot{\boldsymbol{H}}_{\boldsymbol{x}_\circ}(\mathcal{P},t) = \boldsymbol{m}_{\boldsymbol{x}_\circ}(\mathcal{P},t). \tag{2.41}$$

Equations (2.40) and (2.41) are called *Euler's first* and *second* laws, respectively. Note that from (2.37)

$$\begin{aligned} \dot{\boldsymbol{P}}(\mathcal{P},t) &= \int_{\mathcal{P}_t} \rho\, \ddot{\boldsymbol{x}}\, dv, \\ \dot{\boldsymbol{H}}_{\boldsymbol{x}_0}(\mathcal{P},t) &= \int_{\mathcal{P}_t} \rho(\boldsymbol{x} - \boldsymbol{x}_0) \wedge \ddot{\boldsymbol{x}}\, dv, \end{aligned} \tag{2.42}$$

provided that the mass is conserved.

Since the velocity is not frame indifferent, it is easy to see that both $\boldsymbol{P}$ and $\boldsymbol{H}$ are not frame indifferent and hence Euler's laws depend on the frame of reference. Indeed, the first part of the laws of dynamics assures the existence of inertial frames (an equivalent of Newton's first law), relative to which Euler's laws take the above simple forms – the motion has a preferred frame. In view of the results from Sect. 1.7.1 that acceleration is invariant with respect to Galilean transformation, the class of inertial frames can be characterized by the following proposition:

**Proposition 2.3.1** *Let $\phi$ be an inertial frame and assume that the force $\boldsymbol{f}$ is a frame-indifferent vector. Then a frame $\phi^*$ is an inertial frame if and only if $\phi$ and $\phi^*$ are related by a Galilean transformation.*

As a consequence, we have the following

**Proposition 2.3.2** *The laws of dynamics relative to inertial frames are Galilean invariant.*

In other words, the dynamic laws have the same expressions in any frame related to an inertial frame by a Galilean transformation.

### 2.3.1 Forces and Moments

Forces and moments will be treated as primitive concepts of mechanics. We assume that the total force $\boldsymbol{f}(\mathcal{P},t)$ acting on $\mathcal{P} \subset \mathcal{B}$, consists of two parts: the *body force* $\boldsymbol{f}^b(\mathcal{P},t)$, which acts within the body, and the *contact force* $\boldsymbol{f}^c(\mathcal{P},t)$, which acts on the boundary of the body,

$$\boldsymbol{f}(\mathcal{P},t) = \boldsymbol{f}^b(\mathcal{P},t) + \boldsymbol{f}^c(\mathcal{P},t), \tag{2.43}$$

and they can be represented in the configuration $\mathcal{B}_t$ as

$$\begin{aligned} \boldsymbol{f}^b(\mathcal{P},t) &= \int_{\mathcal{P}_t} \rho\, \boldsymbol{b}\, dv, \\ \boldsymbol{f}^c(\mathcal{P},t) &= \int_{\partial\mathcal{P}_t} \boldsymbol{t}\, da, \end{aligned} \tag{2.44}$$

where $\boldsymbol{b}$ is called the *body force density* (per unit mass), and $\boldsymbol{t}$ is called the *surface traction* (per unit surface area). Both $\boldsymbol{b}$ and $\boldsymbol{t}$ are assumed to be frame-indifferent vector quantities.

We shall also assume that the total moment $\boldsymbol{m}_{\boldsymbol{x}_\circ}(\mathcal{P},t)$ acting on $\mathcal{P}$ with respect to $\boldsymbol{x}_\circ \in \mathcal{E}$ can be represented by

$$\boldsymbol{m}_{\boldsymbol{x}_\circ}(\mathcal{P},t) = \int_{\mathcal{P}_t} (\boldsymbol{x} - \boldsymbol{x}_\circ) \wedge \rho\, \boldsymbol{b}\, dv + \int_{\partial\mathcal{P}_t} (\boldsymbol{x} - \boldsymbol{x}_\circ) \wedge \boldsymbol{t}\, da. \tag{2.45}$$

In other words, only the moments due to forces are considered. Indeed, by adopting the definitions (2.39) and (2.45), we have essentially neglected the effect of any spin of material particles, caused by polarized molecular structures, for example. Materials with microstructures are called *polar materials.* We shall be concerned with non-polar materials only.

Euler's first and second laws, (2.40) and (2.41), can now be written as

$$\begin{aligned} &\frac{d}{dt}\int_{\mathcal{P}_t} \rho\, \dot{\boldsymbol{x}}\, dv = \int_{\mathcal{P}_t} \rho\, \boldsymbol{b}\, dv + \int_{\partial\mathcal{P}_t} \boldsymbol{t}\, da, \\ &\frac{d}{dt}\int_{\mathcal{P}_t} \rho(\boldsymbol{x}-\boldsymbol{x}_\circ)\wedge \dot{\boldsymbol{x}}\, dv = \int_{\mathcal{P}_t} (\boldsymbol{x}-\boldsymbol{x}_\circ)\wedge \rho\, \boldsymbol{b}\, dv + \int_{\partial\mathcal{P}_t} (\boldsymbol{x}-\boldsymbol{x}_\circ)\wedge \boldsymbol{t}\, da. \end{aligned} \tag{2.46}$$

### 2.3.2 Stress Tensor

The body force $\boldsymbol{b}$ is regarded as an external force due to sources outside the body, such as the gravitational force. In saying so, we have tacitly assumed that the interaction between different parts of the body (the internal forces) is manifested through the contact forces alone. Therefore, we shall assume that the body force density $\boldsymbol{b} = \boldsymbol{b}(\boldsymbol{x}, t)$ is a function of position and time only. On the other hand, at an interior point $\boldsymbol{x} \in \partial\mathcal{P}_t$, the surface traction $\boldsymbol{t}$ is the force exerted on the part $\mathcal{P}$ by the remaining part of the body through the surface $\partial\mathcal{P}_t$ at $\boldsymbol{x}$. For two different parts with a common surface point at $\boldsymbol{x}$, the surface tractions $\boldsymbol{t}$ at $\boldsymbol{x}$ on the two different surfaces are, in general, different. In other words, the surface traction $\boldsymbol{t}$ at $\boldsymbol{x}$ depends, in general, on the surface $\partial\mathcal{P}_t$ on which $\boldsymbol{x}$ lies. The basic assumption for simplifying the dependence of the traction on $\partial\mathcal{P}$ in classical continuum mechanics states that the surface tractions on all like-oriented surfaces with a common tangent plane at $\boldsymbol{x}$ are the same (see Fig. 2.4). This is known as *Cauchy's Postulate.*

**Postulate** (Cauchy). *If $\boldsymbol{x} \in \partial\mathcal{P}_t \cap \partial\overline{\mathcal{P}}_t$, and $\partial\mathcal{P}_t$ and $\partial\overline{\mathcal{P}}_t$ have a common oriented normal at $\boldsymbol{x}$, then*

$$\boldsymbol{t}(\boldsymbol{x}, t, \partial\mathcal{P}_t) = \boldsymbol{t}(\boldsymbol{x}, t, \partial\overline{\mathcal{P}}_t). \tag{2.47}$$

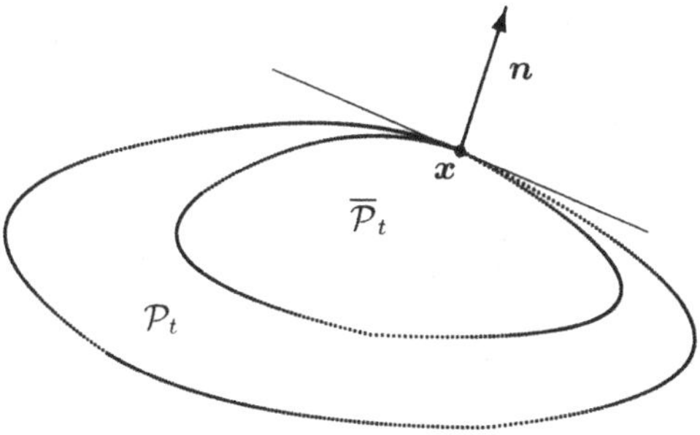


**Fig. 2.4.** Surfaces with common tangent plane

In other words, Cauchy's postulate asserts that $\boldsymbol{t}$ depends on $\partial\mathcal{P}_t$ only through the oriented unit normal $\boldsymbol{n}$ of $\partial\mathcal{P}_t$ at $\boldsymbol{x}$,

$$\boldsymbol{t}(\boldsymbol{x}, t, \partial\mathcal{P}_t) = \boldsymbol{t}(\boldsymbol{x}, t, \boldsymbol{n}). \tag{2.48}$$

An immediate consequence of this postulate is the following theorem.

**Theorem 2.3.3** (Cauchy's lemma). *Suppose that $\boldsymbol{t}(\cdot, \boldsymbol{n})$ is a continuous function of $\boldsymbol{x}$, and $\ddot{\boldsymbol{x}}$, $\boldsymbol{b}$ are bounded in $\mathcal{B}_t$. Then Euler's first law implies that*

$$\boldsymbol{t}(\boldsymbol{x}, -\boldsymbol{n}) = -\boldsymbol{t}(\boldsymbol{x}, \boldsymbol{n}), \tag{2.49}$$

*for any $\boldsymbol{x} \in \mathcal{B}_t$ and any unit vector $\boldsymbol{n} \in V$.*

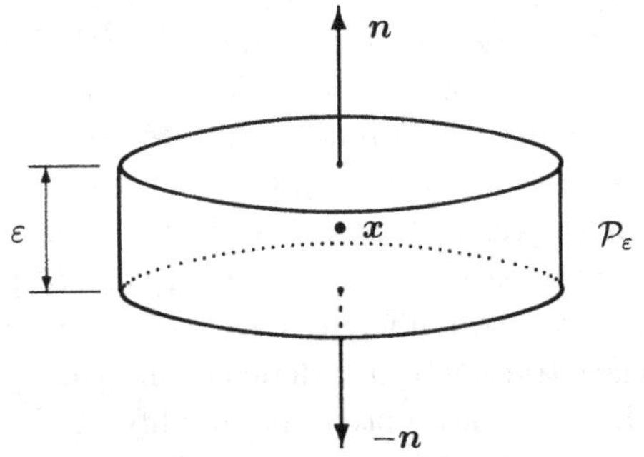


**Fig. 2.5.** Pillbox argument

*Proof.* For any $\boldsymbol{x} \in \mathcal{B}_t$ and any unit vector $\boldsymbol{n} \in V$. We consider a small pillbox $\mathcal{P}_\epsilon \subset \mathcal{B}_t$ of thickness $\epsilon$ and centered at $\boldsymbol{x}$, such that one of its flat surfaces is normal to $\boldsymbol{n}$ (Fig. 2.5). Applying Euler's first law to the pillbox, we obtain

$$\int_{\mathcal{P}_\epsilon} \rho\, \ddot{\boldsymbol{x}}\, dv = \int_{\mathcal{P}_\epsilon} \rho\, \boldsymbol{b}\, dv + \int_{\partial\mathcal{P}_\epsilon} \boldsymbol{t}(\boldsymbol{n})\, da.$$

If we flatten the pillbox down to the middle surface $\mathcal{S}$ then, since $\rho\ddot{\boldsymbol{x}}$ and $\rho\boldsymbol{b}$ are both bounded, the two volume integrals will approach zero, while the surface integral remains finite. Therefore we have

$$\lim_{\epsilon \to 0} \int_{\partial\mathcal{P}_\epsilon} \boldsymbol{t}(\boldsymbol{n})\, da = 0,$$

which by our construction reduces to

$$\int_{\mathcal{S}} \left(\boldsymbol{t}(\boldsymbol{n}) + \boldsymbol{t}(-\boldsymbol{n})\right) da = 0.$$

Since the surface integral along the circular boundary of the pillbox approaches zero as $\epsilon \to 0$, and the two flat ends have exterior normals $\boldsymbol{n}$ and $-\boldsymbol{n}$, respectively. Finally, the continuity of the integrand implies that it must vanish identically at $\boldsymbol{x}$,

$$\boldsymbol{t}(\boldsymbol{x}, \boldsymbol{n}) + \boldsymbol{t}(\boldsymbol{x}, -\boldsymbol{n}) = 0,$$

which proves the lemma. □

This theorem, usually known as *Cauchy's Fundamental Lemma*, an equivalent of Newton's law of action and reaction, ensures the existence of a stress tensor, as shown in the following theorem.

**Theorem 2.3.4** (Cauchy's Theorem). *Under the same conditions assumed in Cauchy's lemma, there exists a second-order tensor field $T$, such that*

$$\boldsymbol{t}(\boldsymbol{x}, t, \boldsymbol{n}) = T(\boldsymbol{x}, t)\, \boldsymbol{n}. \tag{2.50}$$

*Proof.* From (2.48), for fixed $\boldsymbol{x} \in \mathcal{B}_t$, $\boldsymbol{t}$ is a function defined on the set of unit vectors in $V$. We can extend this function to the whole space, $\boldsymbol{t}(\,\cdot\,) : V \to V$, in the following manner,

$$\boldsymbol{t}(\boldsymbol{v}) = \begin{cases} |\boldsymbol{v}|\; \boldsymbol{t}\big(\dfrac{\boldsymbol{v}}{|\boldsymbol{v}|}\big) & \text{if } \boldsymbol{v} \neq 0, \\ 0 & \text{if } \boldsymbol{v} = 0. \end{cases} \tag{2.51}$$

To prove the existence of a tensor field $T$ satisfying (2.50), it is sufficient to show that $\boldsymbol{t}(\,\cdot\,)$ is a linear transformation on $V$. Therefore we have only to show the following two relations:

(1) $\boldsymbol{t}(\alpha \boldsymbol{v}) = \alpha \boldsymbol{t}(\boldsymbol{v}), \quad \forall \alpha \in I\!R, \boldsymbol{v} \in V,$
(2) $\boldsymbol{t}(\boldsymbol{v}_1 + \boldsymbol{v}_2) = \boldsymbol{t}(\boldsymbol{v}_1) + \boldsymbol{t}(\boldsymbol{v}_2), \quad \forall \boldsymbol{v}_1, \boldsymbol{v}_2 \in V.$

Note that if $\alpha = 0$ or $\boldsymbol{v} = 0$, or if $\alpha > 0$ and $\boldsymbol{v} \neq 0$, then (1) follows directly from the definition (2.51). Therefore, we consider the case that $\alpha < 0$ and $\boldsymbol{v} \neq 0$. Since $\alpha = -|\alpha|$, we have, by the use of Cauchy's lemma (2.49)

$$\boldsymbol{t}(\alpha \boldsymbol{v}) = \boldsymbol{t}(|\alpha|\,(-\boldsymbol{v})) = |\alpha|\, \boldsymbol{t}(-\boldsymbol{v}) = -\,|\alpha|\, \boldsymbol{t}(\boldsymbol{v}) = \alpha\, \boldsymbol{t}(\boldsymbol{v}),$$

which proves (1).

To prove (2), since if $\boldsymbol{v}_1$ and $\boldsymbol{v}_2$ are linearly dependent then it reduces to the case (1), we assume that $\boldsymbol{v}_1$ and $\boldsymbol{v}_2$ are linearly independent and let

$$\boldsymbol{v}_3 = -(\boldsymbol{v}_1 + \boldsymbol{v}_2). \tag{2.52}$$

We consider a small triangular block $\mathcal{P} \subset \mathcal{B}_t$ containing $\boldsymbol{x} \in \mathcal{B}_t$, as shown in Fig. 2.6, with its three faces $A_1$, $A_2$, and $A_3$ normal to $\boldsymbol{v}_1$, $\boldsymbol{v}_2$, and $\boldsymbol{v}_3$, respectively, and the two parallel end triangles $A_4$ and $A_5$ at a distance $\delta$

apart. Let the height of the triangle from the side $A_3$ be $\epsilon$ and the areas of $A_i$ be $a_i$, $i = 1, 2, 3$. Then, from the construction, one can easily see that

$$\frac{a_1}{|\boldsymbol{v}_1|} = \frac{a_2}{|\boldsymbol{v}_2|} = \frac{a_3}{|\boldsymbol{v}_3|} = \lambda\epsilon\delta, \tag{2.53}$$

while the area of the end triangle is $\mu\epsilon^2$ and the volume of the block $\mathcal{P}$ is $\mu\epsilon^2\delta$, where $\lambda$ and $\mu$ are some constants.

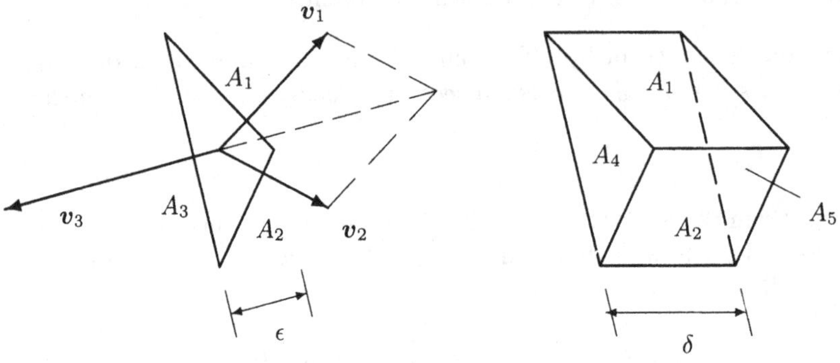


**Fig. 2.6.** A triangular region with chosen normals

If we apply Euler's first law to the block $\mathcal{P}$, we can obtain

$$\frac{1}{\epsilon}\Big(\int_{\mathcal{P}} \rho(\ddot{\boldsymbol{x}} - \boldsymbol{b})\, dv - \int_{A_4 \cup A_5} \boldsymbol{t}(\boldsymbol{n})\, da\Big) = \frac{1}{\epsilon}\sum_{i=1}^{3}\int_{A_i} \boldsymbol{t}(\boldsymbol{n})\, da.$$

Since the integrands are bounded, from the above estimates of areas and volume for $\mathcal{P}$, the left-hand side is of the order of $\epsilon$, while the right-hand side is independent of $\epsilon$. Therefore, we conclude that

$$\lim_{\epsilon\to 0}\frac{1}{\epsilon}\sum_{i=1}^{3}\int_{A_i} \boldsymbol{t}(\boldsymbol{x}, \boldsymbol{n})\, da = 0,$$

and since $\boldsymbol{t}$ is a continuous function of $\boldsymbol{x}$, by the mean value theorem of integral calculus, it follows that

$$\lim_{\epsilon\to 0}\frac{1}{\epsilon}\sum_{i=1}^{3} a_i\, \boldsymbol{t}(\boldsymbol{x}_i, \frac{\boldsymbol{v}_i}{|\boldsymbol{v}_i|}) = 0,$$

where $\boldsymbol{x}_i$, $i = 1, 2, 3$ are some points on the surfaces $A_i$, respectively. From the relation (2.53), this gives

$$\sum_{i=1}^{3} |\boldsymbol{v}_i|\, \boldsymbol{t}(\boldsymbol{x}_i, \frac{\boldsymbol{v}_i}{|\boldsymbol{v}_i|}) = 0.$$

Now, when we shrink the block $\mathcal{P}$ to the point $\boldsymbol{x}$ by taking $\epsilon$ and $\delta$ down to zero, then $\boldsymbol{x}_i \to \boldsymbol{x}$ and the above relation by the use of the definition (2.51), becomes

$$\boldsymbol{t}(\boldsymbol{x}, \boldsymbol{v}_1) + \boldsymbol{t}(\boldsymbol{x}, \boldsymbol{v}_2) + \boldsymbol{t}(\boldsymbol{x}, \boldsymbol{v}_3) = 0,$$

which, from (2.52) and using Cauchy's lemma, proves (2). □

The tensor field $T(\boldsymbol{x}, t)$ in (2.50) is called the *Cauchy stress tensor* or simply *stress tensor*. In the above theorem, the hypothesis of continuity of traction field is often too stringent for the existence of a stress tensor in some applications, for example, in problems of shock waves. However, it has been shown that the theorem remains valid under a much weaker hypothesis of integrability, which would be satisfied in most applications [25, 26].

If, for some unit vector $\boldsymbol{n}$,

$$T\boldsymbol{n} = \sigma\, \boldsymbol{n}, \tag{2.54}$$

then $\sigma$ is called a *principal stress* and $\boldsymbol{n}$ a *principal direction* of $T$. Principal stresses and principal directions are eigenvalues and eigenvectors of $T$. Since $T$ is symmetric, there exist three mutually perpendicular principal directions and three corresponding principal stresses.

The surface traction $\boldsymbol{t}$ on an oriented surface with unit normal $\boldsymbol{n}$ can be decomposed into the sum of a *normal traction*,

$$(\boldsymbol{t} \cdot \boldsymbol{n})\boldsymbol{n} = (\boldsymbol{n} \cdot T\boldsymbol{n})\boldsymbol{n} = (\boldsymbol{n} \otimes \boldsymbol{n})T\boldsymbol{n},$$

and a *shear traction*,

$$\boldsymbol{t} - (\boldsymbol{t} \cdot \boldsymbol{n})\boldsymbol{n} = (\mathit{1} - \boldsymbol{n} \otimes \boldsymbol{n})T\boldsymbol{n}.$$

Let $T_{ij}$ be the components of $T$ relative to a Cartesian coordinate system, then

$T_{ij}$ = the component of the traction $\boldsymbol{t}$ in the direction of $i$-th coordinate line on the $j$-th coordinate surface.

Hence, we call $T_{11}$, $T_{22}$, $T_{33}$ the *normal stress* components and $T_{12}$, $T_{23}$, $T_{31}$ the *shear stress* components on the coordinate surfaces.

**Example 2.3.1** We consider some simple states of stress:

1) *Hydrostatic pressure*
The stress is given by $T = -p\mathit{1}$. In this state of stress, every direction is a principal direction, and $-p$ is the principal stress. $p$ is called the *hydrostatic pressure*. Therefore, the traction on any surface with normal $\boldsymbol{n}$ is the normal traction $-p\,\boldsymbol{n}$.

2) *Pure tension* or *compression*
The stress given by $T = \sigma(\boldsymbol{e} \otimes \boldsymbol{e})$ is called a state of pure tension, or uniaxial tension, in the direction of the unit vector $\boldsymbol{e}$. The traction, $\boldsymbol{t} = \sigma\,\boldsymbol{e}$, on the surface with normal $\boldsymbol{e}$ is a normal traction. It is called a tensile stress if $\sigma$ is positive, and a compressive stress if $\sigma$ is negative.

3) *Pure shear*
The stress given by $T = \tau(\boldsymbol{e} \otimes \boldsymbol{d} + \boldsymbol{d} \otimes \boldsymbol{e})$ is called a state of pure shear in the plane spanned by two vectors $(\boldsymbol{e}, \boldsymbol{d})$. Let $(\boldsymbol{e}^*, \boldsymbol{d}^*)$ be the pair of their dual vectors. Then the tractions on the surface planes parallel to $\boldsymbol{e}$ and $\boldsymbol{d}$, respectively are given by (Fig. 2.7)

$$\boldsymbol{t}_1 = T\frac{\boldsymbol{d}^*}{|\boldsymbol{d}^*|} = \frac{\tau}{|\boldsymbol{d}^*|}\,\boldsymbol{e}, \qquad \boldsymbol{t}_2 = T\frac{\boldsymbol{e}^*}{|\boldsymbol{e}^*|} = \frac{\tau}{|\boldsymbol{e}^*|}\,\boldsymbol{d}.$$

They are purely shear tractions and, in addition, they are of the same magnitude. Indeed, one can easily verify that the angle between the vectors $\boldsymbol{e}$ and $\boldsymbol{e}^*$ is equal to that between $\boldsymbol{d}$ and $\boldsymbol{d}^*$, and

$$\cos\theta(\boldsymbol{e}, \boldsymbol{e}^*) = \frac{\boldsymbol{e}\cdot\boldsymbol{e}^*}{|\boldsymbol{e}|\,|\boldsymbol{e}^*|}, \qquad \cos\theta(\boldsymbol{d}, \boldsymbol{d}^*) = \frac{\boldsymbol{d}\cdot\boldsymbol{d}^*}{|\boldsymbol{d}|\,|\boldsymbol{d}^*|},$$

which implies that $|\boldsymbol{e}|\,|\boldsymbol{e}^*| = |\boldsymbol{d}|\,|\boldsymbol{d}^*|$. Hence, it follows that $|\boldsymbol{t}_1| = |\boldsymbol{t}_2|$.

4) *Plane stress*
If, relative to an orthogonal basis $\{\boldsymbol{e}_1, \boldsymbol{e}_2, \boldsymbol{e}_3\}$, the stress components $T_{13}$, $T_{23}$, and $T_{33}$ vanish, it is called a state of plane stress in the plane spanned by $\boldsymbol{e}_1$ and $\boldsymbol{e}_2$. Clearly pure shear and pure tension are plane stresses. □

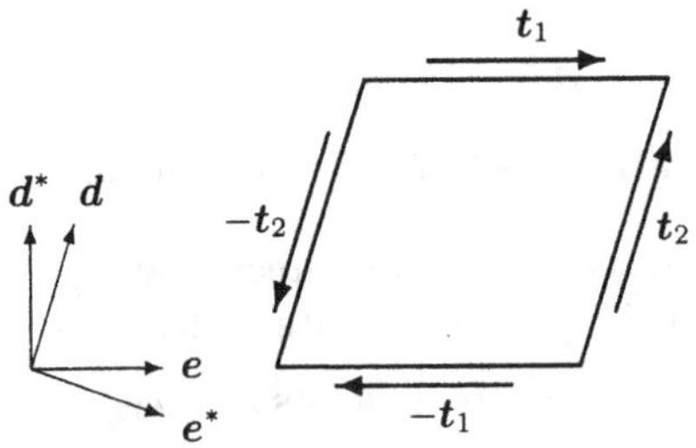


**Fig. 2.7.** Pure shear

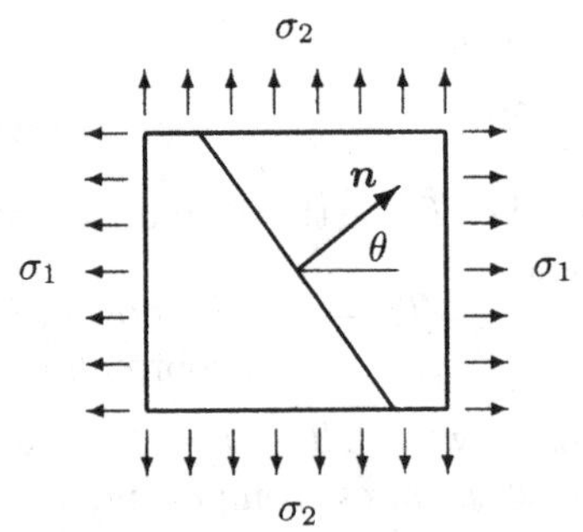


**Fig. 2.8.** Biaxial tension

**Example 2.3.2** *Maximum shear stress in biaxial tension*
Consider a state of stress $T = \sigma_1\boldsymbol{e}_1 \otimes \boldsymbol{e}_1 + \sigma_2\boldsymbol{e}_2 \otimes \boldsymbol{e}_2$ and let $\boldsymbol{n} = \cos\theta\,\boldsymbol{e}_1 + \sin\theta\,\boldsymbol{e}_2$ for $0 \leq \theta \leq \pi$ be the normal of a plane (Fig. 2.8). Then,

the surface traction on the plane is given by

$$\boldsymbol{t} = T\boldsymbol{n} = \sigma_1 \cos\theta\, \boldsymbol{e}_1 + \sigma_2 \sin\theta\, \boldsymbol{e}_2,$$

which can be decomposed into normal and shear tractions. The normal stress,

$$t_n = \boldsymbol{t}\cdot\boldsymbol{n} = \sigma_1 \cos^2\theta + \sigma_2 \sin^2\theta,$$

by the use of trigonometrical identities, can be written as

$$t_n = \frac{\sigma_1 + \sigma_2}{2} + \frac{\sigma_1 - \sigma_2}{2}\cos 2\theta.$$

Therefore, since the value of cosine lies between 1 and $-1$, we conclude that the value of the normal stress lies between the two principal stresses $\sigma_1$ and $\sigma_2$, and the maximum and minimum normal stress acts in the principal direction ($\theta = 0$ and $\theta = \pi/2$).

On the other hand, the shear stress $t_s$ satisfies the relation

$$t_s^2 = \sigma_1^2 \cos^2\theta + \sigma_2^2 \sin^2\theta - (\sigma_1 \cos^2\theta + \sigma_2 \sin^2\theta)^2.$$

Note that, if $\sigma_1 = \sigma_2$, then $t_s = 0$ for any $\theta$. That is, there is no shear stress at all. Therefore, we assume that $\sigma_1 \neq \sigma_2$, in order to find the plane on which the shear stress is maximum. The condition for an extreme value for $|t_s|$ is $dt_s^2/d\theta = 0$, which after a little rearrangement, gives

$$2(\sigma_2 - \sigma_1)\sin\theta\cos\theta\big(\sigma_2 + \sigma_1 - 2(\sigma_1\cos^2\theta + \sigma_2\sin^2\theta)\big) = 0,$$

or, by the use of trigonometrical identities, it becomes

$$(\sigma_2 - \sigma_1)^2 \sin 2\theta \cos 2\theta = 0.$$

Since $\sigma_1 \neq \sigma_2$, there are two solutions to this condition. The first one, $\sin 2\theta = 0$, gives $\theta = 0$ or $\theta = \pi/2$, which leads to $t_s = 0$, the minimum shear stress. The second one, $\cos 2\theta = 0$, gives $\theta = \pi/4$ or $\theta = 3\pi/4$, which leads to

$$t_s = |\sigma_1 - \sigma_2|\,/2.$$

Therefore, we conclude that the maximum shear stress is equal to one half of the difference between the two principal stresses and acts on the plane that bisects the angle between the two principal directions. □

**Exercise 2.3.1** Consider a rectangular block with its sides parallel to the coordinate axes in the reference configuration. Suppose that the stress is given by $T = -p1 + \mu B$. Determine the normal and shear components of the tractions on the surface of the block in the deformed configuration, under the simple shear deformation given by (1.14).

### 2.3.3 Conservation of Linear and Angular Momenta

By Cauchy's theorem, Euler's first law $(2.46)_1$ becomes

$$\frac{d}{dt}\int_{\mathcal{P}_t} \rho\,\dot{\boldsymbol{x}}\,dv = \int_{\mathcal{P}_t} \rho\,\boldsymbol{b}\,dv + \int_{\partial\mathcal{P}_t} T\boldsymbol{n}\,da. \tag{2.55}$$

This equation is often referred to as *Cauchy's first law*. Comparison with the general balance equation (2.1) leads to

$$\psi = \rho\,\dot{\boldsymbol{x}}, \qquad \Phi_\psi = T, \qquad \sigma_\psi = \rho\,\boldsymbol{b},$$

in this case, and hence from (2.13) and (2.15) we obtain the balance equation of linear momentum and its jump condition,

$$\begin{aligned} &\frac{\partial}{\partial t}(\rho\,\dot{\boldsymbol{x}}) + \operatorname{div}(\rho\,\dot{\boldsymbol{x}}\otimes\dot{\boldsymbol{x}} - T) - \rho\,\boldsymbol{b} = 0, \\ &[\![\rho\,\dot{\boldsymbol{x}}(\dot{\boldsymbol{x}}\cdot\boldsymbol{n} - u_n)]\!] - [\![T]\!]\boldsymbol{n} = 0. \end{aligned} \tag{2.56}$$

The first equation, known as the *conservation of linear momentum*, also known as the *equation of motion*, can be rewritten in the following more familiar form by the use of (2.32),

$$\rho\,\ddot{\boldsymbol{x}} - \operatorname{div} T = \rho\,\boldsymbol{b}. \tag{2.57}$$

Similarly, Euler's second law $(2.46)_2$ becomes

$$\frac{d}{dt}\int_{\mathcal{P}_t} \rho(\boldsymbol{x}-\boldsymbol{x}_\circ)\wedge\dot{\boldsymbol{x}}\,dv = \int_{\mathcal{P}_t} (\boldsymbol{x}-\boldsymbol{x}_\circ)\wedge\rho\,\boldsymbol{b}\,dv + \int_{\partial\mathcal{P}_t} (\boldsymbol{x}-\boldsymbol{x}_\circ)\wedge T\boldsymbol{n}\,da. \tag{2.58}$$

This equation is referred to as *Cauchy's second law*, also known as the *conservation of angular momentum*. Comparison with the general balance equation (2.1) leads to

$$\psi = (\boldsymbol{x}-\boldsymbol{x}_\circ)\wedge\rho\,\dot{\boldsymbol{x}}, \quad \Phi_\psi = (\boldsymbol{x}-\boldsymbol{x}_\circ)\wedge T, \quad \sigma_\psi = (\boldsymbol{x}-\boldsymbol{x}_\circ)\wedge\rho\,\boldsymbol{b}.$$

We have the following identity,

$$\operatorname{div}\Big((\boldsymbol{x}-\boldsymbol{x}_\circ)\wedge T\Big) = (\boldsymbol{x}-\boldsymbol{x}_\circ)\wedge\operatorname{div} T + T^T - T.$$

Indeed, in components, the left-hand side gives

$$(p^iT^{jk} - p^jT^{ik})_{,k} = (p^iT^{jk}{}_{,k} - p^jT^{ik}{}_{,k}) + p^i{}_{,k}T^{jk} - p^j{}_{,k}T^{ik},$$

where $\boldsymbol{p} = \boldsymbol{x}-\boldsymbol{x}_\circ$ and therefore $p^i{}_{,k} = \delta^i{}_k$. Hence, from (2.13) and the above identity, after simplification by the use of (2.57), we obtain the following local form of Cauchy's second law,

$$T = T^T. \tag{2.59}$$

In other words, the *symmetry* of the stress tensor for non-polar materials ensures the conservation of angular momentum.

On the other hand, from (2.15), we also obtain the jump condition of angular momentum, which reads

$$[\![(\boldsymbol{x} - \boldsymbol{x}_\circ) \wedge \rho\dot{\boldsymbol{x}}(\dot{\boldsymbol{x}} \cdot \boldsymbol{n} - u_n)]\!] - [\![(\boldsymbol{x} - \boldsymbol{x}_\circ) \wedge T]\!]\boldsymbol{n} = 0.$$

Since $[\![\boldsymbol{x} - \boldsymbol{x}_\circ]\!] = 0$, the above condition is merely a consequence of the jump condition $(2.56)_2$ of linear momentum.

**Exercise 2.3.2** Prove *Bernoulli's Theorem*: Consider a flow with hydrostatic pressure $T = -p1$ and conservative body force $\boldsymbol{b} = -\operatorname{grad}\phi$.
1) If the flow is steady, i.e., $\partial_t \boldsymbol{v} = 0$, then

$$\boldsymbol{v} \cdot \operatorname{grad}\Big(\frac{\boldsymbol{v}^2}{2} + \phi\Big) + \frac{1}{\rho}\boldsymbol{v} \cdot \operatorname{grad} p = 0. \tag{2.60}$$

2) If the flow is steady and irrotational, i.e., $\operatorname{curl}\boldsymbol{v} = 0$, then

$$\operatorname{grad}\Big(\frac{\boldsymbol{v}^2}{2} + \phi\Big) + \frac{1}{\rho}\operatorname{grad} p = 0. \tag{2.61}$$

(Use the relation (1.55).)

## 2.4 Conservation of Energy

For material bodies, besides the mechanical properties associated with the motion, we are also interested in properties associated with heat and temperature. In this section we shall consider the balance of energy. Other thermodynamic properties will be discussed later in Chap. 5.

We define the *kinetic energy* $K$ of a part $\mathcal{P}$ in the motion $\mathcal{X}_t$ by

$$K(\mathcal{P}, t) = \frac{1}{2}\int_{\mathcal{P}_t} \rho\dot{\boldsymbol{x}} \cdot \dot{\boldsymbol{x}}\, dv, \tag{2.62}$$

and the *mechanical power* $P$ by

$$P(\mathcal{P}, t) = \int_{\partial\mathcal{P}_t} \dot{\boldsymbol{x}} \cdot \boldsymbol{t}\, da + \int_{\mathcal{P}_t} \rho\dot{\boldsymbol{x}} \cdot \boldsymbol{b}\, dv. \tag{2.63}$$

For material bodies, the kinetic energy is only a part of the total energy. We call the other part the *internal energy* $E$, and assume that it can be represented as

$$E(\mathcal{P}, t) = \int_{\mathcal{P}_t} \rho\varepsilon\, dv, \tag{2.64}$$

where $\varepsilon(\boldsymbol{x}, t)$ is called the specific *internal energy density*.

We denote the *energy supply* (or *heat supply*) to the part $\mathcal{P}$ by $Q$, and assume that it can be given by

$$Q(\mathcal{P},t) = \int_{\partial\mathcal{P}_t} h\,da + \int_{\mathcal{P}_t} \rho r\,dv, \tag{2.65}$$

where $h = h(\boldsymbol{x},t,\partial\mathcal{P}_t)$ is called the *contact heat supply* and $r = r(\boldsymbol{x},t)$ is called the *energy supply density* (or *heat supply*) due to external sources, such as radiation.

Both internal energy and energy supply are assumed to be frame-indifferent scalar quantities, and consequently, $\varepsilon$, $h$, and $r$ are all frame-indifferent scalars. Obviously, kinetic energy is not frame indifferent.

The *conservation of energy*, also known as the *first law of thermodynamics*, can be stated as follows:

**Conservation of energy.** *Relative to an inertial frame,*

$$\dot{E}(\mathcal{P},t) + \dot{K}(\mathcal{P},t) = P(\mathcal{P},t) + Q(\mathcal{P},t), \tag{2.66}$$

*for all $\mathcal{P} \subset \mathcal{B}$. That is, the rate of change of the total energy of the body is due partly to the mechanical power and partly to the energy supply added to the body.*

The counterpart of Cauchy's postulate and Cauchy's theorem for contact heat supply is the *Fourier–Stokes heat flux principle*, which can be stated as the following relation,

$$h(\boldsymbol{x},t,\partial\mathcal{P}_t) = h(\boldsymbol{x},t,\boldsymbol{n}) = -\boldsymbol{q}(\boldsymbol{x},t)\cdot\boldsymbol{n}, \tag{2.67}$$

where $\boldsymbol{q}(\boldsymbol{x},t)$ is called the *heat flux* vector (or *energy flux*), and $\boldsymbol{n}$ is the exterior unit normal of $\partial\mathcal{P}_t$ at $\boldsymbol{x}$. The above statement can be proved in a similar manner.

With (2.67) and the definitions (2.62) through (2.65), the equation of energy balance (2.66) becomes

$$\frac{d}{dt}\int_{\mathcal{P}_t} (\rho\varepsilon + \frac{\rho}{2}\,\dot{\boldsymbol{x}}\cdot\dot{\boldsymbol{x}})\,dv = \int_{\partial\mathcal{P}_t} (\dot{\boldsymbol{x}}\cdot T\boldsymbol{n} - \boldsymbol{q}\cdot\boldsymbol{n})\,da + \int_{\mathcal{P}_t} (\rho\,\dot{\boldsymbol{x}}\cdot\boldsymbol{b} + \rho\,r)\,dv. \tag{2.68}$$

On comparison with the general balance equation (2.1), we have

$$\psi = (\rho\,\varepsilon + \frac{\rho}{2}\,\dot{\boldsymbol{x}}\cdot\dot{\boldsymbol{x}}), \qquad \Phi_\psi = T\,\dot{\boldsymbol{x}} - \boldsymbol{q}, \qquad \sigma_\psi = \rho\,(\dot{\boldsymbol{x}}\cdot\boldsymbol{b} + r),$$

and hence we have the following local balance equation of total energy and its jump condition,

$$\begin{aligned} &\frac{\partial}{\partial t}(\rho\varepsilon + \frac{\rho}{2}\dot{\boldsymbol{x}}\cdot\dot{\boldsymbol{x}}) + \operatorname{div}\big((\rho\varepsilon + \frac{\rho}{2}\dot{\boldsymbol{x}}\cdot\dot{\boldsymbol{x}})\dot{\boldsymbol{x}} + \boldsymbol{q} - T\dot{\boldsymbol{x}}\big) - \rho(r + \dot{\boldsymbol{x}}\cdot\boldsymbol{b}) = 0, \\ &[\![(\rho\,\varepsilon + \frac{\rho}{2}\,\dot{\boldsymbol{x}}\cdot\dot{\boldsymbol{x}})(\dot{\boldsymbol{x}}\cdot\boldsymbol{n} - u_n)]\!] + [\![\boldsymbol{q} - T\,\dot{\boldsymbol{x}}]\!]\cdot\boldsymbol{n} = 0. \end{aligned} \tag{2.69}$$

On the other hand, if we take the inner product of the velocity $\dot{\boldsymbol{x}}$ and the equation of motion (2.57), we obtain

$$\rho \frac{d}{dt}\left(\frac{1}{2}\dot{\boldsymbol{x}}\cdot\dot{\boldsymbol{x}}\right) = \operatorname{div}(T\dot{\boldsymbol{x}}) + \rho\,\dot{\boldsymbol{x}}\cdot\boldsymbol{b} - T\cdot\operatorname{grad}\dot{\boldsymbol{x}}. \tag{2.70}$$

This equation can be regarded as the *balance of kinetic energy.* Note that the first two terms on the right-hand side are mechanical powers as defined in (2.63), the last term ($T\cdot\operatorname{grad}\dot{\boldsymbol{x}}$), which represents the work done due to the internal stress, will be called the *stress power.*

The energy equation $(2.69)_1$ takes a much simpler form after being subtracted from (2.70),

$$\frac{\partial\rho\varepsilon}{\partial t} + \operatorname{div}(\rho\varepsilon\dot{\boldsymbol{x}} + \boldsymbol{q}) - T\cdot\operatorname{grad}\dot{\boldsymbol{x}} - \rho\,r = 0, \tag{2.71}$$

or

$$\rho\,\dot{\varepsilon} + \operatorname{div}\boldsymbol{q} = T\cdot\operatorname{grad}\dot{\boldsymbol{x}} + \rho\,r. \tag{2.72}$$

This equation is also referred to as the *balance of internal energy.*

Comparison of (2.71) with (2.13) shows that one can get the balance equation for the internal energy in integral form (2.1) with

$$\psi = \rho\,\varepsilon, \qquad \Phi_\psi = -\boldsymbol{q}, \qquad \sigma_\psi = T\cdot\operatorname{grad}\dot{\boldsymbol{x}} + \rho\,r. \tag{2.73}$$

There are two terms in the supply of internal energy, namely, the external energy supply $\rho r$ and the stress power ($T\cdot\operatorname{grad}\dot{\boldsymbol{x}}$), which can be regarded as the *internal production* of energy. The presence of the internal production due to the motion of the body shows that the internal energy is not a conserved quantity in general. By the same token, the kinetic energy is not a conserved quantity either.

Furthermore, we have to point out that the jump condition (2.15) can not be applied to the balance of internal energy, because the supply term contains the velocity gradient, which will become unbounded at the singular surface where the velocity suffers a jump discontinuity. Therefore, there is no jump condition for the internal energy alone.

Another important quantity, associated with the energy and the heat conduction, is the *temperature*, although it does not even appear in the energy equation explicitly. The temperature $\theta(\boldsymbol{x}, t)$ is a measure of how warm a body is. Like the mass measure of the body, the temperature will be treated as a primitive frame-independent scalar quantity and is assumed to be positive for convenience.[1]

[1] In classical thermostatics, the (absolute) temperature is often regarded as a derived concept based on certain hypothesis on physical experiences, instead of the feeling of hot and cold.

## 2.5 Summary of Basic Equations

We can summarize the balance equations of the previous sections in Table 2.1, from which we can write down immediately the corresponding field equations and jump conditions by the use of (2.13) and (2.15).

*Balance of mass*

$$\frac{\partial \rho}{\partial t} + \operatorname{div}(\rho\, \dot{\boldsymbol{x}}) = 0,$$
$$[\![\rho(\dot{\boldsymbol{x}} \cdot \boldsymbol{n} - u_n)]\!] = 0. \tag{2.74}$$

*Balance of linear momentum*

$$\frac{\partial \rho\, \dot{\boldsymbol{x}}}{\partial t} + \operatorname{div}(\rho\, \dot{\boldsymbol{x}} \otimes \dot{\boldsymbol{x}} - T) = \rho\, \boldsymbol{b},$$
$$[\![\rho\, \dot{\boldsymbol{x}}(\dot{\boldsymbol{x}} \cdot \boldsymbol{n} - u_n)]\!] - [\![T]\!]\boldsymbol{n} = 0. \tag{2.75}$$

*Balance of angular momentum*

$$T = T^T,$$
$$[\![(\boldsymbol{x} - \boldsymbol{x}_\circ) \wedge (\rho\, \dot{\boldsymbol{x}}(\dot{\boldsymbol{x}} \cdot \boldsymbol{n} - u_n) - T\boldsymbol{n})]\!] = 0. \tag{2.76}$$

*Balance of energy*

$$\frac{\partial}{\partial t}\Big(\rho\,\varepsilon + \frac{\rho}{2}\,\dot{\boldsymbol{x}} \cdot \dot{\boldsymbol{x}}\Big) + \operatorname{div}\Big(\big(\rho\,\varepsilon + \frac{\rho}{2}\,\dot{\boldsymbol{x}} \cdot \dot{\boldsymbol{x}}\big)\dot{\boldsymbol{x}} + \boldsymbol{q} - T\,\dot{\boldsymbol{x}}\Big) = \rho\, r + \rho\, \dot{\boldsymbol{x}} \cdot \boldsymbol{b},$$
$$[\![\rho(\varepsilon + \frac{1}{2}\,\dot{\boldsymbol{x}} \cdot \dot{\boldsymbol{x}})(\dot{\boldsymbol{x}} \cdot \boldsymbol{n} - u_n)]\!] + [\![\boldsymbol{q} - T\dot{\boldsymbol{x}}]\!] \cdot \boldsymbol{n} = 0. \tag{2.77}$$

*Balance of internal energy*

$$\frac{\partial \rho\varepsilon}{\partial t} + \operatorname{div}(\rho\varepsilon\dot{\boldsymbol{x}} + \boldsymbol{q}) - T \cdot \operatorname{grad} \dot{\boldsymbol{x}} = \rho\, r. \tag{2.78}$$

The balance of angular momentum reduces to the symmetry of the stress tensor (2.76) by the use of $(2.75)_1$, while its jump condition is obviously a consequence of $(2.75)_2$. Moreover, since the balance of kinetic energy is a consequence of the balance of linear momentum, its field equation has not been given above. The balance of internal energy, of course, follows from the balance of energy and linear momentum. Since it is simpler, the balance of internal energy is often used in place of the balance of total energy. Moreover, we have already pointed out that the jump condition (2.15) can not be applied to the balance of internal energy nor to the balance of kinetic energy due to the presence of the internal production ($T \cdot \operatorname{grad} \dot{\boldsymbol{x}}$). We also remark that since

**Table 2.1.** Quantities for balance equations

| | $\psi$ | $\Phi_\psi$ | $\sigma_\psi$ |
|---|---|---|---|
| Mass | $\rho$ | $0$ | $0$ |
| L. Momentum | $\rho\dot{\boldsymbol{x}}$ | $T$ | $\rho\boldsymbol{b}$ |
| A. Momentum | $(\boldsymbol{x}-\boldsymbol{x}_\circ)\wedge\rho\dot{\boldsymbol{x}}$ | $(\boldsymbol{x}-\boldsymbol{x}_\circ)\wedge T$ | $(\boldsymbol{x}-\boldsymbol{x}_\circ)\wedge\rho\boldsymbol{b}$ |
| Energy | $\rho\varepsilon+\frac{\rho}{2}\dot{\boldsymbol{x}}\cdot\dot{\boldsymbol{x}}$ | $-\boldsymbol{q}+T\dot{\boldsymbol{x}}$ | $\rho r+\rho\dot{\boldsymbol{x}}\cdot\boldsymbol{b}$ |
| Int. Energy | $\rho\varepsilon$ | $-\boldsymbol{q}$ | $\rho r+T\cdot\operatorname{grad}\dot{\boldsymbol{x}}$ |
| Kin. Energy | $\frac{\rho}{2}\dot{\boldsymbol{x}}\cdot\dot{\boldsymbol{x}}$ | $T\dot{\boldsymbol{x}}$ | $\rho\dot{\boldsymbol{x}}\cdot\boldsymbol{b}-T\cdot\operatorname{grad}\dot{\boldsymbol{x}}$ |

both the balance equations of internal energy and kinetic energy contain the internal production ($T\cdot\operatorname{grad}\dot{\boldsymbol{x}}$), internal energy and kinetic energy are not conservative quantities. On the other hand, mass, linear momentum, angular momentum, and total energy are all conservative quantities. Their supplies are due to external sources only.

By the use of material time derivative (1.29), the field equations can also be written (see (2.36)) as follows:

$$\begin{aligned}&\dot\rho+\rho\operatorname{div}\boldsymbol{v}=0,\\&\rho\dot{\boldsymbol{v}}-\operatorname{div}T=\rho\boldsymbol{b},\\&\rho\dot\varepsilon+\operatorname{div}\boldsymbol{q}-T\cdot\operatorname{grad}\boldsymbol{v}=\rho r,\end{aligned}\tag{2.79}$$

and the jump conditions in terms of local speed $U$ of the singular surface become

$$\begin{aligned}&[\![\rho U]\!]=0,\\&[\![\rho U\boldsymbol{v}]\!]+[\![T]\!]\boldsymbol{n}=0,\\&[\![\rho U(\varepsilon+\frac{1}{2}v^2)]\!]+[\![T\boldsymbol{v}-\boldsymbol{q}]\!]\boldsymbol{n}=0,\end{aligned}\tag{2.80}$$

where $v^2$ stands for $\boldsymbol{v}\cdot\boldsymbol{v}$.

**Exercise 2.5.1** Suppose that the stress is in a hydrostatic state, $T=-p\mathit{1}$, and $\boldsymbol{q}=0$ (perfect fluid). Show that the following jump conditions hold,

$$\begin{aligned}&[\![p+\rho U^2]\!]=0,\\&[\![\boldsymbol{v}-(\boldsymbol{v}\cdot\boldsymbol{n})\boldsymbol{n}]\!]=0,\\&[\![\varepsilon+\frac{p}{\rho}+\frac{1}{2}U^2]\!]=0.\end{aligned}\tag{2.81}$$

**Exercise 2.5.2** Suppose that the tangential component of the velocity is continuous at the singular surface. Show that the energy jump condition can be written as

$$[\![\boldsymbol{q}]\!]\cdot\boldsymbol{n} = m\,[\![\varepsilon - \frac{1}{\rho}(\boldsymbol{n}\cdot T\boldsymbol{n}) + \frac{1}{2}U^2]\!], \tag{2.82}$$

where $m = (\rho U)^-$ is the mass flux at the surface.

### 2.5.1 Basic Equations in Material Coordinates

It is sometimes more convenient to rewrite the basic equations in material description relative to a reference configuration $\kappa$. They can easily be obtained from $(2.22)_1$,

$$\begin{aligned} &\rho = |J|^{-1}\rho_\kappa, \\ &\rho_\kappa\ddot{\boldsymbol{x}} = \operatorname{Div} T_\kappa + \rho_\kappa \boldsymbol{b}, \\ &\rho_\kappa\dot{\varepsilon} + \operatorname{Div}\boldsymbol{q}_\kappa = T_\kappa\cdot\dot{F} + \rho_\kappa r, \end{aligned} \tag{2.83}$$

where the following definitions have been introduced according to (2.19):

$$T_\kappa = |J|TF^{-T}, \qquad \boldsymbol{q}_\kappa = |J|F^{-1}\boldsymbol{q}. \tag{2.84}$$

$T_\kappa$ is called the *Piola–Kirchhoff stress tensor* and $\boldsymbol{q}_\kappa$ is called the *material heat flux*. Note that unlike the Cauchy stress tensor $T$, the Piola–Kirchhoff stress tensor $T_\kappa$ is not symmetric and it must satisfy

$$T_\kappa F^T = F T_\kappa^T. \tag{2.85}$$

The definition has been introduced according to the relation (1.9), which gives the relation,

$$\int_S T\boldsymbol{n}\,da = \int_{S_\kappa} T_\kappa \boldsymbol{n}_\kappa da_\kappa. \tag{2.86}$$

In other words, $T\boldsymbol{n}$ is the surface traction per unit area in the current configuration, while $T_\kappa\boldsymbol{n}_\kappa$ is the surface traction measured per unit area in the reference configuration. Similarly, $\boldsymbol{q}\cdot\boldsymbol{n}$ and $\boldsymbol{q}_\kappa\cdot\boldsymbol{n}_\kappa$ are the contact heat supplies per unit area in the current and the reference configurations respectively.

We can also write the jump conditions in the reference configuration by the use of $(2.22)_2$,

$$\begin{aligned} &[\![\rho_\kappa]\!]\,U_\kappa = 0, \\ &[\![\rho_\kappa\dot{\boldsymbol{x}}]\!]\,U_\kappa + [\![T_\kappa]\!]\,\boldsymbol{n}_\kappa = 0, \\ &[\![\rho_\kappa(\varepsilon + \frac{1}{2}\dot{\boldsymbol{x}}\cdot\dot{\boldsymbol{x}})]\!]\,U_\kappa + [\![T_\kappa^T\dot{\boldsymbol{x}} - \boldsymbol{q}_\kappa]\!]\cdot\boldsymbol{n}_\kappa = 0, \end{aligned} \tag{2.87}$$

where $\boldsymbol{n}_\kappa$ and $U_\kappa$ are the unit normal and the normal speed at the singular surface in the reference configuration.

**Exercise 2.5.3** By the use of the kinematic compatibility condition (2.26), show that the jump conditions (2.87) can be written as

$$\begin{aligned} &[\![\rho_\kappa]\!] = 0, \\ &[\![T_\kappa \boldsymbol{n}_\kappa]\!] = \rho_\kappa U_\kappa^2 [\![F\boldsymbol{n}_\kappa]\!], \\ &[\![\boldsymbol{q}_\kappa]\!] \cdot \boldsymbol{n}_\kappa = U_\kappa \big( [\![\rho_\kappa \varepsilon]\!] - \langle T_\kappa \boldsymbol{n}_\kappa \rangle \cdot [\![F\boldsymbol{n}_\kappa]\!] \big), \end{aligned} \tag{2.88}$$

where we have introduced the notation $\langle A \rangle = \frac{1}{2}(A^+ + A^-)$ for the mean value of $A$ over the singular surface.

### 2.5.2 Boundary Conditions of a Material Body

The boundary of a material body is a material surface. Therefore, at the boundary $U^\pm = 0$, the jump conditions $(2.80)_{2,3}$ of linear momentum and energy become

$$\begin{aligned} &[\![T]\!]\boldsymbol{n} = 0, \\ &[\![\boldsymbol{q}]\!] \cdot \boldsymbol{n} = [\![\boldsymbol{v} \cdot T\boldsymbol{n}]\!]. \end{aligned}$$

Suppose that the body is acted on the boundary by a force $\boldsymbol{f}$ per unit area, then it requires the stress at the boundary to satisfy the following condition,

$$T\boldsymbol{n} = \boldsymbol{f}, \tag{2.89}$$

and the heat flux to satisfy

$$[\![\boldsymbol{q}]\!] \cdot \boldsymbol{n} = [\![\boldsymbol{v}]\!] \cdot \boldsymbol{f}.$$

In particular, if either the boundary is fixed ($\boldsymbol{v} = 0$) or free ($\boldsymbol{f} = 0$) then the normal component of heat flux must be continuous at the boundary,

$$[\![\boldsymbol{q}]\!] \cdot \boldsymbol{n} = 0. \tag{2.90}$$

Therefore, if the body has a fixed adiabatic (meaning thermally isolated) boundary then the normal component of the heat flux must vanish at the boundary.

Since the deformed configuration is usually unknown for traction boundary value problems of a solid body, the above conditions, which involved the unknown boundary, are sometimes inconvenient.

Of course, we can also express the boundary conditions in the reference configuration. From the jump conditions $(2.87)_{2,3}$, we have

$$T_\kappa \boldsymbol{n}_\kappa = \boldsymbol{f}_\kappa, \tag{2.91}$$

and in a fixed adiabatic boundary,

$$\boldsymbol{q}_\kappa \cdot \boldsymbol{n}_\kappa = 0, \tag{2.92}$$

where $\boldsymbol{f}_\kappa$ is the force per unit area acting on $\partial\mathcal{B}_\kappa$.

The relations (2.91) and (2.92) are the usual boundary conditions in elasticity and heat conduction.

## 2.6 Field Equations in Arbitrary Frames

In the previous sections, we have laid down the basic laws relative to an inertial frame. They can easily be extended to an arbitrary frame of reference from the transformation properties under a change of frame considered in Sect. 1.7.

The field equations relative to the inertial frame $\phi$ can be expressed in the following form (see (2.79)):

$$\begin{aligned} &\dot{\rho} + \rho \operatorname{div} \dot{\boldsymbol{x}} = 0, \\ &\rho \ddot{\boldsymbol{x}} - \operatorname{div} T = \rho \boldsymbol{b}, \\ &\rho \dot{\varepsilon} + \operatorname{div} \boldsymbol{q} - T \cdot \operatorname{grad} \dot{\boldsymbol{x}} = \rho r. \end{aligned} \tag{2.93}$$

Let $\phi$ be an inertial frame and $\phi^*$ be an arbitrary frame. The change of frame from $\phi$ to $\phi^*$, $(\boldsymbol{x}, t) \mapsto (\boldsymbol{x}^*, t^*)$ is given by a Euclidean transformation,

$$\begin{aligned} \boldsymbol{x}^* &= Q(t)(\boldsymbol{x} - \boldsymbol{x}_\circ) + c(t), \\ t^* &= t + a. \end{aligned} \tag{2.94}$$

We have required that the density $\rho$, the forces $\boldsymbol{t}$ (therefore the stress tensor $T$) and $\boldsymbol{b}$, the internal energy $\varepsilon$, and its supplies $h$ (therefore the heat flux $\boldsymbol{q}$) and $r$ are all frame-indifferent quantities, i.e.,

$$\rho^* = \rho, \qquad \varepsilon^* = \varepsilon, \qquad r^* = r,$$

$$\boldsymbol{q}^* = Q\,\boldsymbol{q}, \qquad \boldsymbol{b}^* = Q\,\boldsymbol{b},$$

$$T^* = QTQ^T.$$

We have also shown that the velocity and the acceleration are not frame indifferent with respect to the Euclidean transformations. They satisfy the following transformation relations, (see (1.66) and (1.69))

$$\begin{aligned} \dot{\boldsymbol{x}}^* - Q\,\dot{\boldsymbol{x}} &= \dot{\boldsymbol{c}} + \Omega(\boldsymbol{x}^* - \boldsymbol{c}), \\ \ddot{\boldsymbol{x}}^* - Q\,\ddot{\boldsymbol{x}} &= \ddot{\boldsymbol{c}} + 2\Omega(\dot{\boldsymbol{x}}^* - \dot{\boldsymbol{c}}) + (\dot{\Omega} - \Omega^2)(\boldsymbol{x}^* - \boldsymbol{c}), \end{aligned} \tag{2.95}$$

where $\Omega = \dot{Q}Q^T$ is the spin tensor of the arbitrary frame $\phi^*$ relative to the inertial frame $\phi$, and for the velocity gradient (see (1.77))

$$(\operatorname{grad} \dot{\boldsymbol{x}})^* = Q(\operatorname{grad} \dot{\boldsymbol{x}})Q^T + \Omega,$$

which implies that the divergence of velocity is an objective scalar quantity and so is the stress power, because the stress tensor is symmetric,

$$(\operatorname{div} \dot{\boldsymbol{x}})^* = \operatorname{div} \dot{\boldsymbol{x}}, \qquad (T \cdot \operatorname{grad} \dot{\boldsymbol{x}})^* = T \cdot \operatorname{grad} \dot{\boldsymbol{x}}.$$

Moreover, we can also show that the spatial gradient of an objective field is objective, in particular (see (1.80)),

$$(\operatorname{div} \boldsymbol{q})^* = \operatorname{div} \boldsymbol{q}, \qquad (\operatorname{div} T)^* = Q \operatorname{div} T,$$

and the material time derivative of an objective scalar field is also objective (not true for vector and tensor fields, see (1.81)), therefore,

$$(\dot{\rho})^* = \dot{\rho}, \qquad (\dot{\varepsilon})^* = \dot{\varepsilon}.$$

From the above transformation properties, it follows immediately that the balance of mass and internal energy are objective scalar equations so that they are *Euclidean-invariant*, namely, they are valid in arbitrary frames. The balance of linear momentum, on the other hand, is only *Galilean invariant* because it contains the acceleration, which is frame indifferent with respect to Galilean transformations only. In an arbitrary frame $\phi^*$, it takes the following form,

$$\rho^*(\ddot{\boldsymbol{x}}^* - \boldsymbol{i}^*) - (\operatorname{div} T)^* = \rho^* \boldsymbol{b}^*, \tag{2.96}$$

where

$$\boldsymbol{i}^* = \ddot{\boldsymbol{c}} + 2\Omega(\dot{\boldsymbol{x}}^* - \dot{\boldsymbol{c}}) + (\dot{\Omega} - \Omega^2)(\boldsymbol{x}^* - \boldsymbol{c}), \tag{2.97}$$

stands for the right-hand side of $(2.95)_2$. Equation (2.96) is obtained from $(2.93)_2$ by multiplying by $Q$ and applying the above transformation properties.

Just as the acceleration, the vector field $\boldsymbol{i}^*$ is also not frame indifferent with respect to Euclidean transformations. In fact, $\boldsymbol{i}^* = 0$, if $\phi^*$ is an inertial frame ($\Omega = 0$ and $\ddot{\boldsymbol{c}} = 0$). Nevertheless, we can show that the balance of linear momentum in the form (2.96) is Euclidean invariant. Indeed, we have the following proposition.

**Proposition 2.6.1** *The quantity* $(\ddot{\boldsymbol{x}}^* - \boldsymbol{i}^*)$ *is an objective vector field, i.e., it is invariant under any Euclidean transformation.*

*Proof.* Let $\phi'$ be an arbitrary frame and change of frame from $\phi^*$ to $\phi'$, $(\boldsymbol{x}^*, t^*) \mapsto (\boldsymbol{x}', t')$ be given by a Euclidean transformation,

$$\begin{aligned} \boldsymbol{x}' &= \hat{Q}(t^*)(\boldsymbol{x}^* - \boldsymbol{x}^*_\circ) + \hat{c}(t^*), \\ t' &= t^* + \hat{a}. \end{aligned} \tag{2.98}$$

combining (2.94) and (2.98), we obtain the change of frame from the inertial frame $\phi$ to the frame $\phi'$,

$$\begin{aligned} \boldsymbol{x}' &= Q'(t)(\boldsymbol{x} - \boldsymbol{x}_\circ) + c'(t), \\ t' &= t + a', \end{aligned}$$

where

$$\begin{aligned} a' &= a + \hat{a}, \\ Q'(t) &= \hat{Q}(t+a)Q(t), \\ c'(t) &= \hat{Q}(t+a)\big(c(t) - \boldsymbol{x}_\circ^*\big) + \hat{c}(t+a). \end{aligned}$$

Therefore, in the reference frame $\phi$, we have

$$\ddot{\boldsymbol{x}}' - Q'\,\ddot{\boldsymbol{x}} = \ddot{\boldsymbol{c}}' + 2\Omega'(\dot{\boldsymbol{x}}' - \dot{\boldsymbol{c}}') + (\dot{\Omega}' - \Omega'^2)(\boldsymbol{x}' - \boldsymbol{c}'),$$

where $\Omega' = \dot{Q}'Q'^T$. Denoting the right-hand side by $\boldsymbol{i}'$, we obtain

$$\ddot{\boldsymbol{x}}' - \boldsymbol{i}' = Q'\,\ddot{\boldsymbol{x}} = \hat{Q}\,(Q\,\ddot{\boldsymbol{x}}).$$

On the other hand, from $(2.95)_2$ and (2.97), we have

$$\ddot{\boldsymbol{x}}^* - \boldsymbol{i}^* = Q\,\ddot{\boldsymbol{x}}.$$

Therefore, it follows that

$$\ddot{\boldsymbol{x}}' - \boldsymbol{i}' = \hat{Q}\,(\ddot{\boldsymbol{x}}^* - \boldsymbol{i}^*),$$

which proves that the vector field $(\ddot{\boldsymbol{x}} - \boldsymbol{i})$ is objective under any Euclidean transformation (2.98). □

Dropping the subscript $*$ for simplicity, we can rewrite the field equations in an arbitrary frame.

**Field equations.** Relative to an arbitrary frame $\phi$, the balance of mass, linear momentum, and energy take the following forms:

$$\begin{aligned} &\dot{\rho} + \rho\,\mathrm{div}\,\dot{\boldsymbol{x}} = 0, \\ &\rho\,\ddot{\boldsymbol{x}} - \mathrm{div}\,T = \rho\,(\boldsymbol{b} + \boldsymbol{i}), \\ &\rho\,\dot{\varepsilon} + \mathrm{div}\,\boldsymbol{q} - T\cdot\mathrm{grad}\,\dot{\boldsymbol{x}} = \rho\,r. \end{aligned} \tag{2.99}$$

or

$$\begin{aligned} &\frac{\partial\rho}{\partial t} + \mathrm{div}(\rho\,\dot{\boldsymbol{x}}) = 0, \\ &\frac{\partial}{\partial t}(\rho\,\dot{\boldsymbol{x}}) + \mathrm{div}(\rho\,\dot{\boldsymbol{x}}\otimes\dot{\boldsymbol{x}} - T) = \rho\,(\boldsymbol{b} + \boldsymbol{i}), \\ &\frac{\partial}{\partial t}\Big(\rho\varepsilon + \frac{\rho}{2}\,\dot{\boldsymbol{x}}\cdot\dot{\boldsymbol{x}}\Big) + \mathrm{div}\Big(\big(\rho\varepsilon + \frac{\rho}{2}\,\dot{\boldsymbol{x}}\cdot\dot{\boldsymbol{x}}\big)\dot{\boldsymbol{x}} + \boldsymbol{q} - T\,\dot{\boldsymbol{x}}\Big) = \rho\,r - \rho\,\dot{\boldsymbol{x}}\cdot(\boldsymbol{b} + \boldsymbol{i}), \end{aligned}$$

where

$$\boldsymbol{i} = \ddot{\boldsymbol{c}} + 2\Omega(\dot{\boldsymbol{x}} - \dot{\boldsymbol{c}}) + (\dot{\Omega} - \Omega^2)(\boldsymbol{x} - \boldsymbol{c}), \tag{2.100}$$

and $\Omega$ is the spin tensor and $\boldsymbol{c}$ is the translation of the frame $\phi$ relative to an inertial frame. Moreover, these equations are Euclidean invariant.

We can interpret the vector field $\boldsymbol{i}$ as a force field due to the motion of the frame relative to an inertial frame and call it the *inertial body force*, which consists of a number of contributions, usually known as

| | |
|---|---|
| $2\Omega(\dot{\boldsymbol{x}} - \dot{\boldsymbol{c}})$ | Coriolis force, |
| $-\Omega^2(\boldsymbol{x} - \boldsymbol{c})$ | centrifugal force, |
| $\dot{\Omega}(\boldsymbol{x} - \boldsymbol{c})$ | Euler force, |
| $\ddot{\boldsymbol{c}}$ | inertial force of relative translation. |

By comparison of (2.98) and (2.93), we have seen that the field equations relative to an arbitrary frame have exactly the same forms as that relative to an inertial frame, except the body force $\boldsymbol{b}$ is replaced by $(\boldsymbol{b} + \boldsymbol{i})$, which we shall call the *apparent body force* Therefore, from now on, we shall simply write $\boldsymbol{b}$ for $(\boldsymbol{b} + \boldsymbol{i})$ in any frame of reference, and keep in mind that when the inertial force $\boldsymbol{i}$ does not vanish it must be regarded as an (apparent) part of the body force $\boldsymbol{b}$.

Finally, we remark that the corresponding integral laws of balance also have the same forms in an inertial frame as in an arbitrary frame by replacing the body force with the apparent body force. Since the inertial force is bounded, provided that the velocity is bounded, therefore, the jump conditions for mass, linear momentum and energy of previous sections are valid in any frame of reference.

# 3. Basic Principles of Constitutive Theories

## 3.1 Constitutive Relation

The balance laws introduced in Chap. 2 are the fundamental equations common to all material bodies. However, these laws are insufficient to fully characterize the behavior of material bodies, because physical experiences have shown that two bodies of exactly the same size and shape, in general, will not have the same behavior when they are subjected to exactly the same environment (external supplies and boundary conditions). For example, the same force will not produce the same elongation on wires of rubber and aluminum of the same diameter and length.

Physically, the behavior of a body is characterized by a description of the fields of

$$\begin{aligned} &\rho(X,t) && \text{the density,} \\ &\chi(X,t) && \text{the motion,} \\ &\theta(X,t) && \text{the temperature,} \end{aligned} \tag{3.1}$$

called the *basic fields*. The field equations are based on the system of equations consisting of balance of mass, momentum and energy, and they alone are not sufficient to determine these fields, because the system (2.79) contains other field quantities, namely, the stress tensor $T(X,t)$, the heat flux $\boldsymbol{q}(X,t)$, and the internal energy $\varepsilon(X,t)$, as well as external supplies: the body force $\boldsymbol{b}(X,t)$ and the energy supply $r(X,t)$. These quantities must be specified in order that (2.79) becomes a system of field equations for the basic fields (3.1).

External supplies will be regarded as known functions, determined from the environment that the body encounters. Therefore, it is usually assumed that material properties are independent of external supplies. While the stress tensor, the heat flux, and the internal energy,

$$\{T(X,t),\ \boldsymbol{q}(X,t),\ \varepsilon(X,t)\}, \tag{3.2}$$

will depend not only on the behavior of the body but also on the kind of material that constitutes the body. Such quantities are called *constitutive* quantities, for which additional hypotheses must be introduced to characterize thermomechanical responses of a particular material body.

The material response of a body generally depends on the past history of its thermomechanical behavior. Therefore, let us introduce the notion of the past history of a function. Let $\psi(\cdot)$ be a function of time. The *history* of $\psi$ *up to time* $t$ is defined by

$$\psi^t(s) = \psi(t - s), \tag{3.3}$$

where $s \in [0, \infty)$ denotes the time–coordinate pointed into the past from the present time $t$. Clearly $s = 0$ corresponds to the present time, therefore

$$\psi^t(0) = \psi(t).$$

Mathematical descriptions of material response are called *constitutive relations* or constitutive equations. We postulate that the history of the behavior up to the present time determines the present response of the body.

**Principle of determinism.** *Let $\mathcal{C}$ denote a constitutive quantity, then the constitutive relation for $\mathcal{C}$ is given by a functional of the form,*

$$\mathcal{C}(X,t) = \underset{\substack{Y \in \mathcal{B} \\ 0 \leq s < \infty}}{\mathcal{F}} (\rho^t(Y,s),\, \chi^t(Y,s),\, \theta^t(Y,s),\, X,\, t). \tag{3.4}$$

We call $\mathcal{F}$ the *constitutive function* or *response function* of $\mathcal{C}$. Note that we have indicated the domains of the argument functions as underscripts in the notation of the functional $\mathcal{F}$. Such a functional allows the description of arbitrary non-local effects of any inhomogeneous body with a perfect memory of the past history.

Constitutive relations can be regarded as mathematical models of material bodies. The validity of a model can be verified by experiments on the results it predicts. On the contrary, some experiments may suggest certain functional dependence of the constitutive function on its variables to within a reasonable satisfaction for certain materials. However, experiments alone are rarely, if ever, sufficient to determine constitutive functions of a material body. On the other hand, there are some universal requirements that a model should obey lest its consequences be contradictory to some well-known physical experiences. Therefore, in search of a correct formulation of a mathematical model, in general, we shall first impose these requirements on the proposed model. The most important universal requirements of this kind are:

- *principle of material objectivity,*
- *material symmetry,*
- *thermodynamic considerations.*

These requirements impose severe restrictions on the model and hence lead to a great simplification for general constitutive relations. The reduction of constitutive relations from very general to more specific and mathematically simpler ones for a given class of materials is the main objective of constitutive theories in continuum mechanics.

In this chapter we shall restrict our attention to the discussion of the principle of material objectivity, which deals with the transformation properties of constitutive functions, and the material symmetry, which characterizes the specific symmetric properties of material particles. Thermodynamic considerations, which govern thermodynamic aspects of material bodies will be postponed until Chap. 5.

## 3.2 Principle of Material Objectivity

In writing (3.4), we have taken for granted that the constitutive quantity $\mathcal{C}$ is an observable quantity and the constitutive relation has been expressed relative to a certain frame of reference. Now we shall examine the consequence of a change of frame upon the constitutive functions.

Let $\phi$ be a frame of reference and $\mathcal{C}(X,t;\phi)$ be the value of the constitutive quantity $\mathcal{C}$ at the material point $X$ and time $t$ in the frame $\phi$. We rewrite the constitutive relation (3.4) in the following form,

$$\mathcal{C}(X,t;\phi) = \mathop{\mathcal{F}_\phi}_{\substack{Y\in\mathcal{B}\\ 0\leq s<\infty}} (\rho^t(Y,s),\, \chi^t(Y,s),\, \theta^t(Y,s),\, X,\, t). \tag{3.5}$$

In other words, the constitutive function depends on the choice of frame, in general, so that we have also indicated the frame $\phi$ on $\mathcal{F}$ as a subscript.

Since any intrinsic property of materials should be independent of observers, we postulate that for any objective constitutive quantity, its constitutive function must be invariant with respect to any change of frame. Mathematically, we postulate

**Principle of material objectivity.** *The response function of an objective (frame indifferent with respect to Euclidean transformations) constitutive quantity $\mathcal{C}$ must be independent of the frame, i.e.,*

$$\mathcal{F}_\phi(\,\cdot\,) = \mathcal{F}_{\phi^*}(\,\cdot\,),$$

*for any frames of reference $\phi$ and $\phi^*$.*

More specifically, suppose that $\mathcal{C}$ is an $\mathcal{S}_n$-valued objective constitutive quantity ($n = 0, 1, 2$ for scalar-, vector-, and tensor-valued, respectively) then, for any change of frame from $\phi$ to $\phi^*$, we have the objectivity property (see (1.63))

$$\mathcal{C}(\phi^*) = \boldsymbol{Q}^*\mathcal{C}(\phi), \tag{3.6}$$

where $\boldsymbol{Q}^*$ is the induced linear transformation on the tensor space $\mathcal{S}_n$ from the change of frame,

$$\begin{aligned} \boldsymbol{x}^* &= Q(t)(\boldsymbol{x} - \boldsymbol{x}_\circ) + \boldsymbol{c}(t), \\ t^* &= t + a. \end{aligned} \tag{3.7}$$

The principle of material objectivity imposed the following conditions:

$$\begin{aligned} &\underset{\substack{Y \in \mathcal{B} \\ 0 \leq s < \infty}}{\mathcal{F}} (\rho^*(Y, t^* - s),\, \chi^*(Y, t^* - s),\, \theta^*(Y, t^* - s),\, X,\, t^*) \\ &\quad = \boldsymbol{Q}^* \underset{\substack{Y \in \mathcal{B} \\ 0 \leq s < \infty}}{\mathcal{F}} (\rho(Y, t - s),\, \chi(Y, t - s),\, \theta(Y, t - s),\, X,\, t), \end{aligned} \tag{3.8}$$

where we have used $\mathcal{F} = \mathcal{F}_\phi = \mathcal{F}_{\phi^*}$. Since both density and temperature fields are objective scalars, we have

$$\begin{aligned} \rho^*(Y, t^* - s) &= \rho(Y, t - s), \\ \theta^*(Y, t^* - s) &= \theta(Y, t - s), \end{aligned}$$

and the above condition of material objectivity can be written as

$$\begin{aligned} &\underset{\substack{Y \in \mathcal{B} \\ 0 \leq s < \infty}}{\mathcal{F}} (\rho(Y, t - s),\, \chi^*(Y, t^* - s),\, \theta(Y, t - s),\, X,\, t^*) \\ &\quad = \boldsymbol{Q}^* \underset{\substack{Y \in \mathcal{B} \\ 0 \leq s < \infty}}{\mathcal{F}} (\rho(Y, t - s),\, \chi(Y, t - s),\, \theta(Y, t - s),\, X,\, t), \end{aligned} \tag{3.9}$$

where

$$\chi^*(Y, t^* - s) = Q(t - s)(\chi(Y, t - s) - \boldsymbol{x}_\circ) + \boldsymbol{c}(t - s).$$

This equation is the restriction imposed on any constitutive function $\mathcal{F}$ by the requirement of material objectivity. Note that it can be regarded as a condition expressed in terms of any change of frame $*$ relative to an arbitrary frame of reference and hence it is not necessary to mention the frame $\phi$ explicitly.

The principle of material objectivity is often referred to as the *principle of material frame indifference.* We emphasize that for a non-objective quantity, for example, the total energy density $(\varepsilon + v^2/2)$, the condition (3.9) need not hold. Therefore, to apply the principle of material objectivity, it is essential to replace a non-objective constitutive quantity by an objective one, which can usually be done by adding, subtracting or multiplying some non-objective terms. In our later discussions, we shall always assume that $\mathcal{C}$ is objective.

Since (3.9) holds for any change of frame, an immediate consequence follows from choosing $Q(t) = \mathit{1}$ and $\boldsymbol{c}(t) = \boldsymbol{x}_\circ$, such that

$$\boldsymbol{x}^* = \boldsymbol{x}, \qquad t^* = t + a,$$

for arbitrary constant $a \in I\!R$. Clearly, the induced linear transformation on any tensor space in this case is

$$\boldsymbol{Q}^* = 1.$$

Therefore, the condition (3.9) implies that

$$\mathcal{F}(\rho^t, \chi^t, \theta^t, X, t + a) = \mathcal{F}(\rho^t, \chi^t, \theta^t, X, t).$$

Since this is true for arbitrary $a$, taking the derivative of this equation with respect to $a$, we conclude that

$$\frac{\partial \mathcal{F}}{\partial t} = 0,$$

i.e., $\mathcal{F}$ can not depend on $t$ explicitly. Therefore, we can rewrite the constitutive relation (3.4) as

$$\mathcal{C}(X,t) = \underset{\substack{Y \in \mathcal{B} \\ 0 \leq s < \infty}}{\mathcal{F}} (\rho^t(Y,s),\, \chi^t(Y,s),\, \theta^t(Y,s),\, X), \tag{3.10}$$

and the condition (3.9) becomes

$$\mathcal{F}(\rho^t, (\chi^t)^*, \theta^t, X) = \boldsymbol{Q}^* \mathcal{F}(\rho^t, \chi^t, \theta^t, X), \tag{3.11}$$

where

$$(\chi^t)^* = Q^t(\chi^t - \boldsymbol{x}_\circ) + \boldsymbol{c}^t.$$

We shall often omit the domain and the dependence of the argument functions in $\mathcal{F}$ as indicated in (3.9) for simplicity.

**Remark.** Comparing the arguments of the two sides of the above equation, we note that the condition (3.11) only imposes restrictions on the dependence of the response function upon the histories of motion, and nothing on that of density and temperature fields. In view of this, it leads to a different interpretation of the condition of material objectivity. From (1.65), for a motion $\chi$ and a change of frame $*$, we have

$$\begin{aligned} \chi^*(Y,t^*) &= Q(t)(\chi(Y,t) - \boldsymbol{x}_\circ) + \boldsymbol{c}(t), \\ t^* &= t + a. \end{aligned}$$

If we define a new motion $\chi_*$ associated with the motion $\chi$ by

$$\chi_*(Y,t) = Q(t)(\chi(Y,t) - \boldsymbol{x}_\circ) + \boldsymbol{c}(t), \tag{3.12}$$

then the condition (3.11) can be written as

$$\mathcal{F}(\rho^t, \chi_*^t, \theta^t, X) = \boldsymbol{Q}^* \mathcal{F}(\rho^t, \chi^t, \theta^t, X). \tag{3.13}$$

Two motions $\chi$ and $\chi_*$ related by (3.12) are called *equivalent motions*. Equivalent motions are motions that differ by an arbitrary rigid transformation. We say that a scalar $s$, a vector $\boldsymbol{u}$, or a tensor $T$ is an *invariant* quantity under equivalent motions, if it satisfies

$$\begin{aligned} s(\chi_*) &= s(\chi), \\ \boldsymbol{u}(\chi_*) &= Q(t)\,\boldsymbol{u}(\chi), \\ T(\chi_*) &= Q(t)\,T(\chi)\,Q(t)^T, \end{aligned}$$

where $\chi_*$ and $\chi$ are related by a rigid transformation given by (3.12). Physically, an invariant quantity is a quantity that accompanies the rotation of the rigid transformation. In this sense, the condition (3.13) is sometimes invoked in postulating the principle of material objectivity, especially when only mechanical theories are concerned.

**Principle of material objectivity** (another version). *The response function of an objective constitutive quantity must be invariant for any equivalent motions.*

From this point of view, no change of frame of reference is involved. These two versions of the principle of material objectivity are equivalent, nevertheless, we shall retain to our original interpretation of material objectivity. □

### 3.2.1 In Referential Description

Let $\kappa$ be a reference configuration of the body $\mathcal{B}$. For a motion $\chi$, we have

$$\chi(X,t) = \chi_\kappa(\boldsymbol{X},t), \qquad \boldsymbol{X} = \kappa(X).$$

Similarly, we can also write

$$\rho(X,t) = \rho_\kappa(\boldsymbol{X},t), \qquad \theta(X,t) = \theta_\kappa(\boldsymbol{X},t).$$

We could have abused the notation by writing $\rho(X,t) = \rho(\boldsymbol{X},t)$ and $\theta(X,t) = \theta(\boldsymbol{X},t)$ as we have done usually, but for clarity, more precise notations will be used in this section.

In terms of referential description, we can rewrite the constitutive relation (3.10) relative to $\kappa$ at a point $\boldsymbol{X} \in \mathcal{B}_\kappa$ and time t as

$$\mathcal{C}(\boldsymbol{X},t) = \underset{\substack{\boldsymbol{Y}\in\mathcal{B}_\kappa \\ 0\leq s<\infty}}{\mathcal{F}_\kappa}\,(\rho^t_\kappa(\boldsymbol{Y},s),\, \chi^t_\kappa(\boldsymbol{Y},s),\, \theta^t_\kappa(\boldsymbol{Y},s),\, \boldsymbol{X}), \tag{3.14}$$

where $\mathcal{F}_\kappa$ is related to $\mathcal{F}_\phi$ of (3.5) by

$$\begin{aligned}\mathcal{F}_\kappa(\rho_\kappa^t(\boldsymbol{Y},s),\ \chi_\kappa^t(\boldsymbol{Y},s),\ \theta_\kappa^t(\boldsymbol{Y},s),\boldsymbol{X})\\ = \mathcal{F}_\phi(\rho_\kappa^t(\boldsymbol{Y},s),\ \chi_\kappa^t(\boldsymbol{Y},s),\ \theta_\kappa^t(\boldsymbol{Y},s),\ \kappa^{-1}(\boldsymbol{X})).\end{aligned} \tag{3.15}$$

$\mathcal{F}_\kappa$ is called the constitutive function of $\mathcal{C}$ relative to the configuration $\kappa$. The change of reference configuration for the constitutive function will be considered later in the discussion of material symmetry.

To express the condition of material objectivity in the referential description, let $\phi$ and $\phi^*$ be two arbitrary frames, and denote the reference configuration in these frames by $\kappa$ and $\kappa^*$, respectively,

$$\boldsymbol{X} = \kappa(X), \qquad \boldsymbol{X}^* = \kappa^*(X).$$

From (1.73), the map

$$\gamma = \kappa^* \circ \kappa^{-1} : \mathcal{B}_\kappa \to \mathcal{B}_{\kappa^*} \tag{3.16}$$

is given by a fixed rigid transformation,

$$\boldsymbol{X}^* = K(\boldsymbol{X} - \boldsymbol{x}_\circ) + \boldsymbol{c}_\circ, \tag{3.17}$$

where $K$ is a constant orthogonal transformation due to the change of frame.

For an observable quantity $\mathcal{C}$ given by (3.14), the objectivity property (3.6) for change of frame now takes the form,

$$\begin{aligned}\underset{\substack{\boldsymbol{Y}^*\in\mathcal{B}_{\kappa^*}\\ 0\le s<\infty}}{\mathcal{F}_{\kappa^*}}(\rho_{\kappa^*}^*(\boldsymbol{Y}^*,t^*-s),\ \chi_{\kappa^*}^*(\boldsymbol{Y}^*,t^*-s),\ \theta_{\kappa^*}^*(\boldsymbol{Y}^*,t^*-s),\ \boldsymbol{X}^*)\\ = \boldsymbol{Q}^*\ \underset{\substack{\boldsymbol{Y}\in\mathcal{B}_\kappa\\ 0\le s<\infty}}{\mathcal{F}_\kappa}(\rho_\kappa(\boldsymbol{Y},t-s),\ \chi_\kappa(\boldsymbol{Y},t-s),\ \theta_\kappa(\boldsymbol{Y},t-s),\ \boldsymbol{X}).\end{aligned} \tag{3.18}$$

We can also express the left-hand side in terms of the constitutive function $\mathcal{F}_\kappa$ by the principle of material objectivity. Indeed, from the relation (3.15), without writing out all the arguments, we have

$$\begin{aligned}&\mathcal{F}_{\kappa^*}(\chi_{\kappa^*}^*(\boldsymbol{Y}^*,t^*-s),\ \boldsymbol{X}^*)\\ &\quad= \mathcal{F}_{\phi^*}(\chi_{\kappa^*}^*(\boldsymbol{Y}^*,t^*-s),\ \kappa^{*-1}(\boldsymbol{X}^*))\\ &\quad= \mathcal{F}_\phi(\chi_{\kappa^*}^*(\gamma(\boldsymbol{Y}),t^*-s),\ \kappa^{*-1}(\gamma(\boldsymbol{X})))\\ &\quad= \mathcal{F}_\kappa(\chi_{\kappa^*}^*(\gamma(\boldsymbol{Y}),t^*-s),\ \boldsymbol{X}),\end{aligned}$$

where we have applied the principle of material objectivity that $\mathcal{F}_\phi = \mathcal{F}_{\phi^*}$. The arguments for the density and the temperature are similar, and since they are objective scalar quantities, we have

$$\begin{aligned}\rho_{\kappa^*}^*(\gamma(\boldsymbol{Y}),t^*) = \rho_{\kappa^*}^*(\boldsymbol{Y}^*,t^*) = \rho_\kappa(\boldsymbol{Y},t),\\ \theta_{\kappa^*}^*(\gamma(\boldsymbol{Y}),t^*) = \theta_{\kappa^*}^*(\boldsymbol{Y}^*,t^*) = \theta_\kappa(\boldsymbol{Y},t).\end{aligned} \tag{3.19}$$

Therefore, from (3.18), we have the following condition for material objectivity in the reference configuration,

$$\begin{aligned}\underset{\substack{Y\in\mathcal{B}_\kappa\\0\le s<\infty}}{\mathcal{F}_\kappa}\,(\rho(\boldsymbol{Y},t-s),\ \chi^*_{\kappa^*}(\gamma(\boldsymbol{Y}),t^*-s),\ \theta(\boldsymbol{Y},t-s),\ \boldsymbol{X})\\ = \boldsymbol{Q}^*\ \underset{\substack{Y\in\mathcal{B}_\kappa\\0\le s<\infty}}{\mathcal{F}_\kappa}\,(\rho(\boldsymbol{Y},t-s),\ \chi_\kappa(\boldsymbol{Y},t-s),\ \theta(\boldsymbol{Y},t-s),\ \boldsymbol{X}),\end{aligned} \tag{3.20}$$

where

$$\chi^*_{\kappa^*}(\gamma(\boldsymbol{Y}),t^*) = Q(t)(\chi_\kappa(\boldsymbol{Y},t)-\boldsymbol{x}_\circ)+\boldsymbol{c}(t). \tag{3.21}$$

Note that the condition (3.20) is expressed in the reference configuration $\kappa$ in the frame $\phi$ only.

In the above condition, we have again adopted the usual notation by writing $\rho(\boldsymbol{X},t)$ for $\rho_\kappa(\boldsymbol{X},t)$, $\theta(\boldsymbol{X},t)$ for $\theta_\kappa(\boldsymbol{X},t)$. Similar to the condition (3.11), the condition of material objectivity (3.20) imposes restrictions on the dependence of response function upon the history of motion only and not on the histories of density and temperature of the body. We can write the condition (3.20) simply as

$$\mathcal{F}_\kappa(\rho^t,\ (\chi^t_\kappa)^*\circ\gamma,\ \theta^t,\ \boldsymbol{X}) = \boldsymbol{Q}^*\mathcal{F}_\kappa(\rho^t,\ \chi^t_\kappa,\ \theta^t,\ \boldsymbol{X}), \tag{3.22}$$

where

$$(\chi^t_\kappa)^*\circ\gamma = Q^t(\chi^t_\kappa-\boldsymbol{x}_\circ)+\boldsymbol{c}^t.$$

### 3.2.2 An Example: a Particular Class of Materials

Besides general consequences imposed by the principle of material objectivity, to be considered further in later sections, one can also apply the principle of material objectivity directly to constitutive equations of a particular class of materials, as illustrated in the following example.

Suppose that a given material is characterized by the constitutive equation for the stress tensor in the following form:

$$T = \mathcal{T}(\rho,\boldsymbol{v},L),$$

where $\boldsymbol{v}$ is the velocity and $L = \operatorname{grad}\boldsymbol{v}$, then the condition of material objectivity implies that

$$\mathcal{T}(\rho^*,\boldsymbol{v}^*,L^*) = Q\mathcal{T}(\rho,\boldsymbol{v},L)Q^T,$$

for any change of frame given by (1.57). From (1.66) and (1.77) and $\rho^* = \rho$, since the density is an objective scalar, the above condition becomes

$$\mathcal{T}(\rho, Q\boldsymbol{v}+\dot{Q}(\boldsymbol{x}-\boldsymbol{x}_\circ)+\dot{\boldsymbol{c}}, QLQ^T+\dot{Q}Q^T) = Q\mathcal{T}(\rho,\boldsymbol{v},L)Q^T,$$

for any orthogonal tensor $Q(t)$ and any vector $\boldsymbol{c}(t)$. In particular, if we choose

$Q(t) = \mathit{1}$, then we have

$$\mathcal{T}(\rho, \boldsymbol{v} + \dot{\boldsymbol{c}}, L) = \mathcal{T}(\rho, \boldsymbol{v}, L).$$

Since $\boldsymbol{c}(t)$ is arbitrary, it implies that $\mathcal{T}$ can not depend on the velocity $\boldsymbol{v}$. Hence, from $L = D + W$, where $D$ is the stretching tensor and $W$ the skew-symmetric spin tensor, the objectivity condition becomes

$$\mathcal{T}(\rho, QDQ^T + QWQ^T + \dot{Q}Q^T) = Q\mathcal{T}(\rho, L)Q^T,$$

for any orthogonal tensor $Q(t)$. For any given skew-symmetric tensor $W$, we can define an orthogonal tensor $Q(t)$ such that

$$Q(t_0) = \mathit{1}, \qquad \dot{Q}(t_0) = -W,$$

and hence evaluation of the objectivity condition at $t = t_0$ leads to

$$\mathcal{T}(\rho, L) = \mathcal{T}(\rho, D),$$

which implies that $\mathcal{T}$ can not depend on the spin tensor $W$. Therefore, the constitutive equation for the stress tensor must reduce to

$$T = \mathcal{T}(\rho, D).$$

In addition, the objectivity condition still requires the function $\mathcal{T}$ to satisfy the following relation,

$$\mathcal{T}(\rho, QDQ^T) = Q\,\mathcal{T}(\rho, D)\,Q^T, \tag{3.23}$$

for any orthogonal transformation $Q$. A representation of the function $\mathcal{T}$ satisfying this condition will be given later.

**Exercise 3.2.1** Let the exponential of a matrix $A$ be defined by the series

$$\exp(A) = \mathit{1} + A + \frac{A^2}{2!} + \frac{A^3}{3!} + \cdots.$$

Show that

1) if $AB = BA$, then $\exp(A)\exp(B) = \exp(A + B)$;
2) if $A$ is skew-symmetric, then $\exp(A)$ is orthogonal;
3) for any skew-symmetric tensor $W$, the tensor function $Q(t) = \exp(-(t - t_0)W)$ satisfies the following differential equation,

$$\dot{Q} + QW = 0 \quad \text{and} \quad Q(t_0) = \mathit{1}.$$

**Exercise 3.2.2** Consider a material (a linear viscous fluid) with the stress tensor, relative to a Cartesian coordinate system, represented by

$$T_{ij} = -p\delta_{ij} + V_{ijkl}D_{kl},$$

where $p$ and the fourth-order tensor $V_{ijkl}$ are functions of the density $\rho$. Show that the objectivity condition implies that for any orthogonal tensor $Q$,

$$V_{ijkl} = Q_{im}Q_{jn}Q_{kp}Q_{lq}V_{mnpq}. \tag{3.24}$$

Such a tensor is usually called a fourth-order isotropic tensor.

## 3.3 Simple Material Bodies

Let $\rho_\kappa(\boldsymbol{Y})$ be the density field in the reference configuration $\kappa$, then it determines the density field $\rho(\boldsymbol{Y},t)$ from the deformation field $\chi_\kappa(\boldsymbol{Y},t)$ according to the balance of mass (2.34). Therefore, we can replace $\rho^t(\boldsymbol{Y},s)$ by $\rho_\kappa(\boldsymbol{Y})$ as the independent variable in the constitutive equation (3.14), which can then be written as

$$\mathcal{C}(\boldsymbol{X},t) = \underset{\substack{\boldsymbol{Y}\in\mathcal{B}_\kappa \\ 0\leq s<\infty}}{\mathcal{F}_\kappa} (\rho_\kappa(\boldsymbol{Y}),\, \chi^t_\kappa(\boldsymbol{Y},s),\, \theta^t(\boldsymbol{Y},s),\, \boldsymbol{X}). \tag{3.25}$$

For constitutive equations of this form, thermomechanical histories of any part of the body can affect the response at any point of the body. In most applications, such a non-local property is rarely relevant and may only cause unnecessary complications. Therefore, it is usually assumed that only thermomechanical histories in an arbitrary small neighborhood of $\boldsymbol{X}$ affect the material response at the point $\boldsymbol{X}$, and hence they can be approximated at $\boldsymbol{X}$ by Taylor series up to certain order. In particular, if only the first approximation is concerned, we have

$$\begin{aligned}
\rho_\kappa(\boldsymbol{Y}) &= \rho_\kappa(\boldsymbol{X}) + \operatorname{Grad}\rho_\kappa(\boldsymbol{X})(\boldsymbol{Y}-\boldsymbol{X}) + o(2),\\
\chi_\kappa(\boldsymbol{Y},t) &= \chi_\kappa(\boldsymbol{X},t) + F(\boldsymbol{X},t)(\boldsymbol{Y}-\boldsymbol{X}) + o(2),\\
\theta(\boldsymbol{Y},t) &= \theta(\boldsymbol{X},t) + \operatorname{Grad}\theta(\boldsymbol{X},t)(\boldsymbol{Y}-\boldsymbol{X}) + o(2),
\end{aligned}$$

and the constitutive relation (3.25) can be written as

$$\mathcal{C}(\boldsymbol{X},t) = \underset{0\leq s<\infty}{\mathcal{F}_\kappa} (\chi^t_\kappa(\boldsymbol{X},s),\, F^t(\boldsymbol{X},s),\, \theta^t(\boldsymbol{X},s),\, \boldsymbol{g}^t_\kappa(\boldsymbol{X},s),\, \boldsymbol{X}), \tag{3.26}$$

where $\boldsymbol{g}_\kappa = \operatorname{Grad}\theta$ is the temperature gradient with respect to the reference configuration. Note that, since $\rho_\kappa(\boldsymbol{X})$ and $\operatorname{Grad}\rho_\kappa(\boldsymbol{X})$ are merely functions of $\boldsymbol{X}$, their presence has been absorbed into the explicit dependence of the functional $\mathcal{F}_\kappa$ on $\boldsymbol{X}$.

For a change of frame, from (3.21) we have[1]

$$\nabla_{\boldsymbol{X}}(\chi^*_{\kappa^*}(\gamma(\boldsymbol{X}),t^*)) = Q(t)F(\boldsymbol{X},t).$$

Therefore, the condition of material objectivity (3.20) now takes the form,

$$\begin{aligned}
&\underset{0\leq s<\infty}{\mathcal{F}_\kappa} (\boldsymbol{x}^*(\boldsymbol{X},t^*-s),\, Q^t(s)F^t(\boldsymbol{X},s),\, \theta^t(\boldsymbol{X},s),\, \boldsymbol{g}^t_\kappa(\boldsymbol{X},s),\, \boldsymbol{X})\\
&\qquad = \boldsymbol{Q}^*(t)\ \underset{0\leq s<\infty}{\mathcal{F}_\kappa} (\boldsymbol{x}(\boldsymbol{X},t-s),\, F^t(\boldsymbol{X},s),\, \theta^t(\boldsymbol{X},s),\, \boldsymbol{g}^t_\kappa(\boldsymbol{X},s),\, \boldsymbol{X}),
\end{aligned} \tag{3.27}$$

---

[1] Note that the left-hand side is not $F^*(\boldsymbol{X},t^*)$, in general, (see the remark on p. 27). The assumption that the reference configuration is unaffected by the change of frame will give this relation as $\nabla_{\boldsymbol{X}}(\chi^*_\kappa(\boldsymbol{X},t^*)) = F^*(\boldsymbol{X},t^*) = Q(t)F(\boldsymbol{X},t)$, and hence will lead to the same condition of material objectivity.

where

$$\boldsymbol{x}(\boldsymbol{X},t) = \chi_\kappa(\boldsymbol{X},t), \qquad \boldsymbol{x}^*(\boldsymbol{X},t^*) = \chi^*_{\kappa^*}(\gamma(\boldsymbol{X}),t^*),$$

denote the position of the material point $X$ in the two frames, respectively.

We claim that the dependence of constitutive function on the position is not allowed by the principle of material objectivity. Indeed, since (3.27) is valid for any change of frame, in particular, let us consider a change of frame given by

$$\boldsymbol{x}^* = \boldsymbol{x} - \boldsymbol{x}_\circ + \boldsymbol{c}(t), \qquad t^* = t.$$

In this case, $Q(t)$ is an identity and the condition (3.27) becomes

$$\mathcal{F}_\kappa(\boldsymbol{x}(\boldsymbol{X},t-s)+(\boldsymbol{c}(t-s)-\boldsymbol{x}_\circ), F^t, \theta^t, \boldsymbol{g}^t_\kappa, \boldsymbol{X}) = \mathcal{F}_\kappa(\boldsymbol{x}(\boldsymbol{X},t-s), F^t, \theta^t, \boldsymbol{g}^t_\kappa, \boldsymbol{X}),$$

which proves our claim, since $\boldsymbol{c}(t) - \boldsymbol{x}_\circ \in V$ is arbitrary. Therefore, the constitutive relation (3.26) reduces to

$$\mathcal{C}(\boldsymbol{X},t) = \underset{0\leq s<\infty}{\mathcal{F}_\kappa}(F^t(\boldsymbol{X},s),\ \theta^t(\boldsymbol{X},s),\ \boldsymbol{g}^t_\kappa(\boldsymbol{X},s),\ \boldsymbol{X}), \tag{3.28}$$

and the condition of material objectivity (3.27) becomes

$$\begin{aligned} &\underset{0\leq s<\infty}{\mathcal{F}_\kappa}(Q^t(s)F^t(\boldsymbol{X},s),\ \theta^t(\boldsymbol{X},s),\ \boldsymbol{g}^t_\kappa(\boldsymbol{X},s),\ \boldsymbol{X}) \\ &\qquad = \boldsymbol{Q}^*(t)\ \underset{0\leq s<\infty}{\mathcal{F}_\kappa}(F^t(\boldsymbol{X},s),\ \theta^t(\boldsymbol{X},s),\ \boldsymbol{g}^t_\kappa(\boldsymbol{X},s),\ \boldsymbol{X}), \end{aligned} \tag{3.29}$$

or simply

$$\mathcal{F}_\kappa(Q^tF^t,\ \theta^t,\ \boldsymbol{g}^t_\kappa,\ \boldsymbol{X}) = \boldsymbol{Q}^*\mathcal{F}_\kappa(F^t,\ \theta^t,\ \boldsymbol{g}^t_\kappa,\ \boldsymbol{X}).$$

Note that, as we have remarked before, this condition imposes no restrictions on thermal histories, $\theta^t(\boldsymbol{X},s)$ and $\boldsymbol{g}^t_\kappa(\boldsymbol{X},s)$.

A body with its constitutive relation defined by (3.28) is called a *simple material body* (due to Noll).[2] For simple material bodies, the value of the constitutive quantities at the point $\boldsymbol{X}$ depend on the histories of deformation gradient, temperature, and temperature gradient at the point $\boldsymbol{X}$ only. This class of materials may have a perfect temporal memory but only a very limited non-local effect, namely, the neighboring points are taken into account, in the first gradient of deformation and temperature, only as a first approximation. Nevertheless, it turns out that this class is still general enough to include most material bodies of practical interest. *Non-simple material bodies* are by no means unimportant, as an example, in theories of mixtures and porous media [51] the density gradient (which is related to the second gradient of deformation) must be taken into account in constitutive equations in order to have a consistent theory.

[2] Noll, in his memoir [58], considered purely mechanical theory only.

A body is called *homogeneous* if there exists a reference configuration $\kappa$, called a *homogeneous configuration*, such that the constitutive function $\mathcal{F}_\kappa$ does not depend on $\boldsymbol{X}$ explicitly. Therefore for a homogeneous simple material body (3.28) reduces to

$$\mathcal{C}(\boldsymbol{X},t) = \mathop{\mathcal{F}_\kappa}_{0\leq s<\infty} (F^t(\boldsymbol{X},s),\, \theta^t(\boldsymbol{X},s),\, \boldsymbol{g}^t_\kappa(\boldsymbol{X},s)). \tag{3.30}$$

In our later discussions, explicit dependence on $\boldsymbol{X}$ of response functions is mostly irrelevant, therefore we shall omit it from the arguments of response functions for simplicity.

**Exercise 3.3.1** Consider an elastic material with the stress tensor given by $T = \mathcal{T}_\kappa(F)$ relative to a reference configuration $\kappa$. A configuration $\kappa$ is called a *natural* configuration if

$$\mathcal{T}_\kappa(\mathit{1}) = 0,$$

where $\mathit{1}$ is the identity transformation. Show that any rotation has no effect on a natural configuration. In other words, if $\hat{\kappa}$ is another reference configuration that differs from $\kappa$ by a rigid rotation, then $\hat{\kappa}$ is also a natural configuration.

**Exercise 3.3.2** A *hyperelastic* material is defined by the constitutive equation

$$T = \mathcal{T}(F) = \rho(\partial_F\sigma)F^T, \tag{3.31}$$

where $\sigma = \sigma(F)$ is called the *stored energy* function. Show that if $\sigma$ satisfies the objectivity condition, then so does $\mathcal{T}$.

**Exercise 3.3.3** Let $\boldsymbol{g} = \operatorname{grad}\theta$ be the spatial gradient of temperature and the constitutive relation (3.28) be rewritten as

$$\mathcal{C} = \widetilde{\mathcal{F}}_\kappa(F^t,\, \theta^t,\, \boldsymbol{g}^t),$$

then the condition of material objectivity becomes

$$\widetilde{\mathcal{F}}_\kappa(Q^tF^t,\, \theta^t,\, Q^t\boldsymbol{g}^t) = \boldsymbol{Q}^*\widetilde{\mathcal{F}}_\kappa(F^t,\, \theta^t,\, \boldsymbol{g}^t).$$

Note that this condition imposes restrictions on the response function $\widetilde{\mathcal{F}}$ not only on the history of the deformation gradient but also on that of the spatial temperature gradient, since $\boldsymbol{g}_\kappa = F^T\boldsymbol{g}$.

## 3.4 Reduced Constitutive Relations

For simple material bodies, response functions for the constitutive quantities (3.2), the stress tensor, the heat flux vector and the internal energy, are given by

$$
\begin{aligned}
T(\boldsymbol{X},t) &= \mathcal{T}(F^t,\theta^t,\boldsymbol{g}_\kappa^t),\\
\boldsymbol{q}(\boldsymbol{X},t) &= \mathcal{Q}(F^t,\theta^t,\boldsymbol{g}_\kappa^t),\\
\varepsilon(\boldsymbol{X},t) &= \mathcal{E}(F^t,\theta^t,\boldsymbol{g}_\kappa^t).
\end{aligned} \tag{3.32}
$$

Here, we have assumed that all constitutive quantities of a material body depend on the same set of independent variables. This is referred to as the *rule of equipresence* (Truesdell). Even though this rule is not to be considered as a physical principle, it may serve as a starting assumption for the formulation of a constitutive theory before any other requirements are analyzed, so as not to impose unnecessary restrictions or exclude some possible (but perhaps still unknown) cross-effects prematurely.

The constitutive functions (3.32) must satisfy the condition of material objectivity (3.29). By the definition (1.60) of $\boldsymbol{Q}^*$, we have the following conditions,

$$
\begin{aligned}
\mathcal{T}(Q^tF^t,\theta^t,\boldsymbol{g}_\kappa^t) &= Q\,\mathcal{T}(F^t,\theta^t,\boldsymbol{g}_\kappa^t)\,Q^T,\\
\mathcal{Q}(Q^tF^t,\theta^t,\boldsymbol{g}_\kappa^t) &= Q\,\mathcal{Q}(F^t,\theta^t,\boldsymbol{g}_\kappa^t), \qquad \forall\, Q(t)\in\mathcal{O}(V).\\
\mathcal{E}(Q^tF^t,\theta^t,\boldsymbol{g}_\kappa^t) &= \mathcal{E}(F^t,\theta^t,\boldsymbol{g}_\kappa^t),
\end{aligned} \tag{3.33}
$$

These conditions of material objectivity are typical for tensor, vector, and scalar objective constitutive quantities, respectively.

We have seen that the response functions must satisfy the conditions (3.33) imposed by the principle of material objectivity. Such restrictions can be made explicit in a couple of different ways, as we shall see now. Since the condition of material objectivity does not impose any restrictions on thermal histories, for simplicity, we shall omit them and write, typically for the stress tensor

$$
T=\mathcal{T}(F^t), \qquad \mathcal{T}(Q^tF^t)=Q\,\mathcal{T}(F^t)\,Q^T, \qquad \forall Q(t)\in\mathcal{O}(V), \tag{3.34}
$$

and for all non-singular $F(t)\in\mathcal{L}(V)$. Solutions of this condition are given in the following proposition. The extension to other vector or scalar constitutive functions is straightforward.

**Proposition 3.4.1** *For a simple material body, the constitutive function $\mathcal{T}$ satisfies the condition of material objectivity (3.34) if and only if it can be represented by*

$$
\mathcal{T}(F^t)=R\,\mathcal{T}(U^t)\,R^T, \tag{3.35}
$$

*where $F=RU$ is the polar decomposition of $F$. The restriction of $\mathcal{T}$ to the positive symmetric stretch history $U^t$ is arbitrary.*

*Proof.* Suppose that $\mathcal{T}(F^t)$ satisfies (3.34), then by the polar decomposition $F = RU$, we have

$$\mathcal{T}(Q^t R^t U^t) = Q\,\mathcal{T}(F^t) Q^T, \qquad \forall\, Q(t) \in \mathcal{O}(V),$$

and (3.35) follows immediately by choosing $Q = R^T$. Conversely, if (3.35) holds, then for any $Q \in \mathcal{O}(V)$, $QF$ is non-singular and

$$QF = QRU = (QR)U$$

is its polar decomposition. Hence, (3.35) implies that

$$\mathcal{T}(Q^t F^t) = (QR)\,\mathcal{T}(U^t)\,(QR)^T = Q\,(R\,\mathcal{T}(U^t)\,R^T)\,Q^T = Q\,\mathcal{T}(F^t)\,Q^T.$$

Therefore, (3.35) is also sufficient for (3.34). □

From the representation (3.35), we note that the material response depends only on the current value of rotation part of deformation. The past history of the rotation $R^t(s)$ for all $s > 0$ does not influence the material response at all.

**Corollary 3.4.2** *The representation (3.35) can be put into the following form:*

$$\mathcal{T}(F^t) = F\,\mathcal{S}(C^t)\,F^T, \tag{3.36}$$

*where $\mathcal{S}$ is an arbitrary function of the Cauchy–Green tensor $C = F^T F$.*

*Proof.* From (3.35) and $F = RU$, we have

$$F^{-1} T\, F^{-T} = U^{-1} R^T (R\,\mathcal{T}(U^t)\,R^T)\,RU^{-1} = U^{-1}\mathcal{T}(U^t)\,U^{-1} = \mathcal{S}(C^t),$$

where we have redefined the function $\mathcal{T}$ by the new function $\mathcal{S}$ depending on $C = U^2 = F^T F$. □

The constitutive relations (3.35) and (3.36) are the solutions of the condition (3.34). Therefore, they are not subjected to any further restrictions from the objectivity condition. Constitutive relations of this kind are said to be in *reduced forms.*

We can obtain a different version of the reduced constitutive relation (3.36), for quantities in material description introduced in (2.84), namely, the material heat flux vector $\boldsymbol{q}_\kappa = |J| F^{-1} \boldsymbol{q}$ and the (first) Piola–Kirchhoff stress tensor $T_\kappa = |J| T F^{-T}$. In addition, we shall define the *second Piola–Kirchhoff stress tensor* as

$$S_\kappa = F^{-1} T_\kappa = |J| F^{-1} T F^{-T},$$

which, unlike the (first) Piola–Kirchhoff stress tensor, is a symmetric tensor. As a consequence of (3.36) we have the following

**Proposition 3.4.3** *Constitutive relations for the second Piola–Kirchhoff stress tensor $S_\kappa$, the material heat flux $\boldsymbol{q}_\kappa$ and the internal energy density $\varepsilon$ of a simple material body satisfy the condition of material objectivity if and only if they can be represented in the following forms,*

$$\begin{aligned} S_\kappa &= \mathcal{S}_\kappa(C^t, \theta^t, \boldsymbol{g}_\kappa^t), \\ \boldsymbol{q}_\kappa &= \mathcal{Q}_\kappa(C^t, \theta^t, \boldsymbol{g}_\kappa^t), \\ \varepsilon &= \mathcal{E}_\kappa(C^t, \theta^t, \boldsymbol{g}_\kappa^t), \end{aligned} \tag{3.37}$$

*where $C$ is the right Cauchy–Green strain tensor, and $\boldsymbol{g}_\kappa = \operatorname{Grad}\theta$ is the temperature gradient with respect to the reference configuration.*

**Exercise 3.4.1** Consider the following constitutive relation for the stress tensor,

$$T = t_0 1 + t_1 B + t_2 B^2,$$

where $B$ is the left Cauchy–Green tensor, and $t_0$, $t_1$, and $t_2$ are arbitrary functions of $\operatorname{tr} B$, $\operatorname{tr} B^2$, and $\operatorname{tr} B^3$. Verify that this constitutive relation satisfies the objectivity condition and is a special case of (3.35).

**Exercise 3.4.2** Prove that for a simple material body, the stress tensor has a reduced constitutive relation of the form

$$\mathcal{T}(F^t) = R\,\widehat{\mathcal{T}}(R^T C_t^t R;\, C)\, R^T,$$

where $F = RU$, $C = U^2$, $C_t = F_t^T F_t$ and $F_t$ is the relative deformation gradient (see (1.39) and (1.42)).

**Exercise 3.4.3** Verify that for a simple material body, one has the following reduced constitutive relations:

$$\begin{aligned} T &= R\,\mathcal{T}(U^t, \theta^t, \boldsymbol{g}_\kappa^t)\, R^T, \\ \boldsymbol{q} &= R\,\mathcal{Q}(U^t, \theta^t, \boldsymbol{g}_\kappa^t), \\ \varepsilon &= \mathcal{E}(U^t, \theta^t, \boldsymbol{g}_\kappa^t). \end{aligned}$$

## 3.5 Material Symmetry

The constitutive relations of a simple material relative to a reference configuration $\kappa$ from (3.28) can be written in the form,

$$\mathcal{C}(\boldsymbol{x}, t) = \mathop{\mathcal{F}_\kappa}_{0 \le s < \infty} (F^t(\boldsymbol{X}, s), \theta^t(\boldsymbol{X}, s), \boldsymbol{g}_\kappa^t(\boldsymbol{X}, s)), \qquad \boldsymbol{x} = \chi_\kappa(\boldsymbol{X}, t). \tag{3.38}$$

Suppose that $\hat{\kappa}$ is another reference configuration and the constitutive relations relative to $\hat{\kappa}$ can be written as

$$\mathcal{C}(\boldsymbol{x},t) = \underset{0\leq s<\infty}{\mathcal{F}_{\hat{\kappa}}} \left(\widehat{F}^t(\hat{\boldsymbol{X}},s),\, \theta^t(\hat{\boldsymbol{X}},s),\, \hat{\boldsymbol{g}}^t_{\kappa}(\hat{\boldsymbol{X}},s)\right), \qquad \boldsymbol{x} = \chi_{\hat{\kappa}}(\hat{\boldsymbol{X}},t). \tag{3.39}$$

Let $P = \nabla_{\boldsymbol{X}} \hat{\kappa} \circ \kappa^{-1}$, then from (1.10)

$$F = \widehat{F}\, P, \qquad \boldsymbol{g}_{\kappa} = P^T \hat{\boldsymbol{g}}_{\kappa}.$$

By comparing (3.38) and (3.39), we have the following relation between the response functions $\mathcal{F}_{\kappa}$ and $\mathcal{F}_{\hat{\kappa}}$,

$$\mathcal{F}_{\hat{\kappa}}(\widehat{F}^t, \theta^t, \hat{\boldsymbol{g}}^t_{\kappa}) = \mathcal{F}_{\kappa}(\widehat{F}^t P, \theta^t, P^T \hat{\boldsymbol{g}}^t_{\kappa}). \tag{3.40}$$

Therefore, a response function relative to a reference configuration determines that relative to any other reference configuration. Note that $P$ is time independent since the change of reference configuration is.

A material body subjected to the same thermomechanical history at two different configurations $\kappa$ and $\hat{\kappa}$, in general, have different results. However, it may happen that the results are exactly the same if the material possesses a certain symmetry that makes it unable to distinguish $\kappa$ and $\hat{\kappa}$. For example, one can not distinguish the material body with a cubic crystal structure before and after a rotation of 90° about one of its crystal axes.

**Definition.** Two reference configurations $\kappa$ and $\hat{\kappa}$ are said to be *materially indistinguishable* if

$$\mathcal{F}_{\kappa}(\,\cdot\,) = \mathcal{F}_{\hat{\kappa}}(\,\cdot\,). \tag{3.41}$$

If we denote the gradient of the change from $\kappa$ to $\hat{\kappa}$ by $G = \nabla_{\boldsymbol{X}}(\hat{\kappa} \circ \kappa^{-1})$, then by (3.40), the condition (3.41) is equivalent to

$$\mathcal{F}_{\kappa}(F^t, \theta^t, \boldsymbol{g}^t_{\kappa}) = \mathcal{F}_{\kappa}(F^t G, \theta^t, G^T \boldsymbol{g}^t_{\kappa}), \qquad \forall\, (F, \theta, \boldsymbol{g}_{\kappa}). \tag{3.42}$$

We call a transformation $G \in \mathcal{L}(V)$ that satisfies (3.42) a *material symmetry transformation* with respect to $\kappa$.

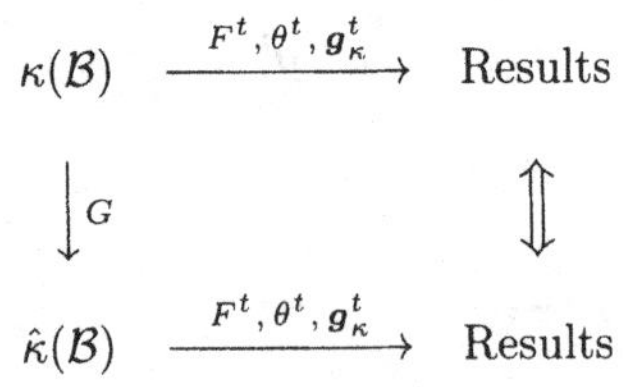


**Fig. 3.1.** Material symmetry transformation

Physically, this means that the response of the material with respect to the configuration $\kappa$ can not be distinguished from that with respect to the configuration $\hat{\kappa}$, obtained from $\kappa$ by a linear transformation $G$, by any possible (or ideal) experiment. The idea is schematically represented in the diagram (Fig. 3.1).

We shall assume that a material symmetry transformation is volume preserving (consequently, by (2.34) also density preserving), since otherwise, a material body could suffer arbitrarily large dilatation with no change in material response - a conclusion that seems physically unacceptable (for another justifications, see [27, 75]). Therefore, we must require that $|\det G| = 1$, i.e., a material transformation $G$ must be a unimodular transformation, $G \in \mathcal{U}(V)$.

**Proposition 3.5.1** *The set of all material symmetry transformations with respect to* $\kappa$, *denoted by* $\mathcal{G}_\kappa$, *is a subgroup of the unimodular group,*

$$\mathcal{G}_\kappa \subseteq \mathcal{U}(V). \tag{3.43}$$

*Proof.* We need only to show that $\mathcal{G}_\kappa$ is a group:
1) It is obvious that $1 \in \mathcal{G}_\kappa$.
2) For any $G_1, G_2 \in \mathcal{G}_\kappa$, by definition, we have

$$\begin{aligned}\mathcal{F}_\kappa(F^t, \theta^t, \boldsymbol{g}_\kappa^t) &= \mathcal{F}_\kappa(F^t G_1, \theta^t, G_1^T \boldsymbol{g}_\kappa^t)\\ &= \mathcal{F}_\kappa((F^t G_1)G_2,\, \theta^t,\, G_2^T(G_1^T \boldsymbol{g}_\kappa^t)) = \mathcal{F}_\kappa(F^t(G_1G_2), \theta^t, (G_1G_2)^T \boldsymbol{g}_\kappa^t),\end{aligned}$$

which implies that $G_1 G_2 \in \mathcal{G}_\kappa$.
3) For any $G \in \mathcal{G}_\kappa$, since $|\det G| = 1$, $G^{-1}$ exists, and

$$\mathcal{F}_\kappa(F^t G^{-1}, \theta^t, G^{-T} \boldsymbol{g}_\kappa^t) = \mathcal{F}_\kappa((F^t G^{-1})G, \theta^t, G^T(G^{-T} \boldsymbol{g}_\kappa^t)) = \mathcal{F}_\kappa(F^t, \theta^t, \boldsymbol{g}_\kappa^t),$$

which implies $G^{-1} \in \mathcal{G}_\kappa$. Therefore, $\mathcal{G}_\kappa$ is a group. □

We call $\mathcal{G}_\kappa$ the *material symmetry group* of $\mathcal{C}$ with respect to the reference configuration $\kappa$. It is clear that $\mathcal{G}_\kappa$ depends on $\kappa$ as well as the constitutive quantity $\mathcal{C}$. More specifically, we can denote it by $\mathcal{G}_\kappa(\mathcal{C})$. In other words, we may have different symmetry groups for different constitutive quantities of the same material body. The largest group contained in the symmetry groups of all constitutive quantities of $\mathcal{B}$ is called the *material symmetry group* of $\mathcal{B}$.

Suppose that $\hat{\kappa}$ is another reference configuration and $P = \nabla_{\boldsymbol{X}}(\hat{\kappa} \circ \kappa^{-1})$, then for any $G \in \mathcal{G}_\kappa$, we have from (3.40)

$$\begin{aligned}\mathcal{F}_{\hat{\kappa}}(F^t, \theta^t, \boldsymbol{g}_\kappa^t) &= \mathcal{F}_\kappa(F^t P, \theta^t, P^T \boldsymbol{g}_\kappa^t) = \mathcal{F}_\kappa((F^t P)G,\, \theta^t,\, G^T(P^T \boldsymbol{g}_\kappa^t))\\ &= \mathcal{F}_\kappa(F^t(PGP^{-1})P,\, \theta^t,\, P^T(PGP^{-1})^T \boldsymbol{g}_\kappa^t)\\ &= \mathcal{F}_{\hat{\kappa}}(F^t(PGP^{-1}),\, \theta^t,\, (PGP^{-1})^T \boldsymbol{g}_\kappa^t),\end{aligned}$$

which implies that $PGP^{-1} \in \mathcal{G}_{\hat{\kappa}}$. Therefore, we have proved the following proposition.

**Proposition 3.5.2** *For any $\kappa$ and $\hat{\kappa}$, such that $P = \nabla_\kappa(\hat{\kappa} \circ \kappa^{-1})$, the following relation holds,*

$$\mathcal{G}_{\hat{\kappa}} = P \mathcal{G}_\kappa P^{-1}. \tag{3.44}$$

The relation (3.44) is referred to as *Noll's rule,* which establishes the relation between symmetry groups with respect to two different reference configurations. Note that since $P$ is a change of reference configuration it is non-singular, but need not be unimodular. In particular, if $P = \alpha 1$, $\alpha \neq 0$, then $\mathcal{G}_{\hat{\kappa}} = \mathcal{G}_\kappa$, i.e., dilatation ($\alpha > 1$), contraction ($0 < \alpha < 1$), and central inversion ($\alpha = -1$) leave the material symmetry group invariant.

Physical concepts of real materials such as *solids* and *fluids*, are usually characterized by their symmetry properties. For example, one such concept can be interpreted as saying that a solid has a preferred configuration such that any non-rigid deformation from it alters its material response. while for a fluid any deformation that preserves the density should not affect the material response. Based on this concept, we shall give the following definitions of solids and fluids due to Noll [58].

**Definition.** A material is called a *solid* if there exists a reference configuration $\kappa$, such that $\mathcal{G}_\kappa$ is the full orthogonal group or a subgroup of it, i.e.,

$$\mathcal{G}_\kappa \subseteq \mathcal{O}(V). \tag{3.45}$$

Such a configuration is called an *undistorted* configuration for the solid.

**Definition.** A material is called a *fluid* if, for a reference configuration $\kappa$, the symmetry group is the full unimodular group, i.e.,

$$\mathcal{G}_\kappa = \mathcal{U}(V). \tag{3.46}$$

For a fluid, Noll's rule (3.44) implies that

$$\mathcal{G}_\kappa = \mathcal{G}_{\hat{\kappa}} \qquad \forall \kappa, \hat{\kappa}. \tag{3.47}$$

Therefore, a fluid has the same symmetry group with respect to any configuration. In other words, a fluid does not have a preferred configuration.

**Definition.** A material that is neither a fluid nor a solid will be called a *fluid crystal.*

In other words, for a fluid crystal, there does not exist a reference configuration $\kappa$ for which either $\mathcal{G}_\kappa \subseteq \mathcal{O}(V)$ or $\mathcal{G}_\kappa = \mathcal{U}(V)$.

**Exercise 3.5.1** Consider a constitutive equation for the stress tensor given by $T = \mathcal{T}_\kappa(F, \theta, \boldsymbol{g})$. Use the material objectivity condition to show that if $G \in \mathcal{G}_\kappa(T)$, then $-G \in \mathcal{G}_\kappa(T)$. Moreover, does this conclusion remain valid for the symmetry group $\mathcal{G}_\kappa(\boldsymbol{q})$ of the heat flux $\boldsymbol{q} = \mathcal{Q}_\kappa(F, \theta, \boldsymbol{g})$?

**Exercise 3.5.2** A material for which $\mathcal{G}_\kappa = \{1, -1\}$ is a *triclinic* solid. Show that the relation (3.47) also holds for this solid.

**Exercise 3.5.3** Let $\kappa$ and $\hat{\kappa}$ be two undistorted configurations of a solid, and let $P = \nabla_\kappa(\hat{\kappa} \circ \kappa^{-1})$. Suppose that the polar decomposition of $P$ is $P = RU$. Show that

$$\mathcal{G}_{\hat{\kappa}} = R\mathcal{G}_\kappa R^T, \quad \text{and} \quad U = QUQ^T, \quad \forall Q \in \mathcal{G}_\kappa. \tag{3.48}$$

Moreover, if $\mathcal{G}_\kappa = \mathcal{O}(V)$ then $\mathcal{G}_{\hat{\kappa}} = \mathcal{G}_\kappa$ and $U = \alpha 1$ for $\alpha \in I\!R$.

### 3.5.1 Constitutive Equation for a Simple Solid Body

Combining the reduced constitutive relations (3.37) and the symmetry condition (3.42), we have the following theorem that characterizes the most general constitutive relations for simple solid bodies.

**Proposition 3.5.3** *For a simple solid body, the constitutive relations for the second Piola–Kirchhoff stress, the material heat flux and the internal energy, relative to an undistorted configuration $\kappa$ with symmetry group $\mathcal{G}_\kappa$, satisfy the conditions of material objectivity and symmetry if and only if the constitutive functions for $\mathcal{C}_\kappa = \{S_\kappa, \boldsymbol{q}_\kappa, \varepsilon\}$ can be represented by*

$$\mathcal{C}_\kappa = \mathcal{F}_\kappa(C^t, \theta^t, \boldsymbol{g}^t_\kappa).$$

*Moreover, $\mathcal{F}_\kappa = \{\mathcal{S}_\kappa, \mathcal{Q}_\kappa, \mathcal{E}_\kappa\}$ must satisfy the following conditions,*

$$\left.\begin{aligned} &\mathcal{S}_\kappa(QC^tQ^T, \theta^t, Q\boldsymbol{g}^t_\kappa) = Q\,\mathcal{S}_\kappa(C^t, \theta^t, \boldsymbol{g}^t_\kappa)\,Q^T, \\ &\mathcal{Q}_\kappa(QC^tQ^T, \theta^t, Q\boldsymbol{g}^t_\kappa) = Q\,\mathcal{Q}_\kappa(C^t, \theta^t, \boldsymbol{g}^t_\kappa), \\ &\mathcal{E}_\kappa(QC^tQ^T, \theta^t, Q\boldsymbol{g}^t_\kappa) = \mathcal{E}_\kappa(C^t, \theta^t, \boldsymbol{g}^t_\kappa), \end{aligned}\right. \qquad \forall\, Q \in \mathcal{G}_\kappa \subseteq \mathcal{O}(V). \tag{3.49}$$

*Proof.* By the Proposition 3.4.3, we only have to prove the conditions (3.49). We shall prove for the stress tensor $T$, from the definition (2.84) and the reduced constitutive relation (3.37), we have

$$\mathcal{T}(F^t, \theta^t, \boldsymbol{g}^t_\kappa) = |\det F|^{-1} F\,\mathcal{S}_\kappa(C^t, \theta^t, \boldsymbol{g}^t_\kappa)\,F^T.$$

For a solid, we have $Q \in \mathcal{G}_\kappa \subseteq \mathcal{O}(V)$, and by taking $G = Q^T = Q^{-1} \in \mathcal{G}_\kappa$ in the condition (3.42), we obtain, from the above relation, that

$$\begin{aligned} &|\det F|^{-1} F\,\mathcal{S}_\kappa(C^t, \theta^t, \boldsymbol{g}^t_\kappa)\,F^T \\ &\quad = |\det(FQ^T)|^{-1}(FQ^T)\,\mathcal{S}_\kappa((F^tQ^T)^T(F^tQ^T),\, \theta^t,\, Q\boldsymbol{g}^t_\kappa)\,(FQ^T)^T. \end{aligned}$$

Since $|\det Q| = 1$, it becomes

$$\mathcal{S}_\kappa(C^t, \theta^t, \boldsymbol{g}^t_\kappa) = Q^T\mathcal{S}_\kappa(QC^tQ^T, \theta^t, Q\boldsymbol{g}^t_\kappa)\,Q,$$

which proves $(3.49)_1$. The proofs for $\mathcal{Q}_\kappa$ and $\mathcal{E}_\kappa$ are similar. □

Relations of the type (3.49) for constitutive functions are of particular interest in the light of representations to be considered in the next chapter.

We note that, in general, $G \in \mathcal{G}_\kappa$ does *not* imply $G^T \in \mathcal{G}_\kappa$, but for $G \in \mathcal{O}(V) \cap \mathcal{G}_\kappa$, since $G^T = G^{-1}$, $G^T$ must also belong to $\mathcal{G}_\kappa$. Therefore, the condition of material objectivity (3.33) and the material symmetry condition (3.42) imply the following condition,

$$\mathcal{T}(QF^tQ^T, \theta^t, Q\boldsymbol{g}_\kappa^t) = Q\,\mathcal{T}(F^t, \theta^t, \boldsymbol{g}_\kappa^t)\,Q^T,$$

for all $Q \in \mathcal{O}(V) \cap \mathcal{G}_\kappa$. In particular, it is valid for solids. However, unlike the constitutive function $\mathcal{S}_\kappa$ of (3.49), the function $\mathcal{T}$ is still required to satisfy the objectivity condition (3.33).

### 3.5.2 Constitutive Equation for a Simple Fluid

In order to derive the general constitutive relations of simple fluids, let us rewrite the constitutive relation (3.38) in terms of relative description (see Fig. 1.5). From (1.42), the *relative* deformation gradient $F_t$, defined by (1.39), is related to the deformation gradient $F$ by the relation,

$$F_t(\boldsymbol{x}, \tau) = F(\boldsymbol{X}, \tau)\, F(\boldsymbol{X}, t)^{-1}, \qquad \boldsymbol{x} = \chi_\kappa(\boldsymbol{X}, t). \tag{3.50}$$

We can also define a *relative* temperature gradient $\boldsymbol{g}_t$ in a similar manner,

$$\boldsymbol{g}_t(\boldsymbol{x}, \tau) = \nabla_{\boldsymbol{x}} \theta(\chi_\kappa^{-1}(\boldsymbol{x}, t), \tau),$$

which implies that

$$\boldsymbol{g}_t(\boldsymbol{x}, \tau) = F^{-T}(\boldsymbol{X}, t)\, \boldsymbol{g}_\kappa(\boldsymbol{X}, \tau), \tag{3.51}$$

where $\boldsymbol{g}_\kappa = \operatorname{Grad}\theta$ is the referential temperature gradient in the configuration $\kappa$. From (3.50) and (3.51) we have

$$F_t^t(s) = F^t(s) F(t)^{-1}, \qquad \boldsymbol{g}_t^t(s) = F(t)^{-T} \boldsymbol{g}_\kappa^t(s),$$

and

$$F_t(t) = \mathit{1}, \qquad \boldsymbol{g}_t(t) = \boldsymbol{g},$$

where $\boldsymbol{g} = \operatorname{grad}\theta$ is the spatial temperature gradient.

In terms of relative gradient of deformation and temperature, we can rewrite (3.38) as

$$C(\boldsymbol{x}, t) = \underset{0 \le s < \infty}{\widetilde{\mathcal{F}}}\,(F_t^t(\boldsymbol{x}, s),\ \theta^t(\boldsymbol{x}, s),\ \boldsymbol{g}_t^t(\boldsymbol{x}, s);\ F(\boldsymbol{x}, t)). \tag{3.52}$$

The constitutive functions $\widetilde{\mathcal{F}}$ and $\mathcal{F}_\kappa$ of (3.38), from (3.50) and (3.51), are related by

$$\mathcal{F}_\kappa(F^t, \theta^t, \boldsymbol{g}_\kappa^t) = \mathcal{F}_\kappa(F_t^t F,\, \theta^t,\, F^T \boldsymbol{g}_t^t) = \widetilde{\mathcal{F}}(F_t^t, \theta^t, \boldsymbol{g}_t^t; F),$$

from which one can verify that the condition of material objectivity (3.33) and material symmetry (3.42) become

$$\begin{aligned} &\widetilde{\mathcal{F}}(Q^t F_t^t Q^T, \theta^t, Q\boldsymbol{g}_t^t; QF) = \boldsymbol{Q}^* \widetilde{\mathcal{F}}(F_t^t, \theta^t, \boldsymbol{g}_t^t; F); \quad && \forall Q \in \mathcal{O}(V), \\ &\widetilde{\mathcal{F}}(F_t^t, \theta^t, \boldsymbol{g}_t^t; FG) = \widetilde{\mathcal{F}}(F_t^t, \theta^t, \boldsymbol{g}_t^t; F), && \forall G \in \mathcal{G}_\kappa, \end{aligned} \tag{3.53}$$

respectively. To see this, note that we have

$$\mathcal{F}_\kappa(\bar{F}^t, \theta^t, \bar{\boldsymbol{g}}_\kappa^t) = \widetilde{\mathcal{F}}(\bar{F}_t^t, \theta^t, \bar{\boldsymbol{g}}_t^t; \bar{F}),$$

with

$$\bar{F}_t^t = \bar{F}^t \bar{F}^{-1}, \qquad \bar{\boldsymbol{g}}_t^t = \bar{F}^{-T} \bar{\boldsymbol{g}}_\kappa^t.$$

By taking first $\bar{F} = QF$ and $\bar{\boldsymbol{g}}_\kappa = \boldsymbol{g}_\kappa$ for $Q \in \mathcal{O}(V)$, we have

$$\begin{aligned} \bar{F}_t^t &= Q^t F^t F^{-1} Q^T = Q^t F_t^t Q^T, \\ \bar{\boldsymbol{g}}_t^t &= Q F^{-T} \boldsymbol{g}_\kappa^t = Q\boldsymbol{g}_t^t, \end{aligned}$$

which lead to $(3.53)_1$ from (3.33). Then, by taking $\bar{F} = FG$ and $\bar{\boldsymbol{g}}_\kappa = G^T \boldsymbol{g}_\kappa$ for $G \in \mathcal{G}_\kappa$, we have

$$\begin{aligned} \bar{F}_t^t &= F^t G G^{-1} F^{-1} = F^t F^{-1} = F_t^t, \\ \bar{\boldsymbol{g}}_t^t &= F^{-T} G^{-T} G^T \boldsymbol{g}_\kappa^t = F^{-T} \boldsymbol{g}_\kappa^t = \boldsymbol{g}_t^t, \end{aligned}$$

which lead to $(3.53)_2$ from (3.42).

We can obtain general constitutive relations for a simple fluid, with symmetry group $\mathcal{G}_\kappa = \mathcal{U}(V)$, from the conditions (3.53).

**Proposition 3.5.4** *For a simple fluid, the constitutive function satisfies the conditions of objectivity and symmetry (3.53) if and only if it can be represented by*

$$\mathcal{C} = \widehat{\mathcal{F}}(C_t^t, \theta^t, \boldsymbol{g}_t^t; \rho), \tag{3.54}$$

*where $C_t = F_t^T F_t$ is the relative Cauchy–Green strain tensor and $\rho$ is the mass density in the present configuration. Moreover, the constitutive functions $\widehat{\mathcal{F}} = \{\widehat{\mathcal{T}}, \widehat{\mathcal{Q}}, \widehat{\mathcal{E}}\}$ for the stress tensor, heat flux vector, and internal energy must satisfy the following conditions:*

$$\begin{aligned} &\widehat{\mathcal{T}}(QC_t^t Q^T, \theta_t^t, Q\boldsymbol{g}_t^t; \rho) = Q\, \widehat{\mathcal{T}}(C_t^t, \theta_t^t, \boldsymbol{g}_t^t; \rho)\, Q^T, \\ &\widehat{\mathcal{Q}}(QC_t^t Q^T, \theta_t^t, Q\boldsymbol{g}_t^t; \rho) = Q\, \widehat{\mathcal{Q}}(C_t^t, \theta_t^t, \boldsymbol{g}_t^t; \rho), \qquad \forall\, Q \in \mathcal{O}(V). \\ &\widehat{\mathcal{E}}(QC_t^t Q^T, \theta_t^t, Q\boldsymbol{g}_t^t; \rho) = \widehat{\mathcal{E}}(C_t^t, \theta_t^t, \boldsymbol{g}_t^t; \rho), \end{aligned} \tag{3.55}$$

*Proof.* Since for a simple fluid, the symmetry group $\mathcal{G}_\kappa = \mathcal{U}(V)$, by taking the unimodular tensor $G = |\det F|^{1/3} F^{-1}$ the condition $(3.53)_2$ gives

$$\widetilde{\mathcal{F}}(F_t^t, \theta^t, \boldsymbol{g}_t^t; F) = \widetilde{\mathcal{F}}(F_t^t, \theta^t, \boldsymbol{g}_t^t; |\det F|^{1/3} \mathit{1}).$$

Since $|\det F| = \rho_\kappa/\rho$, where $\rho_\kappa$ is the mass density at the reference configuration, we can rewrite the function $\widetilde{\mathcal{F}}$ as

$$\widetilde{\mathcal{F}}(F_t^t, \theta^t, \boldsymbol{g}_t^t; F) = \widehat{\mathcal{F}}(F_t^t, \theta^t, \boldsymbol{g}_t^t; \rho).$$

The condition $(3.53)_1$ for the stress tensor now becomes

$$\widetilde{\mathcal{T}}(F_t^t, \theta^t, \boldsymbol{g}_t^t; F) = Q^T \widetilde{\mathcal{T}}(Q^t F_t^t Q^T, \theta_t^t, Q\boldsymbol{g}_t^t; \rho)\, Q.$$

With the decomposition $F_t^t = R_t^t U_t^t$, taking $Q^t(s) = R_t^t(s)^T$, hence $Q(t) = R_t^t(0) = \mathit{1}$, we obtain (3.54) by replacing $U_t^t$ with $C_t^t = (U_t^t)^2$. Similarly, choosing $Q^t(s) = \bar{Q}(t)\, R_t^t(s)^T$, hence $Q(t) = \bar{Q}(t)$, for any $\bar{Q}(t) \in \mathcal{O}(V)$, we obtain

$$\widehat{\mathcal{T}}(\bar{Q} U_t^t \bar{Q}^T, \theta^t, \bar{Q}\boldsymbol{g}_t^t; \rho) = \bar{Q}\, \widehat{\mathcal{F}}(U_t^t, \theta^t, \boldsymbol{g}_t^t; \rho)\, \bar{Q}^T,$$

which implies $(3.55)_1$. The proof of sufficiency is straightforward. □

From this representation, we conclude that, for a simple fluid, the constitutive dependence on the deformation gradient $F$ reduces to the dependence on its determinant $\det F$ only.

### 3.5.3 Fluid Crystal with an Intrinsic Direction

We shall not consider fluid crystals in more detail in this book, except for the following interesting example.

Consider a fluid crystal with the symmetry group $\mathcal{G}_\kappa$ relative to an orthonormal basis $\{\boldsymbol{e}_1, \boldsymbol{e}_2, \boldsymbol{e}_3\}$ in the reference configuration $\kappa$ given by

$$\mathcal{G}_\kappa = \left\{ \begin{bmatrix} a & d & 0 \\ b & e & 0 \\ c & f & g \end{bmatrix}, \ |g(ae - bd)| = 1 \right\}.$$

One can easily verify that this is a subgroup of $\mathcal{U}(V)$. Moreover, for any $G \in \mathcal{G}_\kappa$, we have

$$G\boldsymbol{e}_3 \,/\!/\, \boldsymbol{e}_3.$$

In other words, this material has an intrinsic direction $\boldsymbol{e}_3$ in the reference configuration $\kappa$. A material with needle-like crystals that can flow in such a way that they are always aligned in one direction (therefore, it is called a fluid crystal) is an example of materials that possess such a symmetry.

Let the constitutive equation for the stress tensor be given by $T = \mathcal{T}(F)$, then the objectivity and the symmetry conditions require that

$$\begin{aligned} \mathcal{T}(QF) &= Q\mathcal{T}(F)Q^T, \quad && \forall Q \in \mathcal{O}(V), \\ \mathcal{T}(FG) &= \mathcal{T}(F), && \forall G \in \mathcal{G}_\kappa. \end{aligned} \tag{3.56}$$

In order to analyze these conditions, first we note that they imply

$$\mathcal{T}(RFR^T) = R\mathcal{T}(F)R^T, \quad \forall R \in \mathcal{O}(V) \cap \mathcal{G}_\kappa. \tag{3.57}$$

In particular, for $F = \alpha\mathit{1}$, it leads to

$$\mathcal{T}(\alpha\mathit{1}) = R\mathcal{T}(\alpha\mathit{1})R^T.$$

That is, the orthogonal tensor $R$ commutes with the symmetric tensor $\mathcal{T}(\alpha\mathit{1})$. By the commutation theorem (see Sect. A.1.8), $R$ must preserve all the characteristic spaces of $\mathcal{T}(\alpha\mathit{1})$. Note that the general form for $R \in \mathcal{O}(V) \cap \mathcal{G}_\kappa$ is

$$R = \pm \begin{bmatrix} \cos\theta & \sin\theta & 0 \\ -\sin\theta & \cos\theta & 0 \\ 0 & 0 & 1 \end{bmatrix},$$

which preserves two characteristic spaces, a one-dimensional one, spanned by $\boldsymbol{e}_3$, and a two-dimensional space, spanned by $\{\boldsymbol{e}_1, \boldsymbol{e}_2\}$. Therefore, by the spectral theorem, $\mathcal{T}(\alpha\mathit{1})$ must be of the form,

$$\begin{aligned} \mathcal{T}(\alpha\mathit{1}) &= \tau(\alpha)(\boldsymbol{e}_1 \otimes \boldsymbol{e}_1 + \boldsymbol{e}_2 \otimes \boldsymbol{e}_2) + \tau'(\alpha)\, \boldsymbol{e}_3 \otimes \boldsymbol{e}_3 \\ &= \tau(\alpha)\mathit{1} + \tau_3(\alpha)\boldsymbol{e}_3 \otimes \boldsymbol{e}_3, \end{aligned} \tag{3.58}$$

where $\tau_3 = \tau' - \tau$.

For an arbitrary $F$, suppose we can decompose $F$ in the form,

$$F = \alpha KG, \quad K \in \mathcal{O}(V), \quad G \in \mathcal{G}_\kappa.$$

Then, since $G\boldsymbol{e}_3 \,/\!/\, \boldsymbol{e}_3$, we have $F\boldsymbol{e}_3 \,/\!/\, K\boldsymbol{e}_3$ and hence

$$K\boldsymbol{e}_3 = \frac{F\boldsymbol{e}_3}{|F\boldsymbol{e}_3|}.$$

Therefore, this decomposition is possible if we define $K$ as a rotation that takes the unit vector $\boldsymbol{e}_3$ into the unit vector in the direction of $F\boldsymbol{e}_3$. Moreover, since $|\det K| = |\det G| = 1$, we must have $\alpha = |\det F|^{1/3}$. Substituting this decomposition of $F$ into (3.57), we obtain

$$\begin{aligned} \mathcal{T}(F) &= R^T\mathcal{T}(\alpha RKGR^T)R = R^T\mathcal{T}(\alpha RK)R \\ &= \mathcal{T}(\alpha K) = K\mathcal{T}(\alpha\mathit{1})K^T \\ &= \tau(\alpha)\mathit{1} + \tau_3(\alpha)K(\boldsymbol{e}_3 \otimes \boldsymbol{e}_3)K^T, \end{aligned}$$

where we have made repeated use of (3.56) and, in the last step, of the relation

(3.58). Since $|\det F| = \rho_\kappa/\rho$, we can write the constitutive equation as

$$\mathcal{T}(F) = -p(\rho)1 + \sigma(\rho)\frac{Fe_3}{|Fe_3|} \otimes \frac{Fe_3}{|Fe_3|}. \tag{3.59}$$

Conversely, one can easily show that if (3.59) holds then the conditions (3.56) are satisfied. Therefore, we have shown that the stress of this fluid crystal consists of a hydrostatic pressure and a pure tension in the deformed intrinsic direction. Moreover, the pressure $p$ and the tensile stress $\sigma$ are functions of density only.

Note that in this example, we are able to reduce the constitutive function for the stress $T = \mathcal{T}(F)$, a symmetric tensor function of a tensor variable (equivalent to six scalar functions of nine variables) to a representation that is explicit to within two scalar functions of one variable only. Representations of this kind are among the most useful results of the constitutive theory. Other representations of constitutive functions for solids and fluids will be discussed in the next chapter.

**Exercise 3.5.4** Relative to an orthonormal basis, let the material symmetry group of a fluid crystal be given by

$$\mathcal{G}_\kappa = \left\{ \begin{bmatrix} a & b & c \\ d & e & f \\ 0 & 0 & g \end{bmatrix}, \; |g(ae - bd)| = 1 \right\}.$$

Show that it is a group and analyze the constitutive properties of this material.

## 3.6 Isotropic Materials

**Definition.** A material body is called *isotropic* if there exists a configuration $\kappa$, such that

$$\mathcal{G}_\kappa \supseteq \mathcal{O}(V). \tag{3.60}$$

Such a configuration is called *undistorted* for the isotropic material body, or simply an *isotropic* configuration.

Physically, we can interpret the above definition as saying that any rotation (orthogonal transformation) from an isotropic configuration does not alter the material response. The following theorem characterizes isotropic materials [59].

**Theorem 3.6.1.** *The orthogonal group is maximal in the unimodular group, i.e., if $\mathcal{G}$ is a group such that*

$$\mathcal{O}(V) \subset \mathcal{G} \subset \mathcal{U}(V),$$

*then either $\mathcal{G} = \mathcal{O}(V)$ or $\mathcal{G} = \mathcal{U}(V)$.*

Therefore, any isotropic material is either a fluid, $\mathcal{G}_\kappa = \mathcal{U}(V)$, or an isotropic solid, $\mathcal{G}_\kappa = \mathcal{O}(V)$. Any other materials are anisotropic. Anisotropic materials include fluid crystals and anisotropic solids.

It is obvious that, for an isotropic solid, an isotropic configuration is also an undistorted configuration, defined earlier in (3.45). Conversely, if $\hat{\kappa}$ is another undistorted configuration, i.e., $\mathcal{G}_{\hat{\kappa}} \subseteq \mathcal{O}(V)$, then from (3.48), it follows that

$$\mathcal{G}_{\hat{\kappa}} = \mathcal{G}_\kappa = \mathcal{O}(V),$$

in other words, for an isotropic solid, an undistorted configuration is also an isotropic configuration. Moreover, by Noll's rule (3.44) and the relation (3.48) two isotropic configurations $\kappa$ and $\hat{\kappa}$ can differ by a dilatation and a rotation only (see Exercise 3.5.3).

In general, only certain particular reference configurations are undistorted for a given solid body. For a distorted configuration of the solid, the symmetry group neither contains the orthogonal group nor is contained within it.

For an anisotropic solid, the symmetry group relative to an undistorted configuration is a proper subgroup of the orthogonal group, $\mathcal{G}_\kappa \subset \mathcal{O}(V)$ and $\mathcal{G}_\kappa \neq \mathcal{O}(V)$. Its constitutive relation is already given in Proposition 3.5.3, no further general reductions can be made unless the symmetry group $\mathcal{G}_\kappa$ is specified.

**Example 3.6.1** *Transversely isotropic and orthotropic solid bodies.*

Relative to an orthonormal basis $\{\boldsymbol{n}_1, \boldsymbol{n}_2, \boldsymbol{n}_3\}$ in a reference configuration $\kappa$, let the material symmetry groups of two anisotropic solids be given by

$$\begin{aligned} \mathcal{G}_1 &= \{Q \in \mathcal{O}(V),\ Q\boldsymbol{n}_1 = \boldsymbol{n}_1\}, \\ \mathcal{G}_2 &= \{Q \in \mathcal{O}(V),\ Q(\boldsymbol{n}_1 \otimes \boldsymbol{n}_1)Q^T = \boldsymbol{n}_1 \otimes \boldsymbol{n}_1, \\ &\qquad Q(\boldsymbol{n}_2 \otimes \boldsymbol{n}_2)Q^T = \boldsymbol{n}_2 \otimes \boldsymbol{n}_2,\ Q(\boldsymbol{n}_3 \otimes \boldsymbol{n}_3)Q^T = \boldsymbol{n}_3 \otimes \boldsymbol{n}_3\}. \end{aligned} \tag{3.61}$$

The group $\mathcal{G}_1$, containing any rotation that preserves the vector $\boldsymbol{n}_1$, is called a *transverse isotropy* group, and a material body having this symmetry is called a transversely isotropic solid, while the group $\mathcal{G}_2$, containing any rotation that preserves three mutually orthogonal directions, is called an *orthotropy* group, and a material possessing such a symmetry is called an orthotropic solid.

These groups are merely examples of anisotropic solids. In particular, we remark that $\mathcal{G}_1$ is not the only group that characterizes the so-called transverse isotropy, other characterizations are possible.

Some representations for the constitutive functions of these material bodies will be discussed in the next chapter. □

### 3.6.1 Constitutive Equation of an Isotropic Material

For isotropic materials, besides the general constitutive equations for solid bodies given in Proposition 3.5.3 for $\mathcal{G}_\kappa = \mathcal{O}(V)$, we also have the following representation.

**Proposition 3.6.2** *For an isotropic simple material body, the constitutive function satisfies the condition of material objectivity and symmetry if and only if it can be represented by*

$$\mathcal{C} = \widehat{\mathcal{F}}(C_t^t, \theta_t^t, \boldsymbol{g}_t^t; B), \tag{3.62}$$

*where $C_t = F_t^T F_t$ and $B = FF^T$. Moreover, the function $\widehat{\mathcal{F}} = \{\widehat{\mathcal{T}}, \widehat{\mathcal{Q}}, \widehat{\mathcal{E}}\}$ for the stress tensor, heat flux vector and internal energy must satisfy the following conditions:*

$$\begin{aligned}
\widehat{\mathcal{T}}(QC_t^tQ^T, \theta_t^t, Q\boldsymbol{g}_t^t; QBQ^T) &= Q\,\widehat{\mathcal{T}}(C_t^t, \theta_t^t, \boldsymbol{g}_t^t; B)\,Q^T, \\
\widehat{\mathcal{Q}}(QC_t^tQ^T, \theta_t^t, Q\boldsymbol{g}_t^t; QBQ^T) &= Q\,\widehat{\mathcal{Q}}(C_t^t, \theta_t^t, \boldsymbol{g}_t^t; B), \qquad \forall\, Q \in \mathcal{O}(V). \\
\widehat{\mathcal{E}}(QC_t^tQ^T, \theta_t^t, Q\boldsymbol{g}_t^t; QBQ^T) &= \widehat{\mathcal{E}}(C_t^t, \theta_t^t, \boldsymbol{g}_t^t; B),
\end{aligned} \tag{3.63}$$

*Proof.* For isotropic materials, $\mathcal{G} = \mathcal{O}(V)$, by taking $F = RU$ and $G = R^T$, the condition $(3.53)_2$ becomes

$$\widetilde{\mathcal{F}}(F_t^t, \theta^t, \boldsymbol{g}_t^t; F) = \widetilde{\mathcal{F}}(F_t^t, \theta^t, \boldsymbol{g}_t^t; RUR^T).$$

Since $V = RUR^T$ and $B = V^2$, the function $\widetilde{\mathcal{F}}$ can be rewritten as

$$\widetilde{\mathcal{F}}(F_t^t, \theta^t, \boldsymbol{g}_t^t; F) = \widehat{\mathcal{F}}(F_t^t, \theta^t, \boldsymbol{g}_t^t; B).$$

The rest of the proof is similar to that of Proposition 3.5.4. □

From this proposition we see that the rotation does not affect the response of isotropic materials at all, which is an expected result, of course. Since fluids and isotropic solids are the only isotropic materials, the above representation (3.62) is essentially a representation of isotropic solid bodies. The representation (3.54) for fluids is a special case of (3.62), in which the dependence on the left Cauchy–Green tensor $B$ reduces to the dependence of $\det B$ only.

**Exercise 3.6.1** Consider a thermoelastic material with the constitutive relation for the stress tensor given by $T = \mathcal{T}(F, \theta, \boldsymbol{g})$. Prove directly from the objectivity and the symmetry conditions that if the material is isotropic then the constitutive relation must reduce to the form

$$T = \widehat{\mathcal{T}}(B, \theta, \boldsymbol{g}),$$

where $\kappa$ is an isotropic configuration and $\widehat{\mathcal{T}}$ is an isotropic function.

## 3.7 Fading Memory

In this section, we shall give a brief discussion of memory effects of simple materials and the linear approximation that gives rise to the Boltzmann–Volterra theory of viscoelasticity.

We shall consider only a mechanical theory. From the general solution (3.35) for simple materials, the stress tensor can be represented by

$$T(t) = \mathcal{T}(F^t) = R(t)\,\mathcal{T}(U^t)\,R(t)^T, \qquad F = RU.$$

The dependence on $\boldsymbol{X}$ is not indicated, for simplicity. We emphasize that in this representation, $\mathcal{T}(F^t)$ and $\mathcal{T}(U^t)$ are the values of the same functional $\mathcal{T}(\cdot)$ evaluated at two different functions $F^t(s)$ and $U^t(s)$ for $s \in [0, \infty)$, respectively.

It is useful to put the functional $\mathcal{T}(U^t)$ in a slightly different form by writing it as a sum of a term at the present value of deformation and a term that vanishes when the body has always been at rest,

$$\mathcal{T}(U^t) = \widehat{\mathcal{T}}(U(t)) + \mathcal{K}(G^t; U(t)), \tag{3.64}$$

where

$$G^t(s) = U^t(s) - U(t), \qquad \text{and} \qquad \mathcal{K}(0; U(t)) = 0.$$

$G^t(s)$ is the past history of the digression from the present state. We shall now make an additional physical assumption that the memory fades in time. In order to do this we shall first introduce the concept of an influence function that is used to characterize the memory of past histories the material can recall.[3]

**Definition.** We call $h(s)$, defined for $0 \le s < \infty$, an *influence function* if it is monotonically decreasing and

$$h(s) > 0, \qquad \int_0^\infty h(s)\, ds < \infty. \tag{3.65}$$

The influence function is material dependent. $h(s) = (s+1)^{-p}$ for $p > 1$, and $h(s) = e^{-\beta s}$ for $\beta > 0$, where both $p$ and $\beta$ are material parameters, are some examples of such functions. We can then define a norm to measure the recollection of past histories.

**Definition.** The norm $\|G\|$ of a history $G(s)$ relative to an influence function $h(s)$ is defined as

$$\|G\| = \left( \int_0^\infty |G(s)|^2 h(s)\, ds \right)^{1/2}, \tag{3.66}$$

where $|G|$ is the norm of the tensor $G$, i.e., $|G|^2 = \operatorname{tr} G^T G$.

[3] See [10, 71] and for more general concept of fading memory see [9, 73].

The value of $\|G\|$ will be referred to as the recollection of $G$. The influence function $h(s)$ is regarded as the weight in computing the norm of $G(s)$. Since $h(s)$ decreases monotonically, the values of $G(s)$ for small $s$ (recent past) have a greater weight than the values for large $s$ (distant past). Note that by (3.65) all the constant histories, have finite recollections. A constant history $A^c(s)$ is a history such that $A^c(s) = A(0)$ for $0 \leq s < \infty$.

A deformation history that has always been nearly at a constant history, or that may have suffered some large departures from a constant history only in the distant past, has a small recollection of past digressions in the norm (3.66). These are those circumstances that should not much affect the stress in a material with fading memory, as we expect from our experience.

The set of all histories with finite recollections forms a Hilbert space $I\!H$ with the inner product defined by

$$(G, H)_{I\!H} = \int_0^\infty \operatorname{tr}(G(s)^T H(s))\, h(s)\, ds \tag{3.67}$$

for any $G, H \in I\!H$.

We call a material defined by the constitutive equation

$$T(t) = \mathcal{T}(F^t) = R(t)\, \mathcal{T}(U^t)\, R(t)^T, \qquad F = RU, \tag{3.68}$$

where

$$\mathcal{T}(U^t) = \widehat{\mathcal{T}}(U(t)) + \mathcal{K}(G^t; U(t)), \qquad G^t(s) = U^t(s) - U(t), \tag{3.69}$$

a material with fading memory, if both the function $\widehat{\mathcal{T}}$ and the functional $\mathcal{K}$ are smooth in some proper sense in the usual norm in $\mathcal{L}(V)$ and the fading memory norm (3.66) for past histories in $I\!H$.

### 3.7.1 Linear Viscoelasticity

We need the notion of differentiability in the history space $I\!H$.

**Definition.** A functional $\mathcal{F}$ defined on $I\!H$ is said to be *Fréchet differentiable* at the zero history if there is a continuous linear functional $\delta\mathcal{F}$ such that

$$\mathcal{F}(G) = \mathcal{F}(0) + \delta\mathcal{F}(G) + O(\|G\|),$$

where the remainder is of the order $O(\|G\|)$ in the sense that

$$\lim_{\|G\| \to 0} \frac{O(\|G\|)}{\|G\|} = 0.$$

In other words, it is a second-order term in $\|G\|$, and it will be denoted simply by $o(2)$.

If we assume that the functional $\mathcal{K}$ of (3.69) is Fréchet differentiable, we have

$$\mathcal{K}(G;U) = \delta\mathcal{K}(G;U) + o(2),$$

because $\mathcal{K}(0,U) = 0$. A theorem of the theory of Hilbert spaces (Riesz representation theorem, e.g. see [5]) states that every continuous linear functional may be represented by an inner product. In the present case, with the inner product (3.67) by allowing for the fact that the functional $\mathcal{K}$ is tensor-valued rather than scalar-valued, it follows that

$$\delta\mathcal{K}(G;U) = \int_0^\infty \widehat{\boldsymbol{K}}(s;U)[G(s)]h(s)\,ds.$$

Here, the kernel $\widehat{\boldsymbol{K}} \in Sym(V)\otimes Sym(V)$ is a linear transformation of the space of symmetric tensors into itself, namely in component forms, $(\widehat{\boldsymbol{K}}[G])_{ij} = \widehat{K}_{ijkl}G_{kl}$, and hence, for any indices $i$ and $j$ fixed, $\widehat{K}_{ij\,..}$ must belong to $\mathbb{H}$. It follows that $\widehat{\boldsymbol{K}}$ must have the property:

$$\int_0^\infty |\widehat{\boldsymbol{K}}(s;U)|^2 h(s)\,ds < \infty.$$

We then have the following linear approximation,

$$\mathcal{K}(G^t;U) = \int_0^\infty \boldsymbol{K}(s;U)[G^t(s)]\,ds + o(2), \tag{3.70}$$

where $\boldsymbol{K}(s;U) = \widehat{\boldsymbol{K}}(s;U)h(s)$ have the property:

$$\int_0^\infty |\boldsymbol{K}(s;U)|^2 h(s)^{-1}\,ds < \infty. \tag{3.71}$$

By introducing the stress *relaxation function* $\boldsymbol{M}$,

$$\boldsymbol{M}(t,U) = -\int_t^\infty \boldsymbol{K}(s;U)\,ds. \qquad \dot{\boldsymbol{M}}(s,U) = \boldsymbol{K}(s;U), \tag{3.72}$$

and substituting (3.70) into (3.69), we obtain

$$\mathcal{T}(U^t) = \widehat{\mathcal{T}}(U(t)) + \int_0^\infty \dot{\boldsymbol{M}}(s;U(t))[U(t-s) - U(t)]\,ds + o(2),$$

which, upon integration by parts, gives

$$\mathcal{T}(U^t) = \widehat{\mathcal{T}}(U(t)) + \int_0^\infty \boldsymbol{M}(s;U(t))[\dot{U}(t-s)]\,ds + o(2). \tag{3.73}$$

The constitutive equation, $T = \mathcal{T}(F^t)$, given by the above integral equation up to the first gradient in history takes the following form,

$$R(t)^T \mathcal{T}(F^t) R(t) = \hat{\mathcal{T}}(U(t)) + \int_0^\infty \boldsymbol{M}(s; U(t))[\dot{U}(t-s)]\, ds. \tag{3.74}$$

This constitutive equation defines the so-called *finite linear viscoelasticity*, which is a theory of finite deformation with linear dependence on the history of strain rate.

Note that the relaxation function $\boldsymbol{M} \in Sym(V) \otimes Sym(V)$ is material dependent and must satisfy the condition relative to the influence function $h(s)$ of the fading memory norm (3.66),

$$\lim_{s\to\infty} \boldsymbol{M}(s; U) = 0, \qquad \int_0^\infty |\dot{\boldsymbol{M}}(s; U)|^2 h(s)^{-1}\, ds < \infty. \tag{3.75}$$

### 3.7.2 Boltzmann–Volterra Theory of Viscoelasticity

We further consider the case of small deformations. The displacement gradient $H = F - 1$ is assumed to be an infinitesimal quantity with $|H(t)|$ of the order $\epsilon \ll 1$ for all time $t$. We use the order symbol $o(n)$ in the sense of the usual norm as well as the fading memory norm,

$$|o(n)| < k\epsilon^n, \qquad \|o(n)\| < k'\epsilon^n,$$

for some constants, $k$ and $k'$, as a consequence of (3.65) and (3.66). We have from (1.22)

$$U = 1 + \widetilde{E} + o(2), \qquad R = 1 + \widetilde{R} + o(2),$$

where $\widetilde{E}$ and $\widetilde{R}$ are the infinitesimal strain and rotation tensors, respectively.

The linear approximation for the function $\hat{\mathcal{T}}(U)$ can be written as

$$\hat{\mathcal{T}}(U) = T_0 + \boldsymbol{L}[\widetilde{E}] + o(2), \tag{3.76}$$

where

$$T_0 = \hat{\mathcal{T}}(1)$$

is called the *residual stress*, i.e., the stress the material sustains in the reference configuration, and

$$\boldsymbol{L} = \partial_U \hat{\mathcal{T}}(1),$$

is a fourth-order tensor in $Sym(V) \otimes Sym(V)$. Here, of course, we have assumed that the function $\hat{\mathcal{T}}(U)$ is differentiable in the usual sense.

Moreover, we have

$$\dot{U} = \dot{\widetilde{E}} + o(2),$$
$$\boldsymbol{M}(s, U) = \boldsymbol{M}(s, 1) + \partial_U \boldsymbol{M}(s, 1)[\widetilde{E}] + o(2) = \boldsymbol{M}(s) + o(1).$$

Finally, putting (3.76) and (3.73) into (3.68) and neglecting the higher-order terms, we obtain

$$T(t) = T_0 + \widetilde{R}(t)T_0 - T_0\widetilde{R}(t) + \boldsymbol{L}[\widetilde{E}(t)] + \int_0^\infty \boldsymbol{M}(s)[\dot{\widetilde{E}}(t-s)]\,ds. \quad (3.77)$$

We remark that the stress relaxation function $\boldsymbol{M}$ depends not only on the material property but also on the configuration that has been taken as the reference. The special case $\boldsymbol{M}(s) \equiv 0$ gives

$$T(t) = T_0 + \widetilde{R}(t)T_0 - T_0\widetilde{R}(t) + \boldsymbol{L}[\widetilde{E}(t)],$$

which corresponds to the theory of infinitesimal elastic deformation superposed on a large deformation, taking as the reference configuration with the non-vanishing stress field $T_0$.

When the reference configuration is a natural state, i.e., $T_0 = 0$, (3.77) reduces to

$$T(t) = \boldsymbol{L}[\widetilde{E}(t)] + \int_0^\infty \boldsymbol{M}(s)[\dot{\widetilde{E}}(t-s)]\,ds. \quad (3.78)$$

This is the classical equation of linear viscoelasticity of the Boltzmann-Volterra theory.

The special case $\boldsymbol{M}(s) \equiv 0$ corresponds to the classical theory of linear elasticity.

$$T = \boldsymbol{L}[\widetilde{E}].$$

We call $\boldsymbol{L}$ the *elasticity* tensor of the material. Since both the elasticity tensor $\boldsymbol{L}$ and the stress relaxation function $\boldsymbol{M}$ are fourth-order tensors in $Sym(V) \otimes Sym(V)$, their components satisfy the following symmetry conditions:

$$\begin{aligned} L_{ijkl} &= L_{jikl} = L_{ijlk}, \\ M_{ijkl} &= M_{jikl} = M_{ijlk}. \end{aligned} \quad (3.79)$$

### 3.7.3 Linear Viscoelasticity of Rate Type

From the functional $\mathcal{T}(U^t)$ given in (3.73) – by simply replacing $U^t$ with $F^t$ – the constitutive equation for the stress tensor, $T(t) = \mathcal{T}(F^t)$, of finite linear viscoelasticity can also be expressed as

$$T(t) = \widehat{\mathcal{T}}(F(t)) + \int_0^\infty \boldsymbol{M}(s; F(t))[\dot{F}(t-s)]\,ds,$$

or

$$T(t) = \widehat{\mathcal{T}}(F(t)) + \int_{-\infty}^t \boldsymbol{M}(t-s; F(t))[\dot{F}(s)]\,ds. \quad (3.80)$$

This is a constitutive equation of integral type with stress relaxation function $\boldsymbol{M}(s, U)$ and the influence function $h(s)$ of fading memory. Both are material functions.

In this section, we shall consider a material class characterized by the following functions,

$$h(s) = e^{-s/\tau}, \qquad \boldsymbol{M}(s, U) = \boldsymbol{M}(U)h(s), \tag{3.81}$$

where $\tau$ is called the relaxation time of the material. One can easily check that the conditions (3.75) are satisfied.

Taking the time derivative of (3.80), we obtain the following equation,

$$\dot{T}(t) = \Big(\partial_F \widehat{T}(F) + \boldsymbol{M}(F)\Big)[\dot{F}] - \frac{1}{\tau}\int_{-\infty}^{t} \boldsymbol{M}(t-s; F(t))[\dot{F}(s)]\, ds,$$

where we have neglected the second-order terms in $\dot{F}$. By the use of (3.80) again, it becomes

$$\dot{T} - \boldsymbol{L}'(F)[\dot{F}] = -\frac{1}{\tau}\Big(T - \widehat{T}(F)\Big), \tag{3.82}$$

where $\boldsymbol{L}'(F) = \partial_F \widehat{T}(F) + \boldsymbol{M}(F)$ is called the *dynamic* elasticity tensor. This is a different kind of constitutive equation that contains the stress rate. The history dependence of the material response becomes implicit in the solution of the first-order differential equation for the stress. The material characterized by this equation is called a linear viscoelastic material of rate type. In the one-dimensional case, such a material model has been proposed in [28].

For small deformations, from the previous section, (3.82) reduces to

$$\dot{T} - \boldsymbol{L}'[\dot{\widetilde{E}}] = -\frac{1}{\tau}\Big(T - \boldsymbol{L}[\widetilde{E}]\Big), \tag{3.83}$$

where the dynamic elasticity tensor is given by $\boldsymbol{L}' = \boldsymbol{L} + \boldsymbol{M}(0)$. It has the same symmetry properties as the elasticity tensor $\boldsymbol{L}$ in (3.79). The dynamic elasticity tensor characterizes the instantaneous response of the material subject to a sudden change of strain. In the one-dimensional case, $\boldsymbol{M}(0)$ is usually assumed to be positive, and hence $\boldsymbol{L}' > \boldsymbol{L}$.

### 3.7.4 Remark on Objectivity of Linear Elasticity

We remark that although the constitutive equation (3.68) satisfies the principle of material objectivity, the constitutive equation of linear elasticity derived from it does not satisfy this principle because, in such a theory, arbitrary change of frame is not allowed if the deformation gradients are to be small before and after the change of frame.

Let $H = F - 1$ and $H^* = F^* - 1$ be the displacement gradients in the two frames differed by an orthogonal transformation $Q$, and we have $F^* = QF$, then

$$H^* = Q(H + 1) - 1.$$

If we take $Q$ as the rotation of 180° around the $\boldsymbol{e}_1$ axis,

$$Q = \boldsymbol{e}_1 \otimes \boldsymbol{e}_1 - \boldsymbol{e}_2 \otimes \boldsymbol{e}_2 - \boldsymbol{e}_3 \otimes \boldsymbol{e}_3, \tag{3.84}$$

we have

$$H^* = QH - 2(\boldsymbol{e}_2 \otimes \boldsymbol{e}_2 + \boldsymbol{e}_3 \otimes \boldsymbol{e}_3),$$

in which the second term on the right-hand side is no longer a small quantity. The displacement gradient $H^*$ will remain of $o(1)$ only if $Q$ is an infinitesimal rotation, i.e., $Q = 1 + R$ for $R = -R^T = o(1)$.

For linear elasticity, the stress depends linearly on the infinitesimal strain tensor,

$$T(F) = \boldsymbol{L}[\widetilde{E}],$$

where

$$\widetilde{E} = \frac{1}{2}((F + 1) + (F^T + 1)) = \frac{1}{2}(F + F^T) - 1.$$

The material objectivity requires that

$$T(QF) = Q\,T(F)Q^T,$$

for any $Q \in \mathcal{O}(V)$ and $F \in \mathit{Inv}(V)$. In particular, for $F = 1$, it requires that

$$T(Q) = 0,$$

because $T(1) = \boldsymbol{L}[0] = 0$. However, if we take $Q$ as the rotation given in (3.84), then

$$\begin{aligned} T(Q) &= \boldsymbol{L}\left[\frac{1}{2}(Q + Q^T) - 1\right] = \boldsymbol{L}[Q - 1] \\ &= -2\boldsymbol{L}[\boldsymbol{e}_2 \otimes \boldsymbol{e}_2 + \boldsymbol{e}_3 \otimes \boldsymbol{e}_3] \neq 0. \end{aligned}$$

Hence the condition of material objectivity is not satisfied.

Therefore the linear elasticity can only be regarded as an approximation of a physically meaningful theory, and it should not be adopted as a physical model as far as large deformations are concerned.

# 4. Representation of Constitutive Functions

## 4.1 Materials of Grade $n$

We have seen that the constitutive functions have to satisfy both the condition of material objecivity and the condition of material symmetry. In particular, with respect to an undistorted configuration, the constitutive functions of an isotropic material are isotropic functions. The main difficulty in finding representations for constitutive functions lies in the fact that they generally depend on the past values of thermomechanical histories. Fortunately, in most practical problems, long-range memory effects are mostly irrelevant. In other words, for most materials only short memories are of practical interest, in the sense that we can assume that the history be approximated by a Taylor series expansion,

$$\begin{aligned} h^t(s) &= h^t(0) + \frac{\partial h^t(s)}{\partial s}\bigg|_{s=0} s + \frac{1}{2}\frac{\partial^2 h^t(s)}{\partial s^2}\bigg|_{s=0} s^2 + \cdots \\ &\approx \sum_{n=0}^{n} \frac{1}{k!}\frac{\partial^k h^t(s)}{\partial s^k}\bigg|_{s=0} s^k, \end{aligned} \tag{4.1}$$

and consequently, the dependence of the constitutive functions on the history $h^t(s)$ reduces to the dependence on the derivatives of $h^t(s)$ up to the $n$-th-order at the present time, $s = 0$. A material body with such an infinitesimal memory will be called a *material of grade n* relative to the constitutive variable $h$.

Since we have

$$\begin{aligned} C_t^t(s) &\approx C_t^t(0) + \frac{\partial C_t^t(s)}{\partial s}\bigg|_{s=0} s + \cdots + \frac{1}{n!}\frac{\partial^n C_t^t(s)}{\partial s^n}\bigg|_{s=0} s^n \\ &\approx 1 - A_1 s + \cdots + \frac{(-1)^n}{n!} A_n s^n, \end{aligned}$$

where $A_i$ is the Rivlin–Ericksen tensor defined in (1.53), for a fluid, with infinitesimal memory of grade $n$ and grade 0 relative to mechanical and thermal histories, respectively, from (3.54), we can write the constitutive relation in the form

$$\mathcal{C} = \mathcal{F}(\rho, \theta, \boldsymbol{g}, A_1, \cdots, A_n). \tag{4.2}$$

In particular, for $n = 1$, we have

$$\mathcal{C} = \mathcal{F}(\rho, \theta, \boldsymbol{g}, D), \tag{4.3}$$

where we have used the relation $A_1 = 2D$ from (1.54). We call this material a *viscous heat-conducting fluid.*

For an isotropic solid of grade 0, from (3.62), we have

$$\mathcal{C} = \mathcal{F}_\kappa(\theta, \boldsymbol{g}, B), \tag{4.4}$$

where $\kappa$ is an undistorted configuration. We call this material an *isotropic thermoelastic solid.*

Moreover, the above functions are all isotropic functions, namely

$$\begin{aligned} &\mathcal{T}(\theta, Q\boldsymbol{g}, QBQ^T) = Q\,\mathcal{T}(\theta, \boldsymbol{g}, B)\,Q^T, \\ &\mathcal{Q}(\theta, Q\boldsymbol{g}, QBQ^T) = Q\,\mathcal{Q}(\theta, \boldsymbol{g}, B), \qquad \forall Q \in \mathcal{O}(V), \\ &\mathcal{E}(\theta, Q\boldsymbol{g}, QBQ^T) = \mathcal{E}(\theta, \boldsymbol{g}, B), \end{aligned} \tag{4.5}$$

for isotropic thermoelastic solids, and similar relations for viscous heat-conducting fluids. Functions that satisfy such relations are called *tensor*, *vector*, and *scalar isotropic functions*, respectively.

The conditions (4.5) impose restrictions on the generality of the constitutive functions on their arguments. General solutions in the form of representation for isotropic functions are well known and will be given in the following sections.

## 4.2 Isotropic Functions

We shall give some representations of isotropic functions in this section. Let $\phi$, $\boldsymbol{h}$ and $S$ be scalar-, vector- and tensor-valued functions defined on $I\!R \times V \times \mathcal{L}(V)$, respectively.

**Definition.** We say that $\phi$, $\boldsymbol{h}$, and $S$ are *scalar-*, *vector-*, and *tensor-valued isotropic functions*, respectively, if for any $s \in I\!R$, $\boldsymbol{v} \in V$, $A \in \mathcal{L}(V)$, they satisfy the following conditions:

$$\begin{aligned} &\phi(s, Q\boldsymbol{v}, QAQ^T) = \phi(s, \boldsymbol{v}, A), \\ &\boldsymbol{h}(s, Q\boldsymbol{v}, QAQ^T) = Q\,\boldsymbol{h}(s, \boldsymbol{v}, A), \qquad \forall Q \in \mathcal{O}(V). \\ &S(s, Q\boldsymbol{v}, QAQ^T) = Q\,S(s, \boldsymbol{v}, A)\,Q^T, \end{aligned} \tag{4.6}$$

Isotropic functions are also called *isotropic invariants.* The definition can easily be extended to any number of scalar, vector, and tensor variables.

From the definition, obviously the conditions (4.6) impose no restrictions on the scalar variables of an isotropic function, and hence scalar variables are irrelevant as far as the representations of isotropic invariants are concerned. Some general solutions of the condition (4.6) are given in the following representation theorems.

**Theorem 4.2.1.** *Let $\phi$, $\boldsymbol{h}$, and $S$ be isotropic scalar-, vector-, and symmetric tensor-valued functions of a vector variable $\boldsymbol{v}$, respectively, then it is necessary and sufficient that they have the following representations,*

$$\begin{aligned} &1) \quad \phi(\boldsymbol{v}) = f(\boldsymbol{v}\cdot\boldsymbol{v}),\\ &2) \quad \boldsymbol{h}(\boldsymbol{v}) = h(\boldsymbol{v}\cdot\boldsymbol{v})\boldsymbol{v},\\ &3) \quad S(\boldsymbol{v}) = s_0(\boldsymbol{v}\cdot\boldsymbol{v})\mathit{1} + s_1(\boldsymbol{v}\cdot\boldsymbol{v})\boldsymbol{v}\otimes\boldsymbol{v}, \end{aligned} \tag{4.7}$$

*where $f$, $h$, $s_0$, and $s_1$ are arbitrary scalar functions.*

*Proof.* The sufficiency can easily be checked. We shall only show that the above representations are necessary.

To prove (1) it is sufficient to show that if $\phi$ is a scalar isotropic function, then $\phi(\boldsymbol{v}) = \phi(\boldsymbol{u})$ whenever

$$\boldsymbol{v}\cdot\boldsymbol{v} = \boldsymbol{u}\cdot\boldsymbol{u}.$$

In fact, since two vectors of the same length can always be brought into each other by a rotation, there exists an orthogonal tensor $Q$ such that $\boldsymbol{u} = Q\boldsymbol{v}$. Therefore, by the condition of isotropy,

$$\phi(\boldsymbol{u}) = \phi(Q\boldsymbol{v}) = \phi(\boldsymbol{v}),$$

which proves (1).

For the vector isotropic function $\boldsymbol{h}(\boldsymbol{v})$, if $\boldsymbol{v} = 0$, then we must have

$$\boldsymbol{h}(0) = Q\boldsymbol{h}(0), \qquad \forall Q \in \mathcal{O}(V),$$

which implies that $\boldsymbol{h}(0) = 0$ and hence (2) is satisfied. Now, suppose that $\boldsymbol{v} \neq 0$, then we can write the vector $\boldsymbol{h}$ as

$$\boldsymbol{h}(\boldsymbol{v}) = \alpha(\boldsymbol{v})\boldsymbol{v} + \beta(\boldsymbol{v})\boldsymbol{u}, \tag{4.8}$$

where $\boldsymbol{u}$ is some vector orthogonal to $\boldsymbol{v}$. Since $\boldsymbol{h}$ is isotropic we have

$$\boldsymbol{h}(Q\boldsymbol{v}) = Q\boldsymbol{h}(\boldsymbol{v}), \qquad \forall Q \in \mathcal{O}(V). \tag{4.9}$$

If we consider a rotation $Q$ of 180° about the vector $\boldsymbol{v}$, clearly, $Q\boldsymbol{v} = \boldsymbol{v}$ and $Q\boldsymbol{u} = -\boldsymbol{u}$, then it follows from (4.8) and (4.9) that the second term in

(4.8) must vanish. Moreover, from (4.9) the function $\alpha(\boldsymbol{v})$ must be a scalar isotropic function, therefore, by item (1) it must reduce to a function of $\boldsymbol{v}\cdot\boldsymbol{v}$. The representation (2) is proved.

Finally, for the tensor isotropic function $S(\boldsymbol{v})$, we consider a vector-valued function $\boldsymbol{g}(\boldsymbol{v})$ defined by

$$\boldsymbol{g}(\boldsymbol{v}) = S(\boldsymbol{v})\boldsymbol{v}.$$

One can easily check that $\boldsymbol{g}(\boldsymbol{v})$ is an isotropic vector function, therefore, by representation (2) we have

$$S(\boldsymbol{v})\boldsymbol{v} = \boldsymbol{g}(\boldsymbol{v}) = \alpha(\boldsymbol{v})\boldsymbol{v}.$$

In other words, $\boldsymbol{v}$ is an eigenvector of the tensor $S$. Since $S$ is a symmetric tensor, by the spectral theorem, there exist two unit vectors $\boldsymbol{u}$ and $\boldsymbol{w}$ such that $S$ can be represented as

$$S(\boldsymbol{v}) = \alpha(\boldsymbol{v})\boldsymbol{v}\otimes\boldsymbol{v} + \beta(\boldsymbol{v})\boldsymbol{u}\otimes\boldsymbol{u} + \gamma(\boldsymbol{v})\boldsymbol{w}\otimes\boldsymbol{w}, \tag{4.10}$$

where $\{\boldsymbol{v},\boldsymbol{u},\boldsymbol{w}\}$ is an orthonormal set of eigenvectors of $S$. Moreover, since

$$S(Q\boldsymbol{v}) = QS(\boldsymbol{v})Q^T, \qquad \forall Q\in\mathcal{O}(V), \tag{4.11}$$

if we choose an orthogonal tensor $Q$, such that

$$Q\boldsymbol{v} = \boldsymbol{v}, \quad Q\boldsymbol{u} = \boldsymbol{w}, \quad Q\boldsymbol{w} = \boldsymbol{u},$$

then (4.10) and (4.11) imply that $\beta(\boldsymbol{v}) = \gamma(\boldsymbol{v})$. Therefore, we can rewrite (4.10) as

$$S(\boldsymbol{v}) = s_0(\boldsymbol{v})\mathit{1} + s_1(\boldsymbol{v})\boldsymbol{v}\otimes\boldsymbol{v}.$$

Moreover, (4.11) implies that $s_0$ and $s_1$ must be scalar isotropic functions. This completes the proof of the theorem. □

**Theorem 4.2.2** (Rivlin and Ericksen). *Let $\phi$, $\boldsymbol{h}$, and $S$ be isotropic scalar-, vector-, and symmetric tensor-valued functions of a symmetric tensor variable $A$, respectively, then it is necessary and sufficient that they have the following representations,*

$$\begin{aligned} &1)\quad \phi(A) = f(a_1,a_2,a_3),\\ &2)\quad \boldsymbol{h}(A) = 0,\\ &3)\quad S(A) = s_0\mathit{1} + s_1A + s_2A^2, \end{aligned} \tag{4.12}$$

*where $f$, as well as $s_0$, $s_1$, and $s_2$, are arbitrary scalar functions of the three eigenvalues $\{a_1,a_2,a_3\}$ of $A$.*

*Proof.* Again, the proof of sufficiency is trivial, and we shall only prove the necessity of the above representations.

First of all, for the vector isotropic function $\boldsymbol{h}$,

$$\boldsymbol{h}(QAQ^T) = Q\boldsymbol{h}(A), \qquad \forall Q \in \mathcal{O}(V).$$

By taking $Q = -\mathit{1}$, it follows that $\boldsymbol{h}(A) = -\boldsymbol{h}(A)$, therefore, it must vanish.

For the scalar isotropic function $\phi(A)$, it suffices to show that $\phi(A) = \phi(B)$, whenever $A$ and $B$ have the same set of eigenvalues.

$$\{a_1, a_2, a_3\} = \{b_1, b_2, b_3\}.$$

Thus, let $A$ and $B$ be two symmetric tensors and assume that their eigenvalues are the same. Then, by the spectral theorem, there exist orthonormal bases $\{\boldsymbol{e}_i\}$ and $\{\boldsymbol{d}_i\}$ such that

$$A = \sum_i a_i \boldsymbol{e}_i \otimes \boldsymbol{e}_i, \qquad B = \sum_i a_i \boldsymbol{d}_i \otimes \boldsymbol{d}_i.$$

Let $Q$ be the orthogonal tensor carrying the basis $\{\boldsymbol{d}_i\}$ into the basis $\{\boldsymbol{e}_i\}$, $Q\boldsymbol{d}_i = \boldsymbol{e}_i$. Then, since

$$Q(\boldsymbol{d}_i \otimes \boldsymbol{d}_i)Q^T = (Q\boldsymbol{d}_i) \otimes (Q\boldsymbol{d}_i) = \boldsymbol{e}_i \otimes \boldsymbol{e}_i,$$

it follows that $QBQ^T = A$. But since $\phi$ is isotropic,

$$\phi(A) = \phi(QBQ^T) = \phi(B),$$

which proves (1).

To show the representation (3) for the symmetric tensor isotropic function $S$, we need the following lemma.

**Lemma.** *Let $S(A)$ be an isotropic tensor-valued function of a symmetric tensor $A$, then every eigenvector of $A$ is an eigenvector of $S(A)$.*

To prove the lemma, let $\{\boldsymbol{e}_1, \boldsymbol{e}_2, \boldsymbol{e}_3\}$ be a principal orthonormal basis of $A$. We consider a rotation $Q$ of 180° about the $\boldsymbol{e}_1$-axis, therefore,

$$Q\boldsymbol{e}_1 = \boldsymbol{e}_1, \quad Q\boldsymbol{e}_2 = -\boldsymbol{e}_2, \quad Q\boldsymbol{e}_3 = -\boldsymbol{e}_3.$$

Clearly, we have $QAQ^T = A$ and since $S(A)$ is isotropic, it follows that

$$QS(A) = S(QAQ^T)Q = S(A)Q.$$

Therefore, we have

$$QS(A)\boldsymbol{e}_1 = S(A)Q\boldsymbol{e}_1 = S(A)\boldsymbol{e}_1,$$

i.e., $S(A)\boldsymbol{e}_1$ remains unchanged under $Q$. By our choice of $Q$, this can happen only if $S(A)\boldsymbol{e}_1$ is in the direction of $\boldsymbol{e}_1$. In other words, $\boldsymbol{e}_1$ is an eigenvector of $S(A)$. For other eigenvectors, we can use similar arguments. Hence the lemma is proved. □

From the lemma, if we express $A$ as

$$A = a_1 \boldsymbol{e}_1 \otimes \boldsymbol{e}_1 + a_2 \boldsymbol{e}_2 \otimes \boldsymbol{e}_2 + a_3 \boldsymbol{e}_3 \otimes \boldsymbol{e}_3, \tag{4.13}$$

then $S(A)$ can be expressed as

$$S(A) = b_1 \boldsymbol{e}_1 \otimes \boldsymbol{e}_1 + b_2 \boldsymbol{e}_2 \otimes \boldsymbol{e}_2 + b_3 \boldsymbol{e}_3 \otimes \boldsymbol{e}_3, \tag{4.14}$$

where $b_1$, $b_2$, and $b_3$ are functions of $A$.

Suppose that the three eigenvalues $a_1$, $a_2$, and $a_3$ are distinct. We consider the following simultaneous equations for $s_0$, $s_1$, and $s_2$,

$$\begin{aligned} s_0 + a_1 s_1 + a_1^2 s_2 &= b_1, \\ s_0 + a_2 s_1 + a_2^2 s_2 &= b_2, \\ s_0 + a_3 s_1 + a_3^2 s_2 &= b_3. \end{aligned}$$

Since the determinant of the coefficient matrix does not vanish,

$$\det \begin{bmatrix} 1 & a_1 & a_1^2 \\ 1 & a_2 & a_2^2 \\ 1 & a_3 & a_3^2 \end{bmatrix} = (a_1 - a_2)(a_2 - a_3)(a_3 - a_1) \neq 0,$$

we can solve for $s_0$, $s_1$, and $s_2$, which are functions of $A$ of course. Then, from (4.13), (4.14) can be written in the form

$$S(A) = s_0(A)\mathit{1} + s_1(A)A + s_2(A)A^2. \tag{4.15}$$

Furthermore, since $S(A)$ is isotropic,

$$S(QAQ^T) = QS(A)Q^T, \qquad \forall Q \in \mathcal{O}(V),$$

it follows from (4.15) that

$$s_0(QAQ^T)\mathit{1} + s_1(QAQ^T)A + s_2(QAQ^T)A^2 = s_0(A)\mathit{1} + s_1(A)A + s_2(A)A^2.$$

On the other hand, one can show that $\{1, A, A^2\}$ is linearly independent, therefore, we conclude that $s_i$ are all scalar isotropic functions,

$$s_i(QAQ^T) = s_i(A), \quad i = 0, 1, 2, \quad \forall Q \in \mathcal{O}(V).$$

Hence, the representation (3) for the case of distinct eigenvalues of $A$ is proved.

With similar arguments one can show that when $A$ has exactly two distinct eigenvalues, $S(A)$ admits the representation (3) with $s_2 = 0$. For the case when $A = a\mathit{1}$, every vector is an eigenvector, and hence by the lemma, we must have $S(A) = s_0\mathit{1}$, which is a special case of the representation (3). □

Recall that the Cayley–Hamilton theorem states that for any tensor $A$,

$$A^3 - I_A A^2 + II_A A - III_A 1 = 0, \tag{4.16}$$

where $I_A$, $II_A$, and $III_A$ are called the *principal invariant* of $A$. They are the coefficients of the characteristic polynomial of $A$, i.e.,

$$\det(\lambda 1 - A) = \lambda^3 - I_A \lambda^2 + II_A \lambda - III_A = 0. \tag{4.17}$$

Since eigenvalues of $A$ are the roots of the characteristic equation (4.17), if $A$ is symmetric and $a_1$, $a_2$, and $a_3$ are three eigenvalues of $A$, distinct or not, then it follows that

$$\begin{aligned} I_A &= a_1 + a_2 + a_3, \\ II_A &= a_1 a_2 + a_2 a_3 + a_3 a_1, \\ III_A &= a_1 a_2 a_3. \end{aligned} \tag{4.18}$$

They can be written in terms of another three invariants of $A$, namely,

$$\begin{aligned} I_A &= \operatorname{tr} A, \\ II_A &= \tfrac{1}{2}\big((\operatorname{tr} A)^2 - \operatorname{tr} A^2\big), \\ III_A &= \tfrac{1}{6}\big((\operatorname{tr} A)^3 - 3 \operatorname{tr} A \operatorname{tr} A^2 + 2 \operatorname{tr} A^3\big), \end{aligned} \tag{4.19}$$

or conversely,

$$\begin{aligned} \operatorname{tr} A &= I_A, \\ \operatorname{tr} A^2 &= I_A^2 - 2 II_A, \\ \operatorname{tr} A^3 &= I_A^3 - 3 I_A II_A + 3 III_A. \end{aligned} \tag{4.20}$$

From the above relations we can see that the three sets $\{a_1, a_2, a_3\}$, $\{I_A, II_A, III_A\}$, and $\{\operatorname{tr} A, \operatorname{tr} A^2, \operatorname{tr} A^3\}$ are equivalent in the sense that they uniquely determine one another. Moreover, they are all invariant under any change from $A$ to $QAQ^T$ for any orthogonal tensor $Q$. Therefore, from Theorem 4.2.2, any scalar isotropic function $\phi$ of $A$ may also be expressed as:

$$\phi(A) = \phi(I_A, II_A, III_A) = \tilde{\phi}(\operatorname{tr} A, \operatorname{tr} A^2, \operatorname{tr} A^3).$$

The proof of the following corollary is straightforward.

**Corollary 4.2.3** *Let $S$ be an isotropic symmetric tensor-valued function of a symmetric tensor variable $A$. If, in addition, the function $S$ is linear in the variable $A$, then it has the following representation:*

$$S(A) = \alpha\,(\operatorname{tr} A)1 + \beta\,A, \tag{4.21}$$

*where $\alpha$ and $\beta$ are independent of $A$.*

As we have seen from the above representation theorems, an isotropic function is quite restricted in its dependence on the independent variables. From a different point of view, we can also derive some such restrictions in the form of differential equations. First, we need the following lemma:

**Lemma 4.2.4** *Let $\mathcal{F}(Q)$ be a function (scalar-, vector-, or tensor-valued) defined on $\mathcal{L}(V)$ and suppose that $\mathcal{F}(Q) = 0$ for any orthogonal $Q \in \mathcal{O}(V)$. Then the gradient of $\mathcal{F}$ at the identity tensor is symmetric, i.e.,*

$$\partial_Q \mathcal{F}(1)[W] = 0 \tag{4.22}$$

*for any skew-symmetric $W \in Skw(V)$.*

*Proof.* For $W = -W^T$ and $0 < \varepsilon \ll 1$, since $(1+\varepsilon W)(1+\varepsilon W)^T = 1-\varepsilon^2 W^2$, the tensor $1 + \varepsilon W$ is orthogonal to within second-order terms $o(2)$ in $\varepsilon$. In other words, there exists an $Q_W \in \mathcal{O}(V)$ such that

$$1 + \varepsilon W = Q_W + o(2).$$

By the assumption we have $\mathcal{F}(Q_W) = 0$ and $\mathcal{F}(1) = 0$. Consequently, by the definition of the gradient (A.46),

$$\begin{aligned} \mathcal{F}(1+\varepsilon W) - \mathcal{F}(1) &= \varepsilon\, \partial_Q \mathcal{F}(1)[W] + o(2), \\ &= \mathcal{F}(Q_W + o(2)) - \mathcal{F}(Q_W) = \partial_Q \mathcal{F}(Q_W)[o(2)] + o(2) = o(2), \end{aligned}$$

which implies that the first-order term $\varepsilon\, \partial_Q \mathcal{F}(1)[W]$ must vanish, and (4.22) is proved. □

**Example 4.2.1** Let $\phi(A, \boldsymbol{v})$ be a scalar isotropic function. Then the function $\phi$ satisfies the following relation,

$$\varepsilon_{imn}\Big(\frac{\partial \phi}{\partial v_m} v_n + \frac{\partial \phi}{\partial A_{mp}} A_{np} + \frac{\partial \phi}{\partial A_{pm}} A_{pn}\Big) = 0, \tag{4.23}$$

relative to a Cartesian coordinate system.

*Proof.* Let

$$\mathcal{F}(Q) = \phi(QAQ^T, Q\boldsymbol{v}) - \phi(A, \boldsymbol{v}) \qquad \forall\, Q \in \mathcal{L}(V).$$

By the chain rule, one can easily verify that

$$\begin{aligned} \partial_Q \mathcal{F}(1)[W] &= \partial_A \phi[(1+W)A(1+W)^T - A] + \partial_{\boldsymbol{v}} \phi[(1+W)\boldsymbol{v} - \boldsymbol{v}] \\ &= \partial_A \phi[WA + AW^T] + \partial_{\boldsymbol{v}} \phi[W\boldsymbol{v}] \\ &= (\partial_A \phi)A^T[W] + (\partial_A \phi)^T A[W] + \partial_{\boldsymbol{v}} \phi \otimes \boldsymbol{v}[W]. \end{aligned}$$

Since $\phi(A, \boldsymbol{v})$ is an isotropic function, we have $\mathcal{F}(Q) = 0$ for any $Q \in \mathcal{O}(V)$, and hence, by the above lemma

$$\partial_Q \mathcal{F}(1) = (\partial_A \phi)A^T + (\partial_A \phi)^T A + \partial_{\boldsymbol{v}} \phi \otimes \boldsymbol{v}$$

must be symmetric. Therefore its skew-symmetric part must vanish, which, in component forms, gives the relation (4.23). □

From Lemma 4.2.4 we can prove the following representation theorem for a scalar isotropic function of vector variables as the solution of a system of first-order partial differential equations.

**Theorem 4.2.5.** *Let $\phi(\boldsymbol{v}, \boldsymbol{u})$ be a scalar isotropic function of two vector variables, then it has the following representation,*

$$\phi(\boldsymbol{v}, \boldsymbol{u}) = f(\boldsymbol{v} \cdot \boldsymbol{v}, \boldsymbol{u} \cdot \boldsymbol{u}, \boldsymbol{v} \cdot \boldsymbol{u}).$$

*Proof.* By the use of Lemma 4.2.4, if we define

$$\mathcal{F}(Q) = \phi(Q\boldsymbol{v}, Q\boldsymbol{u}) - \phi(\boldsymbol{v}, \boldsymbol{u}) \qquad \forall Q \in \mathcal{L}(V),$$

then, similar to the previous example, we have

$$\begin{aligned}
&\frac{\partial \phi}{\partial v_1} v_2 - \frac{\partial \phi}{\partial v_2} v_1 + \frac{\partial \phi}{\partial u_1} u_2 - \frac{\partial \phi}{\partial u_2} u_1 = 0,\\
&\frac{\partial \phi}{\partial v_1} v_3 - \frac{\partial \phi}{\partial v_3} v_1 + \frac{\partial \phi}{\partial u_1} u_3 - \frac{\partial \phi}{\partial u_3} u_1 = 0,\\
&\frac{\partial \phi}{\partial v_2} v_3 - \frac{\partial \phi}{\partial v_3} v_2 + \frac{\partial \phi}{\partial u_2} u_3 - \frac{\partial \phi}{\partial u_3} u_2 = 0.
\end{aligned} \tag{4.24}$$

From the first-order partial differential equation $(4.24)_1$ for $\phi$ with $v_3$ and $u_3$ as parameters, one can easily obtain the general solution (e.g., see [29])

$$\phi = f(c_1, c_2, c_3),$$

where $f$ is an arbitrary function, and

$$\begin{aligned}
c_1 &= v_1^2 + v_2^2 + g_1(v_3, u_3),\\
c_2 &= u_1^2 + u_2^2 + g_2(v_3, u_3),\\
c_3 &= v_1 u_1 + v_2 u_2 + g_3(v_3, u_3),
\end{aligned}$$

are the three integrals of $(4.24)_1$. Substituting into the second equation $(4.24)_2$, one obtains

$$\begin{aligned}
\Big(2v_1 v_3 - v_1 \frac{\partial g_1}{\partial v_3} - u_1 \frac{\partial g_1}{\partial u_3}\Big) \frac{\partial f}{\partial c_1} + \Big(2u_1 u_3 - u_1 \frac{\partial g_2}{\partial u_3} - v_1 \frac{\partial g_2}{\partial v_3}\Big) \frac{\partial f}{\partial c_2}&\\
+ \Big(v_1 u_3 + v_3 u_1 - v_1 \frac{\partial g_3}{\partial v_3} - u_1 \frac{\partial g_3}{\partial u_3}\Big) \frac{\partial f}{\partial c_3} = 0.&
\end{aligned}$$

Since $f$ is arbitrary, so are $\partial f / \partial c_i$, and hence their coefficients in the above equation must vanish independently,

$$\begin{aligned}
v_1 \Big(2v_3 - \frac{\partial g_1}{\partial v_3}\Big) &= u_1 \frac{\partial g_1}{\partial u_3},\\
v_1 \frac{\partial g_2}{\partial v_3} &= u_1 \Big(2u_3 - \frac{\partial g_2}{\partial u_3}\Big),\\
v_1 \Big(u_3 - \frac{\partial g_3}{\partial v_3}\Big) &= u_1 \Big(\frac{\partial g_3}{\partial u_3} - v_3\Big).
\end{aligned}$$

Note that in these equations, the left-hand sides do not depend on $u_1$, while the right-hand sides do not depend on $v_1$, therefore, both sides must vanish. From which one can easily show that

$$g_1 = v_3^2, \qquad g_2 = u_3^2, \qquad g_3 = v_3 u_3.$$

Therefore, it follows that

$$\phi = f(v_1^2 + v_2^2 + v_3^2, u_1^2 + u_2^2 + u_3^2, v_1 u_1 + v_2 u_2 + v_3 u_3),$$

which also satisfies $(4.24)_3$ and the theorem is proved. □

This representation theorem can be extended to isotropic functions of an arbitrary number of vector variables.

Before we consider general representations for scalar, vector, and tensor isotropic functions in the next sections, we shall give an example to illustrate how to obtain a representation for higher-order tensor isotropic functions.

**Example 4.2.2** Let $V_{ijkl}$ be a fourth-order isotropic tensor, i.e.,

$$V_{ijkl} = Q_{im} Q_{jn} Q_{kp} Q_{lq} V_{mnpq},$$

for any orthogonal tensor $Q$ relative to a Cartesian coordinate system (see Exercise 3.2.2), with the additional symmetry property:

$$V_{ijkl} = V_{jikl} = V_{ijlk}.$$

Then it has the following representation:

$$V_{ijkl} = \lambda\, \delta_{ij} \delta_{kl} + \mu\, (\delta_{ik} \delta_{jl} + \delta_{il} \delta_{jk}). \tag{4.25}$$

*Proof.* For any symmetric $A$, let

$$S_{ij} = V_{ijkl}\, A_{kl}.$$

Then it is easy to verify that $S = S(A)$ is an isotropic symmetric tensor-valued function and is linear in its symmetric tensor variable $A$. Therefore, from Corollary 4.2.3, we can represent $S(A)$ as an expression linear in $A$ in the following form,

$$S(A) = \lambda\, (\operatorname{tr} A) \mathit{1} + 2\mu\, A,$$

where $\lambda$ and $\mu$ are two scalar constants. In component form, it gives

$$\begin{aligned} S_{ij} &= \lambda\, \delta_{ij} A_{kk} + 2\mu A_{ij} \\ &= \lambda\, \delta_{ij} \delta_{kl}\, A_{kl} + \mu\, (\delta_{ik} \delta_{jl} + \delta_{il} \delta_{jk})\, A_{kl}, \end{aligned}$$

which proves (4.25) by comparison. □

**Exercise 4.2.1** Verify the relations (4.19) and (4.20).

**Exercise 4.2.2** Suppose that $M(\boldsymbol{v})$ is a third-order tensor isotropic function of a vector variable, i.e., for any $Q \in \mathcal{O}(V)$,

$$M_{ijk}(Q\boldsymbol{v}) = Q_{il}Q_{jm}Q_{kn}\, M_{lmn}(\boldsymbol{v}),$$

relative to a Cartesian coordinate system. Find a representation formula for $M$. (Note that if $S(\boldsymbol{v},\boldsymbol{u})$ is defined by $S_{ij} = M_{ijk}u_k$, then $S$ is a second-order tensor isotropic function of two vector variables).

**Exercise 4.2.3** Suppose that $A \in Sym(V)$ and $(\mathit{1} + A) \in Inv(V)$.
1) Let $S(A) = (\mathit{1} + A)^{-1}$. Show that $S(A)$ is an isotropic function.
2) By the use of representation (4.12), show that (see also Exercise A.1.9)

$$(\mathit{1} + A)^{-1} = s_0 \mathit{1} + s_1 A + s_2 A^2,$$

where

$$\begin{aligned} s_2 &= (1 + I_A + II_A + III_A)^{-1}, \\ s_1 &= -(1 + I_A)s_2, \\ s_0 &= (1 + I_A + II_A)s_2. \end{aligned}$$

### 4.2.1 Isotropic Elastic Materials and Linear Elasticity

For an isotropic material with no memory of past deformation histories, from the reduced constitutive relation (3.62), we have the stress tensor $T = \mathcal{T}(B)$, where $\mathcal{T}$ is an isotropic function of the left Cauchy–Green tensor $B$. By representation (4.12), it follows that

$$\begin{aligned} T &= t_0 \mathit{1} + t_1 B + t_2 B^2, \\ t_i &= t_i(I_B, II_B, III_B), \quad i = 0,1,2, \end{aligned} \tag{4.26}$$

or equivalently, by the use of the identity (4.16)

$$\begin{aligned} T &= s_0 \mathit{1} + s_1 B + s_{-1} B^{-1}, \\ s_i &= s_i(I_B, II_B, III_B), \quad i = 0,1,2, \end{aligned} \tag{4.27}$$

where

$$\begin{aligned} s_0 &= t_0 - II_B t_2, \\ s_1 &= t_1 + I_B t_2, \\ s_{-1} &= III_B t_2. \end{aligned} \tag{4.28}$$

This is the general constitutive equation for an *isotropic elastic material* body with finite deformations.

For small deformations, by assuming that the displacement gradient $H = F - \mathit{1}$ is a small quantity of $o(1)$, we have

$$B - \mathit{1} = 2\widetilde{E} + o(2),$$

where $\widetilde{E} = (H + H^T)/2$ is the infinitesimal strain tensor. A linear approximation of $\mathcal{T}(B)$ at $B = \mathit{1}$ can be written as

$$T = \mathcal{T}(B) = \mathcal{T}(\mathit{1}) + \partial_B \mathcal{T}(\mathit{1})[B - \mathit{1}] + o(2).$$

With the additional assumption that the reference configuration is a natural state, $\mathcal{T}(\mathit{1}) = 0$, and by neglecting the second-order terms, it gives

$$T = \mathbf{L}[\widetilde{E}],$$

where $\mathbf{L}$ is the *elasticity tensor* (see (3.79)) given by

$$\mathbf{L} = 2\,\partial_B \mathcal{T}(\mathit{1}).$$

Using the representation (4.26) and carrying out the gradient explicitly by the use of (A.54), we obtain

$$T = \lambda\,(\operatorname{tr}\widetilde{E})\mathit{1} + 2\mu\,\widetilde{E}, \tag{4.29}$$

where $\lambda$ and $\mu$ are called the *Lamé elastic moduli* and are related to the material parameters $t_0$, $t_1$, and $t_2$ of (4.26) by the following relations:

$$\begin{aligned} &\lambda = 2\left(\frac{\partial t}{\partial I_B} + 2\frac{\partial t}{\partial II_B} + \frac{\partial t}{\partial III_B}\right)\bigg|_{(3,3,1)} \qquad t = t_0 + t_1 + t_2, \\ &\mu = t_1(3,3,1) + 2\,t_2(3,3,1). \end{aligned} \tag{4.30}$$

Equation (4.29) is the constitutive relation of the classical theory of *isotropic linear elasticity*, also known as *Hooke's law.* The two material constants $\lambda$ and $\mu$ can be determined in simple experiments, such as uniaxial tension and simple shear. The Lamé constant $\mu$ is also known as the *shear modulus.*

Two more commonly used material constants in linear elasticity, *Young's modulus* and *Poisson's ratio* are usually introduced in uniaxial tension tests: The state of stress in such a test is given by

$$T = \sigma_1 \boldsymbol{e}_1 \otimes \boldsymbol{e}_1,$$

and hence the linear strain tensor has the following form:

$$\widetilde{E} = \epsilon_1 \boldsymbol{e}_1 \otimes \boldsymbol{e}_1 + \epsilon_2(\boldsymbol{e}_2 \otimes \boldsymbol{e}_2 + \boldsymbol{e}_3 \otimes \boldsymbol{e}_3).$$

Young's modulus E and Poisson's ratio $\nu$ are defined as

$$\mathrm{E} = \frac{\sigma_1}{\epsilon_1}, \qquad \nu = -\frac{\epsilon_2}{\epsilon_1},$$

i.e., E is the stress per unit strain in the longitudinal direction of the uniaxial test, while $\nu$ is the ratio of the transverse compression to the longitudinal extension of the strain. It follows immediately from (4.29) that

$$\mathrm{E} = \frac{\mu(3\lambda + 2\mu)}{\lambda + \mu}, \qquad \nu = \frac{\lambda}{2(\lambda + \mu)}, \tag{4.31}$$

and conversely,

$$\lambda = \frac{\nu\mathrm{E}}{(1 - 2\nu)(1 + \nu)}, \qquad \mu = \frac{\mathrm{E}}{2(1 + \nu)}. \tag{4.32}$$

Hooke's law can also be written in a different form,

$$\widetilde{E} = -\frac{\nu}{\mathrm{E}}(\operatorname{tr} T)\mathit{1} + \frac{1 + \nu}{\mathrm{E}} T, \tag{4.33}$$

expressing the linear strain in terms of the stress.

**Exercise 4.2.4** Verify the relations (4.31), (4.32) and (4.33).

**Exercise 4.2.5** From Hooke's law (4.29) and Cauchy's first law (2.57), derive the following equation of linear elasticity,

$$\rho\, \ddot{\boldsymbol{u}} = (\lambda + \mu)\ \operatorname{grad}(\operatorname{div} \boldsymbol{u}) + \mu\, \nabla^2 \boldsymbol{u} + \rho\, \boldsymbol{b},$$

where $\boldsymbol{u}$ is the displacement vector.

### 4.2.2 Reiner–Rivlin Fluids and Navier–Stokes Fluids

A simple fluid of grade 1 is defined by the constitutive equation

$$T = \mathcal{T}(\rho, D).$$

Since it is an isotropic function, by the representation theorem (4.12) it can be expressed as

$$\begin{aligned} T &= \alpha_0 \mathit{1} + \alpha_1 D + \alpha_2 D^2, \\ \alpha_i &= \alpha_i(\rho, I_D, II_D, III_D), \quad i = 0, 1, 2. \end{aligned} \tag{4.34}$$

A fluid characterized by this constitutive equation is called a *Reiner-Rivlin fluid.* This is the most general form of constitutive equations for simple fluids of grade 1. Even though this seems to be a general model for nonlinear viscosity, it has been pointed out that this model is inadequate to describe some observed nonlinear effects in real fluids (See Sect. 119, [71]). Nevertheless, by the use of (4.21), the special case, for which the viscous stress depends linearly on the stretching tensor $D$, leads to the most well-known model, the *Navier–Stokes fluid.* This is given by the following constitutive equation,

$$T = -p(\rho)\mathit{1} + \lambda(\rho)(\operatorname{tr} D)\mathit{1} + 2\mu(\rho)D, \tag{4.35}$$

where $p$ is the pressure, while $\lambda$ and $\mu$ are called the *coefficients of viscosity.* A Navier–Stokes fluid is also known as a *Newtonian fluid* in fluid mechanics. Note that from Exercise 3.2.2, and the representation (4.25) in Example 4.2.2, also proves (4.35).

We can rewrite (4.35) in a different form,

$$T = -p\mathit{1} + (\lambda + \frac{2}{3}\mu)(\operatorname{tr} D)\mathit{1} + 2\mu\hat{D}, \tag{4.36}$$

where

$$\hat{D} = D - \frac{1}{3}(\operatorname{tr} D)\mathit{1},$$

called the deviatoric part of $D$, is traceless, i.e., $\operatorname{tr}\hat{D} = 0$. The parameters $\mu$ and $(\lambda + 2\mu/3)$ are also known as the *shear* and the *bulk viscosities*, respectively. It is usually assumed that

$$\mu \geq 0, \qquad \lambda + \frac{2}{3}\mu \geq 0. \tag{4.37}$$

These inequalities ensure that that the fluid particles tend to flow in the direction of a shear force (for the proof see Sect. 7.2.2). When the bulk viscosity vanishes identically, it is known as a *Stokes fluid*, a model adequate to describe some real fluids and frequently used in numerical calculations,

$$T = -p\,\mathit{1} + 2\mu\,\hat{D}. \tag{4.38}$$

The deviatoric part $\hat{D}$ in component form is given by

$$\hat{D}_{ij} = \frac{1}{2}\left(\frac{\partial v_i}{\partial x_j} + \frac{\partial v_j}{\partial x_i}\right) - \frac{1}{3}\frac{\partial v_k}{\partial x_k}\delta_{ij}.$$

A Navier–Stokes fluid is governed by the system of equations that consists of the conservation of mass (2.32),

$$\frac{\partial \rho}{\partial t} + \frac{\partial}{\partial x_k}(\rho\, v_k) = 0, \tag{4.39}$$

and the equation of motion (2.57),

$$\begin{aligned} \rho\left(\frac{\partial v_i}{\partial t} + v_k\frac{\partial v_i}{\partial x_k}\right) + \frac{\partial p}{\partial x_i} - \frac{\partial}{\partial x_i}\left(\lambda\frac{\partial v_k}{\partial x_k}\right) \\ - \frac{\partial}{\partial x_k}\left(\mu\frac{\partial v_k}{\partial x_i}\right) - \frac{\partial}{\partial x_k}\left(\mu\frac{\partial v_i}{\partial x_k}\right) = \rho\, b_i. \end{aligned} \tag{4.40}$$

The last equation (4.40) is known as the *Navier Stokes equation.* The pressure $p$ and the viscosities, $\lambda$ and $\mu$, are, in general, functions of the density $\rho$. Equations (4.39) and (4.40) form a system for the fields $(\rho(\boldsymbol{x},t),\ \boldsymbol{v}(\boldsymbol{x},t))$. For Stokes fluids, the governing equations are obtained from above by substitution of $\lambda = -2\mu/3$.

Note that aside from the similarity between the constitutive equations of linear elasticity (4.29) and the Navier–Stokes fluids (4.35), there is a fundamental difference between the two theories. Unlike linear elasticity, which is physically meaningless for finite deformations because it does not satisfy the requirement of the principle of material objectivity (see the remark on Sect. 3.7.4), the Navier Stokes fluids do satisfy such a requirement. Therefore the linearization of the constitutive relation in (4.35) need not be regarded as an approximation of the more general constitutive relation (4.34). Any particular form of a constitutive relation (4.34) characterizes a particular class of simple fluids. Thus, it is conceivable that there are some fluids that obey the constitutive equation (4.35) for *arbitrary* rate of deformation. Indeed, water and air are usually treated as Navier Stokes fluids (with additional thermal effects) in most practical applications with very satisfactory results, even in rapid deformation processes.

### 4.2.3 Elastic Fluids

The simplest constitutive equation used in continuum mechanics is that of a fluid that is inviscid, i.e., non-viscous, and a non-conductor of heat,

$$T = -p(\rho, \theta)\, \mathit{1}, \qquad \boldsymbol{q} = 0. \tag{4.41}$$

This fluid is also an elastic material, i.e., its constitutive function depends on the deformation gradient (through $\det F$) only. Therefore, it is also called an *elastic fluid.*

For an elastic fluid, from the balance equations of mass, linear momentum, and energy, we have the following governing equations in component forms:

$$\begin{aligned} &\frac{\partial \rho}{\partial t} + \frac{\partial}{\partial x_j}(\rho\, v_j) = 0, \\ &\frac{\partial}{\partial t}(\rho\, v_i) + \frac{\partial}{\partial x_j}(\rho\, v_i v_j + p\,\delta_{ij}) = \rho\, b_i, \\ &\frac{\partial}{\partial t}(\rho\,\varepsilon + \frac{1}{2}\rho\, v^2) + \frac{\partial}{\partial x_j}\left(\rho\varepsilon\, v_j + \frac{1}{2}\rho\, v^2 v_j + p v_j\right) = \rho\, r. \end{aligned} \tag{4.42}$$

The internal energy $\varepsilon = \varepsilon(\rho, \theta)$ and the pressure $p = p(\rho, \theta)$ have to be specified, for instance by the ideal gas laws, and the equations become a hyperbolic system for the fields $(\rho(\boldsymbol{x}, t), \boldsymbol{v}(\boldsymbol{x}, t), \theta(\boldsymbol{x}, t))$, usually known as the Euler equations for compressible flows of ideal gases.

## 4.3 Representation of Isotropic Functions

In this section, we shall give a more general discussion on isotropic invariants, and restate the well-known results for the representation of isotropic functions of any number of vector and tensor variables.

Let $\mathcal{L}(V)$ be the space of second-order tensors on $V$. Let

$$\mathcal{D} = V^m \times \mathcal{L}(V)^n = \underbrace{V \times \cdots \times V}_{m \text{ times}} \times \underbrace{\mathcal{L}(V) \times \cdots \times \mathcal{L}(V)}_{n \text{ times}}, \tag{4.43}$$

and

$$\begin{aligned} &\phi : \mathcal{D} \to I\!R, \\ &\boldsymbol{h} : \mathcal{D} \to V, \\ &S : \mathcal{D} \to \mathcal{L}(V). \end{aligned} \tag{4.44}$$

**Definition.** We say that $\phi$, $\boldsymbol{h}$, and $S$ are *scalar*, *vector*, and *tensor invariants* relative to the group $\mathcal{G} \subseteq \mathcal{O}(V)$, respectively, if for any $\vec{\boldsymbol{v}} \in V^m$, $\vec{A} \in \mathcal{L}(V)^n$, we have

$$\begin{aligned} \phi(Q\vec{\boldsymbol{v}}, Q\vec{A}Q^T) &= \phi(\vec{\boldsymbol{v}}, \vec{A}), \\ \boldsymbol{h}(Q\vec{\boldsymbol{v}}, Q\vec{A}Q^T) &= Q\boldsymbol{h}(\vec{\boldsymbol{v}}, \vec{A}), \qquad \forall Q \in \mathcal{G}, \\ S(Q\vec{\boldsymbol{v}}, Q\vec{A}Q^T) &= QS(\vec{\boldsymbol{v}}, \vec{A})Q^T. \end{aligned} \tag{4.45}$$

In the above definition, we have introduced the following abbreviations:

$$\begin{aligned} \vec{\boldsymbol{v}} &= (\boldsymbol{v}_1, \cdots, \boldsymbol{v}_m), \\ \vec{A} &= (A_1, \cdots, A_n), \end{aligned} \tag{4.46}$$

and

$$\begin{aligned} Q\vec{\boldsymbol{v}} &= (Q\boldsymbol{v}_1, \cdots, Q\boldsymbol{v}_m), \\ Q\vec{A}Q^T &= (QA_1Q^T, \cdots, QA_nQ^T), \end{aligned}$$

for $\boldsymbol{v}_i \in V$ and $A_j \in \mathcal{L}(V)$.

If the group $\mathcal{G} = \mathcal{O}(V)$, the invariants are usually called *isotropic invariants* or *isotropic functions*, otherwise, they are called *anisotropic invariants.*

**Example 4.3.1** The following functions are
1) isotropic scalar invariants: $\boldsymbol{v} \cdot \boldsymbol{u}$, $\det A$, $\operatorname{tr}(A^m B^n)$, $A^m \boldsymbol{v} \cdot \boldsymbol{B}^n \boldsymbol{u}$;
2) isotropic vector invariants: $A^m \boldsymbol{v}$, $A^m B^n \boldsymbol{v}$;
3) isotropic tensor invariants: $A^m$, $A^m \boldsymbol{v} \otimes B^n \boldsymbol{v}$;
for any $\boldsymbol{u}, \boldsymbol{v} \in V$ and $A, B \in \mathcal{L}(V)$. □

**Example 4.3.2** The vector product $\boldsymbol{v} \times \boldsymbol{u}$ is not an isotropic vector invariant, but it is a vector invariant relative to the proper orthogonal group $\mathcal{O}^+(V)$. To see this, we have

$$Q^T(Q\boldsymbol{v} \times Q\boldsymbol{u}) \cdot \boldsymbol{w} = Q\boldsymbol{v} \times Q\boldsymbol{u} \cdot Q\boldsymbol{w} = (\det Q)(\boldsymbol{v} \times \boldsymbol{u}) \cdot \boldsymbol{w},$$

for any $\boldsymbol{w} \in V$. Hence we obtain

$$(Q\boldsymbol{v} \times Q\boldsymbol{u}) = (\det Q)\, Q(\boldsymbol{v} \times \boldsymbol{u}).$$

Since $\det Q = \pm 1$ for $Q \in \mathcal{O}(V)$ and $\det Q = 1$ for $Q \in \mathcal{O}^+(V)$, therefore, it is not an isotropic vector invariant, but it is an anisotropic vector invariant relative to the proper orthogonal group $\mathcal{O}^+(V)$.

The same is true for the vector function $\boldsymbol{h}(W) = \langle W \rangle$, where $\langle W \rangle$ denotes the axial vector associated with the skew-symmetric tensor $W$ (see (A.29)), since we can also show

$$\langle QWQ^T \rangle = (\det Q)\, Q\langle W \rangle$$

easily. □

Suppose that $\Upsilon_s$, $\Upsilon_v$, and $\Upsilon_t$ are sets of scalar, vector, and tensor invariants, respectively, then it is obvious that the following functions are isotropic:

$$\begin{aligned} \phi &= \phi(\Upsilon_s), \\ \boldsymbol{h} &= \sum_{\boldsymbol{u}_a \in \Upsilon_v} \phi_a(\Upsilon_s)\, \boldsymbol{u}_a, \\ S &= \sum_{T_b \in \Upsilon_t} \phi_b(\Upsilon_s)\, T_b, \end{aligned} \tag{4.47}$$

where $\phi$, $\phi_a$, and $\phi_b$ are arbitrary functions of the scalar invariants in $\Upsilon_s$.

The purpose of the representation problems is to find the sets of invariants $\Upsilon_s$, $\Upsilon_v$, and $\Upsilon_t$, so that isotropic functions can be expressed in the above forms (4.47). We call $\Upsilon_s$ the set of *basic invariants*, and say that $\Upsilon_v$ and $\Upsilon_t$ are the *generating sets* for isotropic vector, and tensor functions, respectively.

A set of basic invariants or a generating set is called a *functional basis* if it is *irreducible*. By irreducible we mean that elements of the basic invariants are not functionally related. and elements of the generating set are linearly independent with respect to isotropic functions, i.e., no elements can be expressed as a linear combination of other elements with coefficients of scalar isotropic functions.

Functional bases for isotropic functions have been extensively studied in the literature (see [7, 66, 74]). For ready reference, we reproduce the complete results from [74] for isotropic invariants in Table 4.1 through Table 4.4, in which $\boldsymbol{v}$ stands for vector, $A$ for symmetric tensor, $W$ for skew-symmetric tensor.

From these tables one can easily construct functional bases $\Upsilon_s$, $\Upsilon_v$, and $\Upsilon_t$ of invariants for any given domain $\mathcal{D}$ of variables and thus obtain the representation of isotropic functions from (4.47). The representation theorems 4.2.1, 4.2.2, and 4.2.5 proved before are just the simplest cases given by these tables.

In the following examples we shall give the functional bases $\Upsilon_s$, $\Upsilon_v$, and $\Upsilon_t$ for isotropic functions of some different cases of variables.

**Example 4.3.3** For $\mathcal{D} = \{(\boldsymbol{v}_1, \cdots, \boldsymbol{v}_m) \in V^m\}$, we obtain

$$\begin{aligned}
\Upsilon_s &= \{\boldsymbol{v}_1 \cdot \boldsymbol{v}_1, \boldsymbol{v}_1 \cdot \boldsymbol{v}_2, \cdots, \boldsymbol{v}_{n-1} \cdot \boldsymbol{v}_n, \boldsymbol{v}_n \cdot \boldsymbol{v}_n\}, \\
\Upsilon_v &= \{\boldsymbol{v}, \cdots, \boldsymbol{v}_n\}, \\
\Upsilon_t &= \{\mathit{1}, \boldsymbol{v}_1 \otimes \boldsymbol{v}_1, \cdots, \boldsymbol{v}_n \otimes \boldsymbol{v}_n, (\boldsymbol{v}_1 \otimes \boldsymbol{v}_2 + \boldsymbol{v}_2 \otimes \boldsymbol{v}_1), \cdots, \\
&\quad (\boldsymbol{v}_{n-1} \otimes \boldsymbol{v}_n + \boldsymbol{v}_n \otimes \boldsymbol{v}_{n-1})\}.
\end{aligned}$$

The above functional bases for scalar, vector, and symmetric tensor invariants are constructed by taking all *possible* combinations of variables according to the tables. □

The above result for scalar invariants of any number of vector variables – Theorem 4.2.5 is a special one with two vectors – is a well-known result due to Euler in the classical invariant theory.

**Example 4.3.4** For $\mathcal{D} = \{(\boldsymbol{v}, A) \in V \times \mathcal{L}(V), A = A^T\}$, we obtain the following functional bases:

$$\begin{aligned}
\Upsilon_s &= \{\operatorname{tr} A, \operatorname{tr} A^2, \operatorname{tr} A^3, \boldsymbol{v} \cdot \boldsymbol{v}, \boldsymbol{v} \cdot A\boldsymbol{v}, \boldsymbol{v} \cdot A^2\boldsymbol{v}\}, \\
\Upsilon_v &= \{\boldsymbol{v}, A\boldsymbol{v}, A^2\boldsymbol{v}\}, \\
\Upsilon_t &= \{\mathit{1}, A, A^2, \boldsymbol{v} \otimes \boldsymbol{v}, (A\boldsymbol{v} \otimes \boldsymbol{v} + \boldsymbol{v} \otimes A\boldsymbol{v}), A\boldsymbol{v} \otimes A\boldsymbol{v}\}.
\end{aligned}$$

for scalar, vector, and symmetric tensor invariants, respectively. □

If we examine Table 4.1 through Table 4.4, we can find that every scalar invariant is the trace of some tensor, e.g., $\boldsymbol{v} \cdot A\boldsymbol{u} = \operatorname{tr}(\boldsymbol{v} \otimes A\boldsymbol{u})$, and every vector or tensor invariant is some vector or tensor formed from the variables intrinsically. Moreover, scalar invariant elements involving more than four variables, and vector or tensor generator elements involving more than three variables,

**Table 4.1.** Isotropic scalar invariants

| One variable | Invariant elements |
|---|---|
| $\boldsymbol{v}$ | $\boldsymbol{v}\cdot\boldsymbol{v}$ |
| $A$ | $\operatorname{tr}A$, $\operatorname{tr}A^2$, $\operatorname{tr}A^3$ |
| $W$ | $\operatorname{tr}W^2$ |
| **Two variables** | **Invariant elements** |
| $\boldsymbol{v}_1$, $\boldsymbol{v}_2$ | $\boldsymbol{v}_1\cdot\boldsymbol{v}_2$ |
| $\boldsymbol{v}$, $A$ | $\boldsymbol{v}\cdot A\boldsymbol{v}$, $\boldsymbol{v}\cdot A^2\boldsymbol{v}$ |
| $\boldsymbol{v}$, $W$ | $\boldsymbol{v}\cdot W^2\boldsymbol{v}$ |
| $A_1$, $A_2$ | $\operatorname{tr}A_1A_2$, $\operatorname{tr}A_1A_2^2$, $\operatorname{tr}A_2A_1^2$, $\operatorname{tr}A_1^2A_2^2$ |
| $W_1$, $W_2$ | $\operatorname{tr}W_1W_2$ |
| $A$, $W$ | $\operatorname{tr}AW^2$, $\operatorname{tr}A^2W^2$, $\operatorname{tr}AWA^2W^2$ |
| **Three variables** | **Invariant elements** |
| $\boldsymbol{v}_1$, $\boldsymbol{v}_2$, $A$ | $\boldsymbol{v}_1\cdot A\boldsymbol{v}_2$, $\boldsymbol{v}_1\cdot A^2\boldsymbol{v}_2$ |
| $\boldsymbol{v}_1$, $\boldsymbol{v}_2$, $W$ | $\boldsymbol{v}_1\cdot W\boldsymbol{v}_2$, $\boldsymbol{v}_1\cdot W^2\boldsymbol{v}_2$ |
| $\boldsymbol{v}$, $A_1$, $A_2$ | $\boldsymbol{v}\cdot A_1A_2\boldsymbol{v}$ |
| $\boldsymbol{v}$, $W_1$, $W_2$ | $\boldsymbol{v}\cdot W_1W_2\boldsymbol{v}$, $\boldsymbol{v}\cdot W_1W_2^2\boldsymbol{v}$, $\boldsymbol{v}\cdot W_2W_1^2\boldsymbol{v}$ |
| $\boldsymbol{v}$, $A$, $W$ | $\boldsymbol{v}\cdot WA\boldsymbol{v}$, $\boldsymbol{v}\cdot WA^2\boldsymbol{v}$, $\boldsymbol{v}\cdot WAW^2\boldsymbol{v}$ |
| $A_1$, $A_2$, $A_3$ | $\operatorname{tr}A_1A_2A_3$ |
| $W_1$, $W_2$, $W_3$ | $\operatorname{tr}W_1W_2W_3$ |
| $A_1$, $A_2$, $W$ | $\operatorname{tr}A_1A_2W$, $\operatorname{tr}A_1A_2^2W$, $\operatorname{tr}A_2A_1^2W$, $\operatorname{tr}A_1WA_2W^2$ |
| $A$, $W_1$, $W_2$ | $\operatorname{tr}AW_1W_2$, $\operatorname{tr}AW_1W_2^2$, $\operatorname{tr}AW_2W_1^2$ |
| **Four variables** | **Invariant elements** |
| $\boldsymbol{v}_1$, $\boldsymbol{v}_2$, $A_1$, $A_2$ | $\boldsymbol{v}_1\cdot A_1A_2\boldsymbol{v}_2$, $\boldsymbol{v}_1\cdot A_2A_1\boldsymbol{v}_2$ |
| $\boldsymbol{v}_1$, $\boldsymbol{v}_2$, $W_1$, $W_2$ | $\boldsymbol{v}_1\cdot W_1W_2\boldsymbol{v}_2$, $\boldsymbol{v}_1\cdot W_2W_1\boldsymbol{v}_2$ |
| $\boldsymbol{v}_1$, $\boldsymbol{v}_2$, $A$, $W$ | $\boldsymbol{v}_1\cdot AW\boldsymbol{v}_2$, $\boldsymbol{v}_1\cdot WA\boldsymbol{v}_2$ |

**Table 4.2.** Isotropic vector invariants

| One variable | Generator elements |
|---|---|
| $\boldsymbol{v}$ | $\boldsymbol{v}$ |
| $A$ or $W$ | 0 |
| **Two variables** | **Generator elements** |
| $\boldsymbol{v}$, $A$ | $A\boldsymbol{v}$, $A^2\boldsymbol{v}$ |
| $\boldsymbol{v}$, $W$ | $W\boldsymbol{v}$, $W^2\boldsymbol{v}$ |
| **Three variables** | **Generator elements** |
| $\boldsymbol{v}$, $A_1$, $A_2$ | $A_1A_2\boldsymbol{v}$, $A_2A_1\boldsymbol{v}$ |
| $\boldsymbol{v}$, $W_1$, $W_2$ | $W_1W_2\boldsymbol{v}$, $W_2W_1\boldsymbol{v}$ |
| $\boldsymbol{v}$, $A$, $W$ | $AW\boldsymbol{v}$, $WA\boldsymbol{v}$ |

**Table 4.3.** Isotropic symmetric tensor invariants

| No variable | Generator elements |
|---|---|
| 0 | $\mathbf{1}$ |
| **One variable** | **Generator elements** |
| $\boldsymbol{v}$ | $\boldsymbol{v}\otimes\boldsymbol{v}$ |
| $A$ | $A,\ A^2$ |
| $W$ | $W^2$ |
| **Two variables** | **Generator elements** |
| $\boldsymbol{v}_1,\ \boldsymbol{v}_2$ | $\boldsymbol{v}_1\otimes\boldsymbol{v}_2+\boldsymbol{v}_2\otimes\boldsymbol{v}_1$ |
| $\boldsymbol{v},\ A$ | $\boldsymbol{v}\otimes A\boldsymbol{v}+A\boldsymbol{v}\otimes\boldsymbol{v},\ A\boldsymbol{v}\otimes A\boldsymbol{v}$ |
| $\boldsymbol{v},\ W$ | $\boldsymbol{v}\otimes W\boldsymbol{v}+W\boldsymbol{v}\otimes\boldsymbol{v},\ W\boldsymbol{v}\otimes W\boldsymbol{v},$ $W\boldsymbol{v}\otimes W^2\boldsymbol{v}+W^2\boldsymbol{v}\otimes W\boldsymbol{v}$ |
| $A_1,\ A_2$ | $A_1A_2+A_2A_1,\ A_1A_2A_1,\ A_2A_1A_2$ |
| $W_1,\ W_2$ | $W_1W_2+W_2W_1,\ W_1W_2^2-W_2^2W_1,\ W_1^2W_2-W_2W_1^2$ |
| $A,\ W$ | $AW-WA,\ WAW,\ A^2W-WA^2,\ WAW^2-W^2AW$ |
| **Three variables** | **Generator elements** |
| $\boldsymbol{v}_1,\ \boldsymbol{v}_2,\ A$ | $\boldsymbol{v}_1\otimes A\boldsymbol{v}_2+A\boldsymbol{v}_2\otimes\boldsymbol{v}_1,\ \boldsymbol{v}_2\otimes A\boldsymbol{v}_1+A\boldsymbol{v}_1\otimes\boldsymbol{v}_2$ |
| $\boldsymbol{v}_1,\ \boldsymbol{v}_2,\ W$ | $\boldsymbol{v}_1\otimes W\boldsymbol{v}_2+W\boldsymbol{v}_2\otimes\boldsymbol{v}_1,\ \boldsymbol{v}_2\otimes W\boldsymbol{v}_1+W\boldsymbol{v}_1\otimes\boldsymbol{v}_2$ |

**Table 4.4.** Isotropic skew-symmetric tensor invariants

| One variable | Generator elements |
|---|---|
| $\boldsymbol{v}$ or $A$ | 0 |
| $W$ | $W$ |
| **Two variables** | **Generator elements** |
| $\boldsymbol{v}_1,\ \boldsymbol{v}_2$ | $\boldsymbol{v}_1\otimes\boldsymbol{v}_2-\boldsymbol{v}_2\otimes\boldsymbol{v}_1$ |
| $\boldsymbol{v},\ A$ | $\boldsymbol{v}\otimes A\boldsymbol{v}-A\boldsymbol{v}\otimes\boldsymbol{v},\ \boldsymbol{v}\otimes A^2\boldsymbol{v}-A^2\boldsymbol{v}\otimes\boldsymbol{v},$ $A\boldsymbol{v}\otimes A^2\boldsymbol{v}-A^2\boldsymbol{v}\otimes A\boldsymbol{v}$ |
| $\boldsymbol{v},\ W$ | $\boldsymbol{v}\otimes W\boldsymbol{v}-W\boldsymbol{v}\otimes\boldsymbol{v},\ \boldsymbol{v}\otimes W^2\boldsymbol{v}-W^2\boldsymbol{v}\otimes\boldsymbol{v}$ |
| $A_1,\ A_2$ | $A_1A_2-A_2A_1,\ A_1A_2^2-A_2^2A_1,\ A_1^2A_2-A_2A_1^2.$ $A_1A_2A_1^2-A_1^2A_2A_1,\ A_2A_1A_2^2-A_2^2A_1A_2$ |
| $W_1,\ W_2$ | $W_1W_2-W_2W_1$ |
| $A,\ W$ | $AW+WA,\ AW^2-W^2A$ |
| **Three variables** | **Generator elements** |
| $\boldsymbol{v}_1,\ \boldsymbol{v}_2,\ A$ | $\boldsymbol{v}_1\otimes A\boldsymbol{v}_2-A\boldsymbol{v}_2\otimes\boldsymbol{v}_1,\ \boldsymbol{v}_2\otimes A\boldsymbol{v}_1-A\boldsymbol{v}_1\otimes\boldsymbol{v}_2$ |
| $\boldsymbol{v}_1,\ \boldsymbol{v}_2,\ W$ | $\boldsymbol{v}_1\otimes W\boldsymbol{v}_2-W\boldsymbol{v}_2\otimes\boldsymbol{v}_1,\ \boldsymbol{v}_2\otimes W\boldsymbol{v}_1-W\boldsymbol{v}_1\otimes\boldsymbol{v}_2$ |
| $\boldsymbol{v},\ A_1,\ A_2$ | $A_1\boldsymbol{v}\otimes A_2\boldsymbol{v}-A_2\boldsymbol{v}\otimes A_1\boldsymbol{v},\ A_1A_2\boldsymbol{v}\otimes\boldsymbol{v}-\boldsymbol{v}\otimes A_1A_2\boldsymbol{v},$ $A_2A_1\boldsymbol{v}\otimes\boldsymbol{v}-\boldsymbol{v}\otimes A_2A_1\boldsymbol{v}$ |
| $A_1,\ A_2,\ A_3$ | $A_1A_2A_3-A_3A_2A_1,\ A_2A_3A_1-A_1A_3A_2,$ $A_3A_1A_2-A_2A_1A_3$ |

are not present in the tables. In other words, such elements are proved to be redundant. These observations are very helpful in forming various invariants from a given set of variables without consulting the tables. In fact, we can form infinitely many such invariants. However, there exist some identities, such as the Cayley–Hamilton theorem (4.16), which enable us to eliminate many of them and thus we are left with a finite set of invariants. Of course, such eliminations using known identities would not give us an irreducible set of invariants, in general. Nevertheless, since general representations are often too complicated to be useful as far as practical applications are concerned, we often need only certain linear or quadratic representations. Such representations are relatively easy to work out based on the above observation, even in the absence of those tables.

**Example 4.3.5** Consider variables $(\boldsymbol{v}, A)$.
1) The isotropic functions linearized in $A$ can be written as

$$
\begin{aligned}
\phi &= \phi_1 + \phi_2 \,\operatorname{tr} A + \phi_3 \,(\boldsymbol{v}\cdot A\boldsymbol{v}),\\
\boldsymbol{h} &= (h_1 + h_2 \,\operatorname{tr} A + h_3 \,(\boldsymbol{v}\cdot A\boldsymbol{v}))\boldsymbol{v} + h_4 A\boldsymbol{v},\\
S &= (s_1 + s_2 \,\operatorname{tr} A + s_3 \,(\boldsymbol{v}\cdot A\boldsymbol{v}))\mathit{1} + s_4 A\\
&\quad + \,(s_5 + s_6 \,\operatorname{tr} A + s_7 \,(\boldsymbol{v}\cdot A\boldsymbol{v}))\boldsymbol{v}\otimes\boldsymbol{v} + s_8(\boldsymbol{v}\otimes A\boldsymbol{v} + A\boldsymbol{v}\otimes\boldsymbol{v}),
\end{aligned}
$$

where the coefficients $\phi_i$, $h_j$, and $s_k$ are scalar functions of $(\boldsymbol{v}\cdot\boldsymbol{v})$.
2) The isotropic functions linearized in both $\boldsymbol{v}$ and $A$ can be written as

$$
\begin{aligned}
\phi &= \phi_1 + \phi_2 \,\operatorname{tr} A,\\
\boldsymbol{h} &= h_1 \boldsymbol{v},\\
S &= (s_1 + s_2 \,\operatorname{tr} A)\mathit{1} + s_3 A,
\end{aligned}
\tag{4.48}
$$

where the coefficients $\phi_1$ through $s_3$ are independent of $\boldsymbol{v}$ and $A$. □

There is an identity that is sometimes very useful in reducing some complicated invariants. Let $A$ and $B$ be two symmetric tensors, then they must satisfy (for a proof of this identity see [62])

$$
\begin{aligned}
&ABA + A^2B + BA^2\\
&\quad = \Big(\operatorname{tr} A^2B - \operatorname{tr} A \,\operatorname{tr} AB - \frac{1}{2}\operatorname{tr} B\,(\operatorname{tr} A^2 - (\operatorname{tr} A)^2)\Big)\mathit{1}\\
&\qquad + \,(\operatorname{tr} AB - \operatorname{tr} A \,\operatorname{tr} B)A + \frac{1}{2}(\operatorname{tr} A^2 - (\operatorname{tr} A)^2)B\\
&\qquad + \,\operatorname{tr} A\,(AB + BA) + (\operatorname{tr} B)\,A^2.
\end{aligned}
\tag{4.49}
$$

Note that when $B = A$, this identity reduces to (4.16) of the Cayley–Hamilton theorem.

As an example, let us take $B = \boldsymbol{v} \otimes \boldsymbol{v}$, then the left-hand side of (4.49) becomes

$$A\boldsymbol{v} \otimes A\boldsymbol{v} + A^2\boldsymbol{v} \otimes \boldsymbol{v} + \boldsymbol{v} \otimes A^2\boldsymbol{v}.$$

Examining the terms on the right-hand side, one can conclude that the generator element $A\boldsymbol{v} \otimes A\boldsymbol{v}$, for two variables $\boldsymbol{v}$ and $A$ in Table 4.3, can be replaced by the element $(\boldsymbol{v} \otimes A^2\boldsymbol{v} + A^2\boldsymbol{v} \otimes \boldsymbol{v})$.

**Exercise 4.3.1** Consider the variables $(\boldsymbol{u}, \boldsymbol{v}, A)$.
1) Find the general representations of scalar, vector, and symmetric tensor isotropic functions.
2) Find their representations that are linearized in the variables $\boldsymbol{v}$ and $A$.

**Exercise 4.3.2** Show that if $\boldsymbol{h}(A, \boldsymbol{v})$ is a vector isotropic function, then $\partial_{\boldsymbol{v}}\boldsymbol{h}$ is a tensor isotropic function, (1) directly from definition; (2) by using the representation formula.

**Exercise 4.3.3** Derive the explicit relation between $A\boldsymbol{v} \otimes A\boldsymbol{v}$ and $(\boldsymbol{v} \otimes A^2\boldsymbol{v} + A^2\boldsymbol{v} \otimes \boldsymbol{v})$, by the use of (4.49). Therefore, justify the replacement of the generator element $A\boldsymbol{v} \otimes A\boldsymbol{v}$ for two variables $\boldsymbol{v}$ and $A$ by the element $(\boldsymbol{v} \otimes A^2\boldsymbol{v} + A^2\boldsymbol{v} \otimes \boldsymbol{v})$ in Table 4.3.

**Exercise 4.3.4** By the use of (4.49), show that both $(A\boldsymbol{v} \otimes A^2\boldsymbol{v} + A^2\boldsymbol{v} \otimes A\boldsymbol{v})$ and $A^2\boldsymbol{v} \otimes A^2\boldsymbol{v}$ are not needed as generator elements for isotropic symmetric tensor invariants of variables $\boldsymbol{v}$ and $A$.

### 4.3.1 Isotropic Thermoelastic Solids and Viscous Heat-Conducting Fluids

For an isotropic thermoelastic solid defined by (4.4), from Example 4.3.4, we have the following constitutive equations,

$$\begin{aligned} T &= t_0 1 + t_1 B + t_2 B^2 \\ &\quad + t_3 \boldsymbol{g} \otimes \boldsymbol{g} + t_4(B\boldsymbol{g} \otimes \boldsymbol{g} + \boldsymbol{g} \otimes B\boldsymbol{g}) + t_5 B\boldsymbol{g} \otimes B\boldsymbol{g}, \\ \boldsymbol{q} &= k_1 \boldsymbol{g} + k_2 B\boldsymbol{g} + k_3 B^2\boldsymbol{g}, \\ \varepsilon &= \varepsilon(\theta, \operatorname{tr} B, \operatorname{tr} B^2, \operatorname{tr} B^3, \boldsymbol{g}\cdot\boldsymbol{g},\ \boldsymbol{g}\cdot B\boldsymbol{g},\ \boldsymbol{g}\cdot B^2\boldsymbol{g}), \end{aligned} \tag{4.50}$$

where the coefficients $t_i$ and $k_j$, as well as $\varepsilon$, are scalar functions of seven variables indicated in the argument of $\varepsilon$.

Similarly, one can immediately write down the constitutive equations for a viscous heat-conducting fluid defined by (4.3),

$$
\begin{aligned}
T &= \alpha_0 1 + \alpha_1 D + \alpha_2 D^2 \\
&\quad + \alpha_3 \boldsymbol{g} \otimes \boldsymbol{g} + \alpha_4 (D\boldsymbol{g} \otimes \boldsymbol{g} + \boldsymbol{g} \otimes D\boldsymbol{g}) + \alpha_5 D\boldsymbol{g} \otimes D\boldsymbol{g}, \\
\boldsymbol{q} &= \beta_1 \boldsymbol{g} + \beta_2 D\boldsymbol{g} + \beta_3 D^2 \boldsymbol{g}, \\
\varepsilon &= \varepsilon(\rho, \theta, \operatorname{tr} D, \operatorname{tr} D^2, \operatorname{tr} D^3, \boldsymbol{g}\cdot\boldsymbol{g},\ \boldsymbol{g}\cdot D\boldsymbol{g},\ \boldsymbol{g}\cdot D^2\boldsymbol{g}),
\end{aligned}
\tag{4.51}
$$

where the coefficients $\alpha_i$ and $\beta_j$, as well as $\varepsilon$, are scalar functions of eight variables indicated in the arguments of $\varepsilon$.

If we linearize the constitutive equations (4.51) of a viscous heat-conducting fluid in both $D$ and $\boldsymbol{g}$, from Example 4.3.5, we obtain the following representations for the stress tensor and the heat flux,

$$
\begin{aligned}
T &= (-p + \lambda \operatorname{tr} D) 1 + 2\mu D, \\
\boldsymbol{q} &= -\kappa \boldsymbol{g},
\end{aligned}
\tag{4.52}
$$

where the coefficients are functions of $(\rho, \theta)$. These are the classical Navier–Stokes theory and *Fourier's law*, and we shall call it a *Navier–Stokes–Fourier fluid*. The coefficients $\lambda$, $\mu$ are the viscosities and $\kappa$ is the thermal conductivity.

## 4.4 Hemitropic Invariants

For thermoelastic solids, the constitutive relations for the second Piola–Kirchhoff stress tensor, the material heat flux, and the internal energy, relative to an undistorted configuration $\kappa$ with symmetry group $\mathcal{G}_\kappa$, must satisfy the following symmetry conditions, according to (3.49),

$$
\begin{aligned}
\mathcal{S}_\kappa(QCQ^T, \theta, Q\boldsymbol{g}_\kappa) &= Q\,\mathcal{S}_\kappa(C, \theta, \boldsymbol{g}_\kappa)\,Q^T, \\
\mathcal{Q}_\kappa(QCQ^T, \theta, Q\boldsymbol{g}_\kappa) &= Q\,\mathcal{Q}_\kappa(C, \theta, \boldsymbol{g}_\kappa), \qquad \forall Q \in \mathcal{G}_\kappa. \\
\mathcal{E}_\kappa(QCQ^T, \theta, Q\boldsymbol{g}_\kappa) &= \mathcal{E}_\kappa(C, \theta, \boldsymbol{g}_\kappa),
\end{aligned}
\tag{4.53}
$$

In other words, the functions $\mathcal{S}_\kappa$, $\mathcal{Q}_\kappa$, and $\mathcal{E}_\kappa$ are invariants relative to the group $\mathcal{G}_\kappa$, according to the definition (4.45).

If the group $\mathcal{G}_\kappa = \mathcal{O}(V)$, the solid is isotropic and the functions are isotropic. For anisotropic solids in general, $\mathcal{G}_\kappa \neq \mathcal{O}(V)$. In this and the following section we shall consider representation theorems for invariants relative to some subgroups of the orthogonal group $O(V)$.

**Definition.** We call an invariant relative to the proper orthogonal group $\mathcal{O}^+(V)$ a *hemitropic function.*

Since a proper orthogonal transformation preserves the orientation, one can always identify a vector with a skew-symmetric tensor relative to the action of any proper orthogonal transformation (see Sect. A.1.6). Let $\boldsymbol{v}$ be a vector and $W$ be its associated skew-symmetric tensor, $\boldsymbol{v} = \langle W \rangle$. One can easily verify that (see Example 4.3.2), for any $Q \in \mathcal{O}^+(V)$,

$$Q\boldsymbol{v} = \langle QWQ^T \rangle.$$

We have the following representation theorems for hemitropic functions.

**Theorem 4.4.1.** *For any scalar-valued (or tensor-valued) function* $\psi(\vec{\boldsymbol{v}}, \vec{A})$, *define*

$$\widehat{\psi}(\vec{W}, \vec{A}) = \psi(\vec{\boldsymbol{v}}, \vec{A}), \quad \vec{\boldsymbol{v}} = \langle \vec{W} \rangle. \tag{4.54}$$

*Then* $\psi(\vec{\boldsymbol{v}}, \vec{A})$ *is a scalar (or tensor) hemitropic function if and only if* $\widehat{\psi}(\vec{W}, \vec{A})$ *is a scalar (or tensor) isotropic function.*

**Theorem 4.4.2.** *For any vector-valued function* $\boldsymbol{h}(\vec{\boldsymbol{v}}, \vec{A})$, *let* $H$ *be the skew-symmetric tensor-valued function, such that* $\boldsymbol{h} = \langle H \rangle$ *and define*

$$\widehat{H}(\vec{W}, \vec{A}) = H(\vec{\boldsymbol{v}}, \vec{A}), \quad \vec{\boldsymbol{v}} = \langle \vec{W} \rangle. \tag{4.55}$$

*Then* $\boldsymbol{h}(\vec{\boldsymbol{v}}, \vec{A})$ *is a vector hemitropic function if and only if* $\widehat{H}(\vec{W}, \vec{A})$ *is a skew-symmetric tensor isotropic function.*

In these statements we have used the notation introduced in (4.46). The proof of these theorems follows from the simple observation that a scalar or tensor hemitropic function of only tensor variables is also isotropic because the transformation $QAQ^T$ is unchanged if $Q$ is replaced by $-Q$. Based on these theorems, one can obtain representations for any hemitropic functions by simply replacing vectors with their associated skew-symmetric tensors. From Tables 4.1, 4.3, and 4.4 one can immediately obtain irreducible functional bases for scalar, tensor, and vector hemitropic functions.[1] They are given in Tables 4.5, 4.6, and 4.7, respectively.

**Example 4.4.1** Let $\mathcal{D} = \{(\boldsymbol{v}, A) \in V \times \mathcal{L}(V), A = A^T\}$ and $\boldsymbol{h} : \mathcal{D} \to V$ be a vector hemitropic function. Suppose that $\boldsymbol{v} = \langle W \rangle$, then from Table 4.1 and Table 4.4 for variables $W$ and $A$, we have the following irreducible bases for scalar and skew-symmetric tensor isotropic invariants:

$$\begin{aligned} \Upsilon_s &= \{\operatorname{tr} A,\ \operatorname{tr} A^2,\ \operatorname{tr} A^3,\ \operatorname{tr} W^2,\ \operatorname{tr} AW^2,\ \operatorname{tr} A^2W^2,\ \operatorname{tr} AWA^2W^2\}, \\ \Upsilon_t &= \{W,\ AW + WA,\ AW^2 - W^2A\}. \end{aligned}$$

Therefore, by Theorem 4.4.2, we can write the representation of $\boldsymbol{h}(\boldsymbol{v}, A)$ as

$$\boldsymbol{h} = h_1\boldsymbol{v} + h_2A\boldsymbol{v} + h_3\boldsymbol{v} \times A\boldsymbol{v}, \tag{4.56}$$

[1] Integrity bases for hemitropic invariants can be found in [19].

**Table 4.5.** Hemitropic scalar invariants

| One variable | Invariant elements |
|---|---|
| $A$ | $\operatorname{tr} A,\ \operatorname{tr} A^2,\ \operatorname{tr} A^3$ |
| $\boldsymbol{v}$ | $\boldsymbol{v}\cdot\boldsymbol{v}$ |
| **Two variables** | **Invariant elements** |
| $A_1,\ A_2$ | $\operatorname{tr} A_1A_2,\ \operatorname{tr} A_1A_2^2,\ \operatorname{tr} A_2A_1^2,\ \operatorname{tr} A_1^2A_2^2$ |
| $\boldsymbol{v}_1,\ \boldsymbol{v}_2$ | $\boldsymbol{v}_1\cdot\boldsymbol{v}_2$ |
| $A,\ \boldsymbol{v}$ | $\boldsymbol{v}\cdot A\boldsymbol{v},\ \boldsymbol{v}\cdot A^2\boldsymbol{v},\ \boldsymbol{v}\cdot A\boldsymbol{v}\times A^2\boldsymbol{v}$ |
| **Three variables** | **Invariant elements** |
| $A_1,\ A_2,\ A_3$ | $\operatorname{tr} A_1A_2A_3$ |
| $\boldsymbol{v}_1,\ \boldsymbol{v}_2,\ \boldsymbol{v}_3$ | $\boldsymbol{v}_1\cdot\boldsymbol{v}_2\times\boldsymbol{v}_3$ |
| $A_1,\ A_2,\ \boldsymbol{v}$ | $\boldsymbol{v}\cdot\langle A_1A_2\rangle,\ \boldsymbol{v}\cdot\langle A_1A_2^2\rangle,\ \boldsymbol{v}\cdot\langle A_2A_1^2\rangle,\ \boldsymbol{v}\cdot A_1\boldsymbol{v}\times A_2\boldsymbol{v}$ |
| $A,\ \boldsymbol{v}_1,\ \boldsymbol{v}_2$ | $\boldsymbol{v}_1\cdot A\boldsymbol{v}_2,\ \boldsymbol{v}_1\cdot\boldsymbol{v}_2\times A\boldsymbol{v}_1,\ \boldsymbol{v}_2\cdot\boldsymbol{v}_1\times A\boldsymbol{v}_2$ |

**Table 4.6.** Hemitropic vector invariants

| One variable | Generator elements |
|---|---|
| $\boldsymbol{v}$ | $\boldsymbol{v}$ |
| $A$ | 0 |
| **Two variables** | **Generator elements** |
| $A_1,\ A_2$ | $\langle A_1A_2\rangle,\ \langle A_1A_2^2\rangle,\ \langle A_2A_1^2\rangle,\ \langle A_1A_2A_1^2\rangle,\ \langle A_2A_1A_2^2\rangle$ |
| $\boldsymbol{v}_1,\ \boldsymbol{v}_2$ | $\boldsymbol{v}_1\times\boldsymbol{v}_2$ |
| $A,\ \boldsymbol{v}$ | $A\boldsymbol{v},\ \boldsymbol{v}\times A\boldsymbol{v}$ |
| **Three variables** | **Generator elements** |
| $A_1\ A_2,\ A_3$ | $\langle A_1A_2A_3\rangle,\ \langle A_2A_3A_1\rangle,\ \langle A_3A_1A_2\rangle$ |

**Table 4.7.** Hemitropic symmetric tensor invariants

| No variable | Generator elements |
|---|---|
| 0 | $\mathit{1}$ |
| **One variable** | **Generator elements** |
| $A$ | $A,\ A^2$ |
| $\boldsymbol{v}$ | $\boldsymbol{v}\otimes\boldsymbol{v}$ |
| **Two variables** | **Generator elements** |
| $A_1,\ A_2$ | $A_1A_2+A_2A_1,\ A_1A_2A_1,\ A_2A_1A_2$ |
| $\boldsymbol{v}_1,\ \boldsymbol{v}_2$ | $\boldsymbol{v}_1\otimes\boldsymbol{v}_2+\boldsymbol{v}_2\otimes\boldsymbol{v}_1,$ <br> $\boldsymbol{v}_1\otimes(\boldsymbol{v}_1\times\boldsymbol{v}_2)+(\boldsymbol{v}_1\times\boldsymbol{v}_2)\otimes\boldsymbol{v}_1,$ <br> $\boldsymbol{v}_2\otimes(\boldsymbol{v}_1\times\boldsymbol{v}_2)+(\boldsymbol{v}_1\times\boldsymbol{v}_2)\otimes\boldsymbol{v}_2$ |
| $A,\ \boldsymbol{v}=\langle W\rangle$ | $AW-WA,\ WAW,\ A^2W-WA^2,\ WAW^2-W^2AW$ |

where for $i = 1, 2, 3$,

$$h_i = h_i(\operatorname{tr} A,\ \operatorname{tr} A^2,\ \operatorname{tr} A^3,\ \boldsymbol{v}\cdot\boldsymbol{v},\ \boldsymbol{v}\cdot A\boldsymbol{v},\ \boldsymbol{v}\cdot A^2\boldsymbol{v},\ \boldsymbol{v}\cdot A\boldsymbol{v}\times A^2\boldsymbol{v}).$$

The above results can also be read off from Tables 4.5 and 4.6.

In obtaining (4.56), we have changed the skew-symmetric tensor $W$ back to its corresponding vector $\boldsymbol{v}$. For example, with $W_{ij} = \varepsilon_{ijk}v_k$ from (A.30), we have

$$\begin{aligned} \operatorname{tr} W^2 &= W_{ij}W_{ji} = \varepsilon_{ijk}\varepsilon_{jil}v_k v_l = -2\delta_{kl}v_k v_l = -2v_j v_j, \\ \langle AW + WA\rangle_i &= 2\varepsilon_{ijk}A_{jl}W_{lk} = 2\varepsilon_{ijk}\varepsilon_{lkn}A_{jl}v_n \\ &= 2(\delta_{in}\delta_{jl} - \delta_{il}\delta_{jn})A_{jl}v_n = 2A_{jj}v_i - 2A_{ji}v_j, \end{aligned}$$

or

$$\operatorname{tr} W^2 = -2\,\boldsymbol{v}\cdot\boldsymbol{v}, \qquad \langle AW + WA\rangle = 2(\operatorname{tr} A)\boldsymbol{v} - 2\,A\boldsymbol{v}.$$

Therefore, we can replace $\operatorname{tr} W^2$ by $\boldsymbol{v}\cdot\boldsymbol{v}$ and $\langle AW+WA\rangle$ by $A\boldsymbol{v}$, bearing in mind that the remainder of the expression consists of elements already in the list of invariants. □

**Exercise 4.4.1** Find the general representations of scalar, vector, and symmetric tensor hemitropic functions of two vector variables $(\boldsymbol{u}, \boldsymbol{v})$.

## 4.5 Anisotropic Invariants

Many anisotropic solid bodies possess symmetries that can be characterized by certain directions, lines or planes, more specifically, say, characterized by some unit vectors $\vec{\boldsymbol{m}} = (\boldsymbol{m}_1, \cdots, \boldsymbol{m}_a)$, and some tensors $\vec{M} = (M_1, \cdots, M_b)$. Let $\mathcal{G}$ be the group that preserves these characteristics, i.e.,

$$\mathcal{G} = \{Q \in G,\ Q\vec{\boldsymbol{m}} = \vec{\boldsymbol{m}},\ Q\vec{M}Q^T = \vec{M}\}, \tag{4.57}$$

where $G$ is a subgroup of $O(V)$, usually $\mathcal{O}(V)$ itself or $\mathcal{O}^+(V)$. In (4.57) we have also used the notations introduced in (4.46).

Obviously, not every anisotropic solid can be specified by a symmetry group of the type (4.57). However, many of them do, among them, transversely isotropic, orthotropic bodies, and some classes of crystalline solid. For such material bodies, we can obtain representations of invariant functions relative to symmetry group $\mathcal{G}$ in terms of invariant functions relative to the group $G$, which is usually $\mathcal{O}(V)$ or $\mathcal{O}^+(V)$, from the following theorem.

**Theorem 4.5.1.** *Let $f(\vec{v}, \vec{A})$ be a scalar-, vector-, or tensor-valued function and $\mathcal{G}$ be a group of the type (4.57). Then $f(\vec{v}, \vec{A})$ is invariant relative to $\mathcal{G}$ if and only if it can be represented by*

$$f(\vec{v}, \vec{A}) = \hat{f}(\vec{v}, \vec{A}, \vec{m}, \vec{M}), \tag{4.58}$$

*where the function $\hat{f}(\vec{v}, \vec{A}, \vec{m}, \vec{M})$ is invariant relative to $G$.*

*Proof.* We shall prove the case for a scalar-valued function only. The proofs for the other cases are similar. For sufficiency, note that if

$$\hat{f}(Q\vec{v}, Q\vec{A}Q^T, Q\vec{m}, Q\vec{M}Q^T) = \hat{f}(\vec{v}, \vec{A}, \vec{m}, \vec{M}), \qquad \forall\, Q \in G,$$

then for any $Q \in \mathcal{G} \subset G$, we have $Q\vec{m} = \vec{m}$, and $Q\vec{M}Q^T = \vec{M}$, hence

$$\hat{f}(Q\vec{v}, Q\vec{A}Q^T, \vec{m}, \vec{M}) = \hat{f}(\vec{v}, \vec{A}, \vec{m}, \vec{M}), \qquad \forall\, Q \in \mathcal{G},$$

which from (4.58) implies that $f(\vec{v}, \vec{A})$ is invariant relative to $\mathcal{G}$.

Conversely, let $\mathcal{M}$ be the orbit of $(\vec{m}, \vec{M})$ in $G$, that is,

$$\mathcal{M} = \{(Q\vec{m}, Q\vec{M}Q^T),\ \forall\, Q \in G\}.$$

Then for $f(\vec{v}, \vec{A})$ invariant relative to $\mathcal{G}$, we can define a function $\hat{f}$ on $\mathcal{D} \times \mathcal{M}$ by

$$\hat{f}(\vec{v}, \vec{A}, \vec{n}, \vec{N}) = f(R^T\vec{v}, R^T\vec{A}R), \tag{4.59}$$

for any $(\vec{v}, \vec{A}) \in \mathcal{D}$ and $(\vec{n}, \vec{N}) \in \mathcal{M}$, where $R \in G$ is such that

$$R\vec{m} = \vec{n}, \qquad R\vec{M}R^T = \vec{N}. \tag{4.60}$$

In general, $R$ is not uniquely determined by the condition (4.60). However, if $R' \in G$ also satisfies the condition, then

$$R\vec{m} = R'\vec{m}, \qquad R\vec{M}R^T = R'\vec{M}R'^T,$$

which implies that $Q = R^T R' \in \mathcal{G}$, and

$$\begin{aligned} f(R^T\vec{v}, R^T\vec{A}R) &= f(R^T(R'R'^T)\vec{v}, R^T(R'R'^T)\vec{A}(R'R'^T)R) \\ &= f(Q(R'^T\vec{v}), Q(R'^T\vec{A}R')Q^T) = f(R'^T\vec{v}, R'^T\vec{A}R'), \end{aligned}$$

since $f$ is invariant relative to $\mathcal{G}$. Therefore, the definition (4.59) is well defined.

Now, for any $Q \in G$, by definition

$$\hat{f}(Q\vec{v}, Q\vec{A}Q^T, Q\vec{m}, Q\vec{M}Q^T) = f(R^T(Q\vec{v}), R^T(Q\vec{A}Q^T)R), \tag{4.61}$$

where $R\vec{m} = Q\vec{m}$ and $R\vec{M}R^T = Q\vec{M}Q^T$, or equivalently

$$R^TQ\vec{m} = \vec{m}, \qquad (R^TQ)\vec{M}(R^TQ)^T = \vec{M},$$

which implies $R^TQ \in \mathcal{G}$. Therefore, since $f$ is invariant relative to $\mathcal{G}$, (4.61) reduces to

$$\hat{f}(Q\vec{v}, Q\vec{A}Q^T, Q\vec{m}, Q\vec{M}Q^T) = f(\vec{v}, \vec{A}) = \hat{f}(\vec{v}, \vec{A}, \vec{m}, \vec{M}), \qquad \forall\, Q \in G,$$

which proves that $\hat{f}$ is invariant relative to $G$. □

### 4.5.1 Transverse Isotropy and Orthotropy

As examples, we consider the symmetry groups given by (3.61)

$$\begin{aligned} \mathcal{G}_1 &= \{Q \in \mathcal{O}(V),\ Q\boldsymbol{n} = \boldsymbol{n}\}, \\ \mathcal{G}_2 &= \{Q \in \mathcal{O}(V),\ Q(\boldsymbol{n}_1 \otimes \boldsymbol{n}_1)Q^T = \boldsymbol{n}_1 \otimes \boldsymbol{n}_1, \\ &\qquad Q(\boldsymbol{n}_2 \otimes \boldsymbol{n}_2)Q^T = \boldsymbol{n}_2 \otimes \boldsymbol{n}_2,\ Q(\boldsymbol{n}_3 \otimes \boldsymbol{n}_3)Q^T = \boldsymbol{n}_3 \otimes \boldsymbol{n}_3\}, \end{aligned} \tag{4.62}$$

where $\boldsymbol{n}$ is a unit vector, and $\{\boldsymbol{n}_1, \boldsymbol{n}_2, \boldsymbol{n}_3\}$ is a set of orthonormal unit vectors.

The group $\mathcal{G}_1$ has a preferred direction characterized by the unit vector $\boldsymbol{n}$. A material body that has such a preferred direction in an undistorted configuration is usually called a *transversely isotropic* solid.

By the commutation theorem, since $Q(\boldsymbol{n}_1 \otimes \boldsymbol{n}_1)Q^T = \boldsymbol{n}_1 \otimes \boldsymbol{n}_1$, the transformation $Q$ preserves the characteristic space of $\boldsymbol{n}_1 \otimes \boldsymbol{n}_1$, which is the line in the direction of $\boldsymbol{n}_1$. Therefore, the group $\mathcal{G}_2$ preserves three mutually orthogonal lines in the directions of $\{\boldsymbol{n}_1, \boldsymbol{n}_2, \boldsymbol{n}_3\}$. A body with this symmetry relative to an undistorted configuration is usually called an *orthotropic* solid body.[2]

**Example 4.5.1** For a transversely isotropic elastic solid, from (4.53) the second Piola–Kirchhoff stress tensor $S_\kappa(C)$ is a tensor invariant relative to the symmetry group $\mathcal{G}_1$ in the undistorted configuration $\kappa$. By Theorem 4.5.1, we can derive the representation for $S_\kappa$ by considering the isotropic invariants of two variables $C$ and $\boldsymbol{n}$. For this case, from Tables 4.1 and 4.3, we obtain, for the second Piola–Kirchhoff stress tensor

$$\begin{aligned} S_\kappa &= s_0 1 + s_1 C + s_2 C^2 + s_3 \boldsymbol{n} \otimes \boldsymbol{n} \\ &\quad + s_4(\boldsymbol{n} \otimes C\boldsymbol{n} + C\boldsymbol{n} \otimes \boldsymbol{n}) + s_5 C\boldsymbol{n} \otimes C\boldsymbol{n}, \end{aligned} \tag{4.63}$$

where for $i = 0, \cdots, 5$,

$$s_i = s_i(\operatorname{tr} C,\ \operatorname{tr} C^2,\ \operatorname{tr} C^3,\ \boldsymbol{n} \cdot C\boldsymbol{n},\ \boldsymbol{n} \cdot C^2\boldsymbol{n}).$$

The scalar invariant $\boldsymbol{n} \cdot \boldsymbol{n}$ has been eliminated since it is equal to the constant 1.

Therefore, from (4.63) we obtain the representation for the Cauchy stress tensor $T = J^{-1} F S_\kappa F^T$,

$$\begin{aligned} T &= t_0 1 + t_1 B + t_2 B^2 + t_3 F\boldsymbol{n} \otimes F\boldsymbol{n} \\ &\quad + t_4(F\boldsymbol{n} \otimes BF\boldsymbol{n} + BF\boldsymbol{n} \otimes F\boldsymbol{n}) + t_5 BF\boldsymbol{n} \otimes BF\boldsymbol{n}, \end{aligned} \tag{4.64}$$

where for $i = 0, \cdots, 5$

$$t_i = t_i(\operatorname{tr} B,\ \operatorname{tr} B^2,\ \operatorname{tr} B^3,\ F\boldsymbol{n} \cdot F\boldsymbol{n},\ F\boldsymbol{n} \cdot BF\boldsymbol{n}). \tag{4.65}$$

In the above transitions, we have used the relations (4.16) and (1.13) and the fact that $C$ and $B$ have the same principal invariants. □

[2] For more discussions on characterization of some anisotropic material bodies see [38].

**Example 4.5.2** Consider an orthotropic elastic solid body with the symmetry group $\mathcal{G}_2$ defined in (4.62) in an undistorted reference configuration $\kappa$. Similarly, the representation for $S_\kappa(C)$ can be obtained from isotropic invariants of four symmetric tensor variables $C$, $\boldsymbol{n}_1 \otimes \boldsymbol{n}_1$, $\boldsymbol{n}_2 \otimes \boldsymbol{n}_2$, and $\boldsymbol{n}_3 \otimes \boldsymbol{n}_3$. For this case, we have

$$\begin{aligned}\Upsilon_s = \{&\underline{\operatorname{tr} C},\ \underline{\operatorname{tr} C^2},\ \operatorname{tr} C^3,\ \operatorname{tr} C(\boldsymbol{n}_1\otimes\boldsymbol{n}_1),\ \operatorname{tr} C(\boldsymbol{n}_2\otimes\boldsymbol{n}_2),\ \operatorname{tr} C(\boldsymbol{n}_3\otimes\boldsymbol{n}_3),\\ &\operatorname{tr} C^2(\boldsymbol{n}_1\otimes\boldsymbol{n}_1),\ \operatorname{tr} C^2(\boldsymbol{n}_2\otimes\boldsymbol{n}_2),\ \operatorname{tr} C^2(\boldsymbol{n}_3\otimes\boldsymbol{n}_3)\},\end{aligned}$$

and

$$\begin{aligned}\Upsilon_t = \{&\underline{1},\ \underline{C},\ \underline{C^2},\ \boldsymbol{n}_1\otimes\boldsymbol{n}_1,\ \boldsymbol{n}_2\otimes\boldsymbol{n}_2,\ \boldsymbol{n}_3\otimes\boldsymbol{n}_3,\ C(\boldsymbol{n}_1\otimes\boldsymbol{n}_1)+(\boldsymbol{n}_1\otimes\boldsymbol{n}_1)C,\\ &C(\boldsymbol{n}_2\otimes\boldsymbol{n}_2)+(\boldsymbol{n}_2\otimes\boldsymbol{n}_2)C,\ C(\boldsymbol{n}_3\otimes\boldsymbol{n}_3)+(\boldsymbol{n}_3\otimes\boldsymbol{n}_3)C,\\ &C(\boldsymbol{n}_1\otimes\boldsymbol{n}_1)C,\ C(\boldsymbol{n}_2\otimes\boldsymbol{n}_2)C,\ C(\boldsymbol{n}_3\otimes\boldsymbol{n}_3)C\},\end{aligned}$$

where we have already eliminated many redundant elements by inspection, using the identities such as $(\boldsymbol{n}_1 \otimes \boldsymbol{n}_1)^2 = \boldsymbol{n}_1 \otimes \boldsymbol{n}_1$ and $(\boldsymbol{n}_1 \otimes \boldsymbol{n}_1)(\boldsymbol{n}_2 \otimes \boldsymbol{n}_2) = 0$. Furthermore, by the use of the identities:

$$\begin{aligned}&\boldsymbol{n}_1\cdot C\boldsymbol{n}_1. + \boldsymbol{n}_2\cdot C\boldsymbol{n}_2 + \boldsymbol{n}_3\cdot C\boldsymbol{n}_3 = \operatorname{tr} C,\\ &\boldsymbol{n}_1\otimes\boldsymbol{n}_1 + \boldsymbol{n}_2\otimes\boldsymbol{n}_2 + \boldsymbol{n}_3\otimes\boldsymbol{n}_3 = 1,\end{aligned}$$

we can easily see that the underlined elements $\operatorname{tr} C$, $\operatorname{tr} C^2$, $1$, $C$, and $C^2$ are also redundant in the above lists. Therefore, we can express $S_\kappa$ as

$$\begin{aligned}S_\kappa = {}& s_1\boldsymbol{n}_1\otimes\boldsymbol{n}_1 + s_2\boldsymbol{n}_2\otimes\boldsymbol{n}_2 + s_3\boldsymbol{n}_3\otimes\boldsymbol{n}_3\\ &+ s_4(\boldsymbol{n}_1\otimes C\boldsymbol{n}_1 + C\boldsymbol{n}_1\otimes\boldsymbol{n}_1) + s_5(\boldsymbol{n}_2\otimes C\boldsymbol{n}_2 + C\boldsymbol{n}_2\otimes\boldsymbol{n}_2)\\ &+ s_6(\boldsymbol{n}_3\otimes C\boldsymbol{n}_3 + C\boldsymbol{n}_3\otimes\boldsymbol{n}_3)\\ &+ s_7C\boldsymbol{n}_1\otimes C\boldsymbol{n}_1 + s_8C\boldsymbol{n}_2\otimes C\boldsymbol{n}_2 + s_9C\boldsymbol{n}_3\otimes C\boldsymbol{n}_3,\end{aligned} \tag{4.66}$$

where $s_i$ depends on $(\det C,\ \boldsymbol{n}_1\cdot C\boldsymbol{n}_1,\ \boldsymbol{n}_2\cdot C\boldsymbol{n}_2,\ \boldsymbol{n}_3\cdot C\boldsymbol{n}_3,\ \boldsymbol{n}_1\cdot C^2\boldsymbol{n}_1,\ \boldsymbol{n}_2\cdot C^2\boldsymbol{n}_2,\ \boldsymbol{n}_3\cdot C^2\boldsymbol{n}_3)$. We have also replaced $\operatorname{tr} C^3$ by $\det C$ by the use of relation (4.19). One can easily turn (4.66) into a representation for the Cauchy stress tensor $T$. □

**Exercise 4.5.1** Find the general representations of the heat flux,

$$\boldsymbol{q} = \mathcal{Q}(\theta, \boldsymbol{g}), \qquad \boldsymbol{g} = \operatorname{grad}\theta,$$

for a rigid heat conductor with the following different symmetry groups:[3]

1) $\mathcal{G}_1 = \{Q \in \mathcal{O}(V),\ Q\boldsymbol{n} = \boldsymbol{n}\}$,
2) $\mathcal{G}_2 = \{Q \in \mathcal{O}^+(V),\ Q\boldsymbol{n} = \boldsymbol{n}\}$,
3) $\mathcal{G}_3 = \{Q \in \mathcal{O}(V),\ Q(\boldsymbol{n}\otimes\boldsymbol{n})Q^T = \boldsymbol{n}\otimes\boldsymbol{n}\}$,

respectively.

[3] The groups $\mathcal{G}_1$, $\mathcal{G}_2$, and $\mathcal{G}_3$ characterize three different classes of transverse isotropy, see [38].

**Exercise 4.5.2** From (4.66), write out the representation for the Cauchy stress tensor $T$ of an orthotropic elastic body.

### 4.5.2 On Irreducibility of Invariant Sets

From the above examples, we note that the invariant sets obtained in this manner are in general not irreducible sets. In fact, some trivial redundant elements can be eliminated by inspection using known identities. However, elimination of such redundant elements need not render irreducibility of these sets due to the constancy of the added variables $\vec{\boldsymbol{m}}$ and $\vec{M}$. Proof of irreducibility is not trivial in general. We give a simple example below.

**Example 4.5.3** Consider a vector invariant $\boldsymbol{h}(\boldsymbol{v}, A)$ of a symmetric tensor variable $A$ and a vector variable $\boldsymbol{v}$ relative to the group $\mathcal{G}_1$ defined in (4.62). We obtain from Table 4.2,

$$\Upsilon_v = \{\boldsymbol{v},\ \boldsymbol{n},\ A\boldsymbol{v},\ A\boldsymbol{n},\ A^2\boldsymbol{v},\ A^2\boldsymbol{n}\}.$$

Apparently, there are no trivial redundant elements in this list. However, we can show that $A^2\boldsymbol{v}$ is redundant and by removing it we have indeed an irreducible generator set,

$$\Upsilon_v' = \{\boldsymbol{v},\ \boldsymbol{n},\ A\boldsymbol{v},\ A\boldsymbol{n},\ A^2\boldsymbol{n}\}.$$

First, let us prove that $A^2\boldsymbol{v}$ is redundant. Clearly, if $\boldsymbol{v} = \alpha\boldsymbol{n}$, the redundancy is obvious. So let us assume that $\{\boldsymbol{v}, \boldsymbol{n}\}$ is linearly independent and that $\Upsilon_v'$ is not a generator set. In other words, we assume that $A^2\boldsymbol{v}$ can not be expressed as a linear combination of the elements in $\Upsilon_v'$. Since the space $V$ is three-dimensional, the set $\Upsilon_v'$ spans at most a two-dimensional space. Since $\{\boldsymbol{v}, \boldsymbol{n}\}$ is linearly independent, $\Upsilon_v'$ must span the subspace $[\boldsymbol{v}, \boldsymbol{n}]$, where we use this notation for the space generated by the vectors in the square brackets. In particular, we have

$$A\boldsymbol{v} \in [\boldsymbol{v}, \boldsymbol{n}] \quad \text{and} \quad A\boldsymbol{n} \in [\boldsymbol{v}, \boldsymbol{n}].$$

Therefore, we have

$$A\boldsymbol{v} = a\boldsymbol{v} + b\boldsymbol{n},$$

for some $a, b \in I\!R$. Consequently,

$$A^2\boldsymbol{v} = aA\boldsymbol{v} + bA\boldsymbol{n},$$

and hence $A^2\boldsymbol{v}$ must belong to the subspace $[\boldsymbol{v}, \boldsymbol{n}]$, which is a contradiction to our assumption, and the redundancy is proved.

Now let us prove that $\Upsilon_v'$ is irreducible. Note that if $\boldsymbol{v} = \alpha\boldsymbol{n}$, then $\Upsilon_v'$ reduces to the set $\{\boldsymbol{n}, A\boldsymbol{n}, A^2\boldsymbol{n}\}$. Since the space is three-dimensional, this set is irreducible. Now, suppose that $\{\boldsymbol{v}, \boldsymbol{n}\}$ is linearly independent and $\boldsymbol{n}$ is an eigenvector of $A$, then

$$[\boldsymbol{n}, A\boldsymbol{n}, A^2\boldsymbol{n}] = [\boldsymbol{n}].$$

Suppose that $A\boldsymbol{v} \notin [\boldsymbol{v}, \boldsymbol{n}]$, then the set $\{\boldsymbol{v}, \boldsymbol{n}, A\boldsymbol{v}\}$ is linearly independent. In these case, the elements of $\{\boldsymbol{v}, \boldsymbol{n}, A\boldsymbol{v}\}$ are irreducible. Therefore, $\Upsilon_v'$ is an irreducible generator set. □

# 5. Entropy Principle

## 5.1 Entropy Inequality

In this chapter, we shall give an introduction to thermodynamic constitutive theories. We have already mentioned the first law of thermodynamics, the energy balance, in Chap. 2. Now we are going to consider the second law for which the essential quantity is the entropy.

We denote the entropy of a part $\mathcal{P}$ by $H$ and assume that it is given by

$$H(\mathcal{P},t) = \int_{\mathcal{P}_t} \rho\eta \, dv, \tag{5.1}$$

where $\eta(\boldsymbol{x},t)$ is called the specific *entropy density.*

Associated with the change of entropy, there is a quantity called *entropy supply* $\varPhi$. We assume that $\varPhi$ is given by

$$\varPhi(\mathcal{P},t) = -\int_{\partial\mathcal{P}_t} \boldsymbol{\varPhi}\cdot\boldsymbol{n}\, da + \int_{\mathcal{P}_t} \rho s \, dv, \tag{5.2}$$

where we call $\boldsymbol{\varPhi}(\boldsymbol{x},t)$ the *entropy flux* and $s$ the *entropy supply density* due to external sources. It is assumed that $\eta$ and $s$ are objective scalar quantities, while $\boldsymbol{\varPhi}$ is assumed to be an objective vector quantity.

Let the difference between the rate of change of entropy and its supply be called the *entropy production* and denoted by

$$\Sigma(\mathcal{P},t) = \dot{H}(\mathcal{P},t) - \varPhi(\mathcal{P},t). \tag{5.3}$$

Unlike the energy, the entropy is not a conservative quantity. However, it is commonly assumed that *the entropy production is a non-negative quantity.* Such a statement is often referred to as the *second law of thermodynamics*, i.e.,

$$\Sigma(\mathcal{P},t) \geq 0 \qquad \forall\mathcal{P}\subset\mathcal{B},$$

or

$$\frac{d}{dt}\int_{\mathcal{P}_t} \rho\eta\, dv + \int_{\partial\mathcal{P}_t} \boldsymbol{\varPhi}\cdot\boldsymbol{n}\, da - \int_{\mathcal{P}_t} \rho s\, dv \geq 0. \tag{5.4}$$

This relation is called the *entropy inequality.*

Similar to the balance laws, from (2.13) we have the local forms of the entropy inequality,

$$\rho\dot{\eta} + \operatorname{div}\boldsymbol{\Phi} - \rho s \geq 0, \tag{5.5}$$

or

$$\frac{\partial \rho\eta}{\partial t} + \operatorname{div}(\rho\eta\dot{\boldsymbol{x}} + \boldsymbol{\Phi}) - \rho s \geq 0, \tag{5.6}$$

and from (2.15) the entropy jump condition at a singular surface,

$$[\![\rho\eta(\dot{\boldsymbol{x}}\cdot\boldsymbol{n} - u_n)]\!] + [\![\boldsymbol{\Phi}]\!]\cdot\boldsymbol{n} \geq 0. \tag{5.7}$$

In deriving this jump condition, we do not rule out the possibility that there might be an entropy production on the surface and it is assumed to be a non-negative quantity, in general, as well.

In the reference configuration, from (2.22) the entropy inequality and its jump condition can be written as

$$\begin{aligned} &\rho_\kappa\dot{\eta} + \operatorname{Div}\boldsymbol{\Phi}_\kappa - \rho_\kappa s \geq 0, \\ &-\rho_\kappa U_\kappa[\![\eta]\!] + [\![\boldsymbol{\Phi}_\kappa\cdot\boldsymbol{n}_\kappa]\!] \geq 0, \end{aligned} \tag{5.8}$$

where the material entropy flux $\boldsymbol{\Phi}_\kappa = |J|F^{-1}\boldsymbol{\Phi}$.

The entropy inequality in the form (5.4) is quite general. Motivated by the results of classical thermostatics, it is often assumed that the entropy flux and the entropy supply are proportional to the heat flux and the heat supply, respectively, in classical thermodynamic theories. Moreover, both proportional constants are assumed to be the reciprocal of the *absolute temperature* $\theta$,

$$\boldsymbol{\Phi} = \frac{1}{\theta}\boldsymbol{q}, \qquad s = \frac{1}{\theta}r. \tag{5.9}$$

The resulting entropy inequality is called the *Clausius–Duhem inequality*,

$$\int_{\mathcal{P}_t}\rho\dot{\eta}\,dv + \int_{\partial\mathcal{P}_t}\frac{1}{\theta}\boldsymbol{q}\cdot\boldsymbol{n}\,da - \int_{\mathcal{P}_t}\frac{1}{\theta}\rho r\,dv \geq 0, \tag{5.10}$$

and the local form (5.5) becomes

$$\rho\dot{\eta} + \operatorname{div}\frac{\boldsymbol{q}}{\theta} - \rho\frac{r}{\theta} \geq 0. \tag{5.11}$$

The role of the Clausius–Duhem inequality in the development of modern rational thermodynamics will be illustrated in this chapter. However, we should remark that although theories based on the Clausius–Duhem inequality seem to be acceptable in various situations, the assumptions (5.9) are known to be inconsistent with the results from the kinetic theory of gases. We shall postpone the discussion of rational thermodynamics based on the general entropy inequality (5.5) to Chap. 7.

## 5.2 Entropy Principle

One of the principal objectives of continuum mechanics is to determine or predict the behavior of a body once the external causes are specified. Mathematically, this amounts to solving an initial boundary valued problem governed by the balance laws (2.79),

$$\begin{aligned}
&\dot{\rho} + \rho \operatorname{div} \dot{\boldsymbol{x}} = 0, \\
&\rho \ddot{\boldsymbol{x}} - \operatorname{div} T = \rho \boldsymbol{b}, \\
&T = T^T, \\
&\rho \dot{\varepsilon} + \operatorname{div} \boldsymbol{q} - T \cdot \operatorname{grad} \dot{\boldsymbol{x}} = \rho r,
\end{aligned} \tag{5.12}$$

when the external supplies $\boldsymbol{b}$ and $r$ are given. After introducing the constitutive relations for $T$, $\varepsilon$, and $\boldsymbol{q}$, the balance laws (5.12) become a system of partial differential equations for the determination of the fields $\rho$, $\chi$, and $\theta$. We call a set of fields $\{\rho, \chi, \theta, \boldsymbol{b}, r\}$ a *thermodynamic process* of $\mathcal{B}$, if it satisfies the balance laws in $\mathcal{B}$.

On the other hand, the behavior of a body must also obey the second law of thermodynamics, i.e., the thermodynamic process must also satisfy the entropy inequality (5.5),

$$\rho \dot{\eta} + \operatorname{div} \boldsymbol{\Phi} - \rho s \geq 0.$$

Classical constitutive relations were usually proposed based on physical experiences or experimental observations. Thermodynamic processes, obtained as solutions of the balance laws and the proposed constitutive relations, are further required to be consistent with the second law of thermodynamics so as to be regarded as physically admissible processes. This is the usual role of the second law of thermodynamics in the traditional formulation of physical problems.

Following the idea set forth by Coleman and Noll [11], in modern rational thermodynamics, the second law of thermodynamics plays a more important role:

> **Entropy principle.** *It is required that constitutive relations be such that the entropy inequality is satisfied identically for any thermodynamic process.*

From this point of view, the entropy principle imposes restrictions on constitutive functions, just as the principle of material objectivity and the material symmetry do, so that the resulting constitutive functions can be established in a physical model for the material body. Consequently, with such restrictions taken care of, all thermodynamic processes will always be consistent with the second law.

To find the restrictions imposed on constitutive functions by the entropy principle is one of the major objectives in modern continuum thermodynamics. In this chapter, we shall illustrate the procedure for exploiting this requirement for thermoelastic materials.

## 5.3 Thermodynamics of Elastic Materials

We shall now exploit the entropy principle based on the Clausius-Duhem inequality for thermoelastic materials. Following the method of its exploitation suggested by Coleman and Noll [11], we define the *free energy density* $\psi$ by

$$\psi = \varepsilon - \theta\eta, \tag{5.13}$$

and use the energy balance $(5.12)_4$ to eliminate $r$, so that the Clausius–Duhem inequality becomes

$$\rho\dot{\psi} + \rho\eta\dot{\theta} - TF^{-T} \cdot \dot{F} + \frac{1}{\theta}\boldsymbol{q} \cdot \boldsymbol{g} \leq 0, \tag{5.14}$$

where (1.31) has been used.

The constitutive relations for thermoelastic materials are given by

$$\begin{aligned} T &= T(F, \theta, \boldsymbol{g}), & \eta &= \eta(F, \theta, \boldsymbol{g}), \\ \boldsymbol{q} &= \boldsymbol{q}(F, \theta, \boldsymbol{g}), & \psi &= \psi(F, \theta, \boldsymbol{g}). \\ \varepsilon &= \varepsilon(F, \theta, \boldsymbol{g}), \end{aligned} \tag{5.15}$$

We assume that the response functions are smooth, then

$$\dot{\psi} = \frac{\partial \psi}{\partial F} \cdot \dot{F} + \frac{\partial \psi}{\partial \theta}\dot{\theta} + \frac{\partial \psi}{\partial \boldsymbol{g}} \cdot \dot{\boldsymbol{g}},$$

and hence (5.14) becomes

$$\rho(\eta + \frac{\partial \psi}{\partial \theta})\dot{\theta} + \rho\frac{\partial \psi}{\partial \boldsymbol{g}} \cdot \dot{\boldsymbol{g}} - (TF^{-T} - \rho\frac{\partial \psi}{\partial F}) \cdot \dot{F} + \frac{1}{\theta}\boldsymbol{q} \cdot \boldsymbol{g} \leq 0. \tag{5.16}$$

The entropy principle requires that the above inequality must hold for any thermodynamic process $\{\rho, \chi, \theta, \boldsymbol{b}, r\}$. Now we shall evaluate the restrictions imposed by this requirement.

Note that for any given fields $\{\chi, \theta\}$, one can determine the fields $\{\rho, \boldsymbol{b}, r\}$ from the balance laws (5.12), so that $\{\rho, \chi, \theta, \boldsymbol{b}, r\}$ is a thermodynamic process. In particular, for any given values of $\{F, \theta, \boldsymbol{g}, \dot{F}, \dot{\theta}, \dot{\boldsymbol{g}}\}$ at $(\boldsymbol{X}, t)$ such that $\det F \neq 0$ and $\theta > 0$, there exists a thermodynamic process having these values at $(\boldsymbol{X}, t)$. In other words, for any given $\{F, \theta, \boldsymbol{g}\}$ the inequality (5.16) must hold for arbitrary values of $\dot{F}$, $\dot{\theta}$, and $\dot{\boldsymbol{g}}$, in which the inequality is linear. Consequently, their coefficients must vanish identically,

$$\eta + \frac{\partial \psi}{\partial \theta} = 0,$$
$$\frac{\partial \psi}{\partial \boldsymbol{g}} = 0, \tag{5.17}$$
$$TF^{-T} - \rho \frac{\partial \psi}{\partial F} = 0,$$

or equivalently,

$$\begin{aligned} \psi &= \psi(F, \theta), & \varepsilon &= \psi - \theta \frac{\partial \psi}{\partial \theta}, \\ \eta &= -\frac{\partial \psi}{\partial \theta}, & T &= \rho \frac{\partial \psi}{\partial F} F^T. \end{aligned} \tag{5.18}$$

Moreover, we have from $(5.12)_3$ the relation,

$$\frac{\partial \psi}{\partial F} F^T = F \Big(\frac{\partial \psi}{\partial F}\Big)^T, \tag{5.19}$$

and from (5.16) the remaining inequality,

$$\boldsymbol{q}(F, \theta, \boldsymbol{g}) \cdot \boldsymbol{g} \leq 0. \tag{5.20}$$

The above relations represent the most general forms of the constitutive relations for thermoelastic materials. Compared to the constitutive relations (5.15) they are greatly simplified. Indeed, the stress, the internal energy and the entropy are completely determined once the free energy function $\psi(F, \theta)$ is known.

The inequality (5.20) is called *Fourier's inequality*, which states that the heat flux always points in the direction from a hot spot to a cold one. Furthermore, since the left-hand side of (5.20), denoted by $f(F, \theta, \boldsymbol{g})$, is a smooth function of $\boldsymbol{g}$ and it assumes its maximum, namely zero, at $\boldsymbol{g} = 0$, we can conclude that

$$\begin{aligned} \partial_{\boldsymbol{g}} f \big|_{\boldsymbol{g}=0} &= \Big(\boldsymbol{q} + \frac{\partial \boldsymbol{q}}{\partial \boldsymbol{g}}^T \boldsymbol{g}\Big)\Big|_{\boldsymbol{g}=0} = \boldsymbol{q}(F, \theta, 0) = 0, \\ \partial^2_{\boldsymbol{g}} f \big|_{\boldsymbol{g}=0} &= \Big(\frac{\partial \boldsymbol{q}}{\partial \boldsymbol{g}} + \frac{\partial \boldsymbol{q}}{\partial \boldsymbol{g}}^T\Big)\Big|_{\boldsymbol{g}=0} \quad \text{is negative semi-definite,} \end{aligned} \tag{5.21}$$

for any $(F, \theta)$. If we assume the classical *Fourier's law* of heat conduction,

$$\boldsymbol{q} = -K(F, \theta)\boldsymbol{g}, \tag{5.22}$$

where $K$ denotes the *thermal conductivity tensor*, then $(5.21)_2$ implies that the symmetric part of $K(F, \theta)$ is positive semi-definite for any $(F, \theta)$.

From (5.18) we can write the differential of $\psi$ as

$$d\psi = \frac{1}{\rho} T F^{-T} \cdot dF - \eta \, d\theta, \tag{5.23}$$

or equivalently,

$$d\eta = \frac{1}{\theta}\Big( d\varepsilon - \frac{1}{\rho} T F^{-T} \cdot dF \Big). \tag{5.24}$$

This is known as the *Gibbs relation* for elastic materials. In particular, for a thermoelastic fluid, which depends on the deformation gradient $F$ only through its dependence on the density $\rho$, we have

$$T = -p\mathit{1}, \qquad d\rho = -\rho F^{-T} \cdot dF,$$

where $p = p(\rho, \theta)$ is the pressure and the relation (5.24) reduces to the well-known Gibbs relation for fluids in classical thermostatics:

$$d\eta = \frac{1}{\theta}\Big( d\varepsilon - \frac{p}{\rho^2} d\rho \Big) = \frac{1}{\theta}\frac{\partial \varepsilon}{\partial \theta}\, d\theta + \frac{1}{\theta}\Big( \frac{\partial \varepsilon}{\partial \rho} - \frac{p}{\rho^2}\Big) d\rho. \tag{5.25}$$

An immediate consequence of (5.25) is the following integrability condition for $\eta(\rho, \theta)$,

$$\frac{\partial \varepsilon}{\partial \rho} = \frac{p}{\rho^2} - \frac{\theta}{\rho^2}\frac{\partial p}{\partial \theta}. \tag{5.26}$$

Similar integrability conditions can be obtained from (5.23) and (5.24).

From the relation $(5.18)_4$ and the relation (3.31), it follows that in any isothermal process, a thermoelastic material is a hyperelastic material, with free energy function $\psi$ served as the stored energy function,

$$\sigma(F) = \psi(F, \theta)\big|_{\theta = \text{constant}}, \tag{5.27}$$

and

$$T = \rho \frac{\partial \sigma}{\partial F} F^T. \tag{5.28}$$

Note that if we introduce the Piola–Kirchhoff stress tensor $T_\kappa$, the relation between the stored energy function and the stress becomes even simpler,

$$T_\kappa = \rho_\kappa \frac{\partial \sigma}{\partial F}, \tag{5.29}$$

where $\rho_\kappa$ is the density in the reference configuration $\kappa$ and we have used the relations $(2.84)_1$ and (2.34).

One may notice that the momentum and the energy balance impose no restrictions in the evaluation of the inequality (5.16), simply because one argues that the external supplies $\boldsymbol{b}$ and $r$ can be suitably adjusted so as to satisfy the momentum and the energy balance. This argument may sound wishful in the real world[1] but it is usually regarded as acceptable. Nevertheless, we shall see in Chap. 7 that such an argument can be avoided.

[1] This argument has been criticized by Woods [79].

**Exercise 5.3.1** Show that the Gibbs relation (5.25) can also be written in the following forms:

$$\begin{aligned} d\psi &= \frac{p}{\rho^2}\,d\rho - \eta\,d\theta, \\ d(\rho\eta) &= \frac{1}{\theta}\big(d(\rho\varepsilon) - g\,d\rho\big), \end{aligned} \tag{5.30}$$

where $\psi$ is the *free energy* and $g$ is the *free enthalpy* defined as

$$\psi = \varepsilon - \theta\eta, \qquad g = \psi + \frac{p}{\rho}. \tag{5.31}$$

**Exercise 5.3.2** For an isotropic thermoelastic solid, the stress tensor $(5.18)_4$ and the Gibbs relation (5.24) become

$$\begin{aligned} T &= 2\rho\frac{\partial\psi}{\partial B}B, \\ d\eta &= \frac{1}{\theta}\Big(d\varepsilon - \frac{1}{2\rho}TB^{-1}\cdot dB\Big). \end{aligned} \tag{5.32}$$

From the representation theorem (4.26), we can write

$$\begin{aligned} T &= t_0 1 + t_1 B + t_2 B^2, \\ \psi &= \psi(\theta, I_B, II_B, III_B). \end{aligned} \tag{5.33}$$

Show that the parameters $t_i$ are given by

$$\begin{aligned} t_0 &= 2\rho III_B \frac{\partial\psi}{\partial III_B}, \\ t_1 &= 2\rho\left(\frac{\partial\psi}{\partial I_B} + I_B\frac{\partial\psi}{\partial II_B}\right), \\ t_2 &= -2\rho\frac{\partial\psi}{\partial II_B}. \end{aligned} \tag{5.34}$$

### 5.3.1 Linear Thermoelasticity

First, we note that for thermoelastic materials, by the use of the Gibbs relation (5.24),

$$\theta\dot\eta = \dot\varepsilon - \frac{1}{\rho}TF^{-T}\cdot\dot F,$$

and hence the energy equation $(5.12)_4$ can be written as

$$\rho\theta\dot{\eta} + \operatorname{div} \boldsymbol{q} = \rho r. \tag{5.35}$$

For the classical linear theory, the displacement gradient $H$ and the temperature increment $\tilde{\theta} = \theta - \theta_0$ are assumed to be small quantities, where $\theta_0$ is the reference temperature of the body. On the other hand, similar to (3.35) the principle of material objectivity requires that the dependence of the free energy function on the deformation gradient $F$ must reduce to the dependence of the right stretch tensor $U$. Since $U = 1 + E + o(2)$ from (1.22), in the linear theory we have $\psi = \psi(E, \theta)$. Therefore, let us express the function $\psi$ up to the second-order terms in $E$ and $\tilde{\theta}$ in the following form,

$$\psi = \psi_0 - \eta_0\tilde{\theta} + M_{ij}E_{ij} - \frac{1}{2}\frac{c_v}{\theta_0}\tilde{\theta}^2 - \frac{1}{\rho_0}P_{ij}E_{ij}\tilde{\theta} + \frac{1}{2\rho_0}L_{ijkl}E_{ij}E_{kl}. \tag{5.36}$$

The relations $(5.18)_{3,4}$ now take the form,

$$T = \rho_0\frac{\partial\psi}{\partial E}, \qquad \eta = -\frac{\partial\psi}{\partial\tilde{\theta}},$$

and if the reference state is assumed to be a natural state, then $M_{ij}$ in (5.36) must vanish and we obtain

$$\begin{aligned} T_{ij} &= L_{ijkl}E_{kl} - P_{ij}\tilde{\theta}, \\ \eta &= \eta_0 + \frac{c_v}{\theta_0}\tilde{\theta} + \frac{1}{\rho_0}P_{ij}E_{ij}. \end{aligned} \tag{5.37}$$

The fourth-order tensor $\boldsymbol{L}$ is the elasticity tensor (see (3.79)) and $c_v$ is called the *specific heat* (at constant volume) because, from $(5.18)_2$ it follows that

$$c_v = \frac{\partial\varepsilon}{\partial\theta}.$$

For the linear theory, we shall assume in addition that Fourier's law of heat conduction holds, so that the linear expression for the heat flux is given by

$$\boldsymbol{q} = -K\,\boldsymbol{g}, \tag{5.38}$$

where $K$ is the thermal conductivity tensor.

Therefore, we can summarize the constitutive equations of linear thermoelasticity for anisotropic materials in the following:

$$\begin{aligned} T &= \mathbf{L}[\tilde{E}] - P\tilde{\theta}, \\ \boldsymbol{q} &= -K\,\boldsymbol{g}, \\ \eta &= \eta_0 + \frac{c_v}{\theta_0}\tilde{\theta} + \frac{P}{\rho_0}\cdot\tilde{E}. \end{aligned} \tag{5.39}$$

In general, the coefficients are functions of the temperature $\theta_0$ in the reference configuration. Moreover, the coefficients must satisfy the following conditions:

$$\begin{aligned} & L_{ijkl} = L_{jikl} = L_{ijlk} = L_{klij}, \quad P_{ij} = P_{ji}, \\ & K \text{ is positive semi-definite}, \\ & c_v > 0. \end{aligned} \tag{5.40}$$

The first two conditions follow from the symmetry of the stress and the linear strain tensors, and the last inequality is a consequence of thermal stability, which will be discussed later.

If the material is isotropic, then we have, for any orthogonal tensor $Q$,

$$\begin{aligned} & T(QFQ^T, \theta) = Q\,T(F,\theta)\,Q^T, \\ & \boldsymbol{q}(QFQ^T, \theta, Q\boldsymbol{g}) = Q\,\boldsymbol{q}(F,\theta,\boldsymbol{g}). \end{aligned}$$

Since

$$\tilde{E} = \frac{1}{2}(H + H^T) = \frac{1}{2}(F + F^T) - 1,$$

from (5.39) it follows immediately that

$$\begin{aligned} & \mathbf{L}[Q\tilde{E}Q^T] = Q\,\mathbf{L}[\tilde{E}]\,Q^T, \\ & P = QPQ^T, \\ & KQ = QK. \end{aligned}$$

Therefore, we conclude that, in component forms,

$$P_{ij} = \alpha\,\delta_{ij}, \qquad K_{ij} = \kappa\,\delta_{ij},$$

and from the first relation, we have, for any orthogonal tensor $Q$,

$$L_{ijkl} = Q_{im}Q_{jn}Q_{kp}Q_{lq}L_{mnpq}.$$

This implies the following representation (see Example 4.2.2),

$$L_{ijkl} = \lambda\,\delta_{ij}\delta_{kl} + \mu\,(\delta_{ik}\delta_{jl} + \delta_{il}\delta_{jk}),$$

where $\lambda$ and $\mu$ are the Lamé elastic moduli (see also the derivation given in Sect. 4.2.1).

In summary, the constitutive equations of linear thermoelasticity for isotropic materials are given in the following:

$$\begin{aligned} & T = \lambda \operatorname{tr} \tilde{E}\, 1 + 2\mu\,\tilde{E} - \alpha\,\tilde{\theta}\, 1, \\ & \boldsymbol{q} = -\kappa\,\boldsymbol{g}, \\ & \eta = \eta_0 + \frac{c_v}{\theta_0}\,\tilde{\theta} + \frac{\alpha}{\rho_0}\operatorname{tr}\tilde{E}. \end{aligned} \tag{5.41}$$

The field equations for thermoelasticity consist of the momentum equation and the energy equation (or the equivalent equation (5.35)) for the displacement $\boldsymbol{u}(\boldsymbol{x},t)$ and the temperature $\theta(\boldsymbol{x},t)$. In component forms, from (1.24)

$$\tilde{E}_{ij} = \frac{1}{2}\Big(\frac{\partial u_i}{\partial x_j} + \frac{\partial u_j}{\partial x_i}\Big),$$

and the constitutive equations (5.39), we have the following field equations for anisotropic thermoelastic materials:

$$\begin{aligned} &\rho_0 \frac{\partial^2 u_i}{\partial t^2} - \frac{\partial}{\partial x_j}\Big(L_{ijkl}\frac{\partial u_k}{\partial x_l}\Big) + \frac{\partial}{\partial x_j}\Big(P_{ij}(\theta-\theta_0)\Big) = \rho_0 b_i, \\ &\rho_0 c_v \frac{\partial \theta}{\partial t} + \theta_0 P_{ij}\frac{\partial}{\partial t}\Big(\frac{\partial u_i}{\partial x_j}\Big) - \frac{\partial}{\partial x_i}\Big(K_{ij}\frac{\partial \theta}{\partial x_j}\Big) = \rho_0 r. \end{aligned} \tag{5.42}$$

The material coefficients are, in general, functions of the reference temperature. In particular, if the reference temperature is uniform throughout the body so that the material coefficients are constants, then the field equations (5.42) become

$$\begin{aligned} &\rho_0 \frac{\partial^2 u_i}{\partial t^2} - L_{ijkl}\frac{\partial^2 u_k}{\partial x_j \partial x_l} + P_{ij}\frac{\partial \theta}{\partial x_j} = \rho_0 b_i, \\ &\rho_0 c_v \frac{\partial \theta}{\partial t} + \theta_0 P_{ij}\frac{\partial}{\partial t}\Big(\frac{\partial u_i}{\partial x_j}\Big) - K_{ij}\frac{\partial^2 \theta}{\partial x_i \partial x_j} = \rho_0 r, \end{aligned} \tag{5.43}$$

or, if the body is rigid and fixed, then $P_{ij} = 0$ and the only field equation is the classical equation of heat conduction,

$$\rho_0 c_v \frac{\partial \theta}{\partial t} - \frac{\partial}{\partial x_i}\Big(K_{ij}\frac{\partial \theta}{\partial x_j}\Big) = \rho_0 r. \tag{5.44}$$

For isotropic thermoelastic materials, from (5.41) the field equations become

$$\begin{aligned} &\rho_0 \frac{\partial^2 u_i}{\partial t^2} - \frac{\partial}{\partial x_i}\Big(\lambda\frac{\partial u_k}{\partial x_k}\Big) - \frac{\partial}{\partial x_j}\Big(\mu\big(\frac{\partial u_i}{\partial x_j} + \frac{\partial u_j}{\partial x_i}\big)\Big) + \frac{\partial}{\partial x_i}\Big(\alpha(\theta-\theta_0)\Big) = \rho_0 b_i, \\ &\rho_0 c_v \frac{\partial \theta}{\partial t} + \alpha\theta_0 \frac{\partial}{\partial t}\Big(\frac{\partial u_i}{\partial x_i}\Big) - \frac{\partial}{\partial x_i}\Big(\kappa\frac{\partial \theta}{\partial x_i}\Big) = \rho_0 r. \end{aligned}$$

If the material coefficients are constants then the field equations become

$$\begin{aligned} &\rho_0 \frac{\partial^2 u_i}{\partial t^2} - (\lambda+\mu)\frac{\partial^2 u_k}{\partial x_i \partial x_k} - \mu\frac{\partial^2 u_i}{\partial x_k \partial x_k} + \alpha\frac{\partial \theta}{\partial x_i} = \rho_0 b_i, \\ &\rho_0 c_v \frac{\partial \theta}{\partial t} + \alpha\theta_0 \frac{\partial}{\partial t}\Big(\frac{\partial u_k}{\partial x_k}\Big) - \kappa\frac{\partial^2 \theta}{\partial x_k \partial x_k} = \rho_0 r. \end{aligned} \tag{5.45}$$

**Exercise 5.3.3** Suppose that the entropy function $\eta(F, \theta, \boldsymbol{g})$ is invertible in $\theta$ for each fixed $F$ and $\boldsymbol{g}$. Then we can regard $(F, \eta, \boldsymbol{g})$ as the field variables for a thermoelastic material body with constitutive equations of the form

$$\mathcal{C} = \hat{\mathcal{F}}(F, \eta, \boldsymbol{g}).$$

1) Evaluate the consequence of the entropy principle with the Clausius–Duhem inequality using $(F, \eta, \boldsymbol{g})$ as field variables.
2) Show that for an isentropic process, i.e., constant entropy, a thermoelastic material is hyperelastic, and the stored energy function $\sigma$ is given by

$$\sigma(F) = \hat{\varepsilon}(F, \eta)\big|_{\eta=\text{constant}}. \tag{5.46}$$

**Exercise 5.3.4** Evaluate the consequence of the entropy principle with the Clausius–Duhem inequality,
1) for viscous heat-conducting fluids defined by (4.3);
2) for Navier–Stokes–Fourier fluids with constitutive equations (4.52), and show that the viscosities $\lambda$, $\mu$, and the thermal conductivity $\kappa$ satisfy

$$\mu \geq 0, \quad 3\lambda + 2\mu \geq 0, \quad \kappa \geq 0, \tag{5.47}$$

while the free-energy density $\psi$ and the pressure $p$ are related by

$$\psi = \psi(\rho, \theta), \qquad p = \rho^2 \frac{\partial \psi}{\partial \rho}. \tag{5.48}$$

## 5.4 Elastic Materials with Internal Constraints

In the previous section, to obtain the main conclusion (5.17) from the inequality (5.16) by being able to choose arbitrary values of $F$ and $\dot{F}$, we have tacitly assumed that the material body is capable of undergoing any deformation. However, for some material bodies, such an assumption may not be appropriate. For example, an incompressible material body can not undergo any deformation that changes its volume.

If the class of possible deformations for the body is limited, the body is said to be subjected to *internal constraints*. Such materials are called *constrained materials*. In this section, we shall consider mechanical constraints[2] for thermoelastic bodies.

Let $\mathcal{D} = \{F \in \mathcal{L}(V) | \, \mu(F) = 0\}$, where $\mu$ is a scalar-valued function. We say that $\mathcal{D}$ is a *mechanical constraint* if the body admits only local deformations compatible with $\mathcal{D}$. For bodies with such a constraint, the values of $F$

[2] For some discussions on thermomechanical constraints, see [8].

and $\dot{F}$ can no longer be assigned arbitrary, since $F$ must belong to $\mathcal{D}$ and $\dot{F}$ must satisfy the following relation,

$$\dot{\mu}(F) = \frac{\partial \mu}{\partial F} \cdot \dot{F} = 0. \tag{5.49}$$

This relation has a simple geometrical interpretation. If we call $Z = \partial\mu/\partial F$ a vector in $\mathcal{L}(V)$, then the vector $\dot{F}$ must lie in the plane orthogonal to $Z$. Let us denote this plane by $Z^\perp$, then we must have $\dot{F} \in Z^\perp$.

For thermoelastic materials with such a constraint, we can reconsider the consequences of the inequality (5.16),

$$\rho(\eta + \frac{\partial \psi}{\partial \theta})\dot{\theta} + \rho\frac{\partial \psi}{\partial \boldsymbol{g}} \cdot \dot{\boldsymbol{g}} - (TF^{-T} - \rho\frac{\partial \psi}{\partial F}) \cdot \dot{F} + \frac{1}{\theta}\boldsymbol{q} \cdot \boldsymbol{g} \leq 0.$$

For any given $\theta$, $\boldsymbol{g}$, and $F \in \mathcal{D}$, this inequality must still hold for any values of $\dot{\theta}$ and $\dot{\boldsymbol{g}}$. Therefore, the first two relations of (5.17) remain valid, and the inequality reduces to

$$-(TF^{-T} - \rho\frac{\partial \psi}{\partial F}) \cdot \dot{F} + \frac{1}{\theta}\boldsymbol{q} \cdot \boldsymbol{g} \leq 0, \tag{5.50}$$

which must hold also for any $\dot{F} \in Z^\perp$. We claim that this implies

$$TF^{-T} - \rho\frac{\partial \psi}{\partial F} = \lambda\frac{\partial \mu}{\partial F}, \tag{5.51}$$

where $\lambda$ is an arbitrary parameter. In other words, $TF^{-T} - \rho\partial\psi/\partial F$ must be parallel to $Z$. For if it does not, then there exists a vector $\dot{F}$ in the plane $Z^\perp$ such that $(TF^{-T} - \rho\partial\psi/\partial F) \cdot \dot{F} \neq 0$, and it can also be chosen in such a way that the inequality (5.50) is violated. Therefore, we conclude from (5.51) that the stress tensor for a thermoelastic body must be given by

$$T = N + \mathcal{T}(F,\theta), \qquad \mathcal{T}(F,\theta) = \rho\frac{\partial \psi}{\partial F}F^T\Big|_{F \in \mathcal{D}}, \tag{5.52}$$

where

$$N = \lambda\frac{\partial \mu}{\partial F}F^T, \tag{5.53}$$

is called the *reaction stress* due to the constraint.

Note that there is no work done due to the reaction stress. Indeed, the stress power produced by the reaction stress is

$$N \cdot L = \lambda\frac{\partial \mu}{\partial F}F^T \cdot L = \lambda\frac{\partial \mu}{\partial F} \cdot \dot{F} = 0, \tag{5.54}$$

by the use of (5.49) and $L = \dot{F}F^{-1}$. Conversely, if $N \cdot L = 0$ for any $F$ satisfying $\mu(F) = 0$, then one can show that $N$ must be of the form (5.53). Therefore, we can say that the constraint is maintained by the reaction stress, which does not do any work in the actual deformation. This observation is often taken as a postulate:[3]

> *The stress, for a material body with mechanical constraint, is determined by the history of motion only to within an arbitrary tensor $N$ that does no work, i.e., $N \cdot L = 0$, in any motion compatible with the constraint.*

Since internal constraints are intrinsic properties of material bodies, we assume that the constraint satisfies the condition of material objectivity, namely,

$$\mu(QF) = \mu(F), \qquad \forall Q \in \mathcal{O}(V), \tag{5.55}$$

which implies that

$$\mu(F) = \mu(U) = \tilde{\mu}(C).$$

Therefore, (5.53) can be written as

$$N = 2\lambda F \frac{\partial \tilde{\mu}}{\partial C} F^T, \tag{5.56}$$

from which we conclude that the reaction stress $N$ is symmetric and objective like the stress $T$ itself. Moreover, the condition (5.54) can also be written as

$$N \cdot D = 0, \tag{5.57}$$

for any $D$, the symmetric part of $L$, compatible with the constraint. Furthermore, one can show that the reaction stress $N$ is independent of the reference configuration. Indeed, if $\hat{\kappa}$ is another reference configuration, then we have

$$\mu(F_\kappa) = \mu(F_{\hat{\kappa}} P) = \hat{\mu}(F_{\hat{\kappa}}),$$

where $P = \nabla(\hat{\kappa} \circ \kappa^{-1})$ and $F_\kappa = F_{\hat{\kappa}} P$. Hence, the constraint is given by $\hat{\mu}(F_{\hat{\kappa}}) = 0$ in $\hat{\kappa}$ and one can easily show that

$$\frac{\partial \hat{\mu}}{\partial F_{\hat{\kappa}}} = \frac{\partial \mu}{\partial F_\kappa} P^T.$$

By (5.53), the reaction stress $\widehat{N}(F_{\hat{\kappa}})$ relative to $\hat{\kappa}$ is

$$\widehat{N}(F_{\hat{\kappa}}) = \lambda \frac{\partial \hat{\mu}}{\partial F_{\hat{\kappa}}} F_{\hat{\kappa}}^T = \lambda \frac{\partial \mu}{\partial F_\kappa} P^T F_{\hat{\kappa}}^T = \lambda \frac{\partial \mu}{\partial F_\kappa} F_\kappa^T = N(F_\kappa).$$

Since the reaction stress is symmetric, objective, and independent of the reference configuration, the constitutive function $\mathcal{T}(F, \theta)$ in (5.52) is subjected to the same conditions of material objectivity and material symmetry for the stress tensor itself.

---

[3] This postulate is referred to as the *principle of determinism* for constrained simple material bodies [71].

Suppose that the constraint $\mathcal{D}$ is specified by multiple conditions,

$$\mathcal{D} = \{F \in \mathcal{L}(V) \mid \mu^a(F) = 0, a = 1, \cdots, m\}. \tag{5.58}$$

Then with similar arguments leading to (5.53), we can obtain the reaction stress

$$N = \sum_{a=1}^{m} \lambda^a \frac{\partial \mu^a}{\partial F} F^T, \tag{5.59}$$

where $\lambda^a, a = 1, \cdots, m$ are arbitrary parameters. We remark that these parameters can not be determined from the thermomechanical state variables alone, therefore, the reaction stress $N$ is also "undeterminate". Nevertheless, the reaction stress $N$, as a part of the total stress $T$, can be determined by solving the field equations with appropriate boundary conditions.

**Example 5.4.1** *Incompressibility*

We now consider the most important internal constraint – *incompressibility*. This requires that the deformation is volume preserving, which can be described by

$$\mathcal{D} = \{F \in \mathcal{L}(V) \mid |\det F| = 1\} = \mathcal{U}(V), \tag{5.60}$$

or $\mu(F) = |\det F| - 1$. Therefore, from (5.53) and the relation $\partial_F \det F = F^{-T} \det F$, we conclude that the reaction stress $N$ is an arbitrary pressure,

$$N = -p\mathit{1}. \tag{5.61}$$

Consequently, the constitutive equation for an incompressible thermoelastic material body must be of the form,

$$T = -p\mathit{1} + \mathcal{T}(F, \theta), \qquad \forall F \in \mathcal{U}(V), \tag{5.62}$$

where $p$, the undeterminate hydrostatic pressure, is the reaction force of the body to maintain incompressibility. □

**Example 5.4.2** *Rigidity*

A material is called rigid if it admits only rigid motions. i.e.,

$$\chi_\kappa(\boldsymbol{X}, t) = Q(t)(\boldsymbol{X} - \boldsymbol{X}_\circ) + \boldsymbol{c}, \;\; Q \in \mathcal{O}(V).$$

Therefore, the constraint is given by

$$\mathcal{D} = \mathcal{O}(V). \tag{5.63}$$

For any $F \in \mathcal{D}$, we have

$$L = \dot{F}F^{-1}, \qquad F \in \mathcal{O}(V),$$

which is a skew tensor, and hence its symmetric part $D$ must vanish. Therefore, the condition (5.57) is identically satisfied for any tensor $N$. In other words, the stress of a rigid material body is totally unaffected by the motion. The reaction stress $N$ is an arbitrary symmetric tensor. □

**Example 5.4.3** *Inextensibility*

Suppose that the body is inextensible in the direction parallel to a unit vector $\boldsymbol{e}$ in a reference configuration $\kappa$, then the constraint can be expressed by

$$\mathcal{D} = \{F \in \mathcal{L}(V) \mid |F\boldsymbol{e}| = |\boldsymbol{e}|\}. \tag{5.64}$$

In this case, we can take $\mu(F) = F\boldsymbol{e} \cdot F\boldsymbol{e} - \boldsymbol{e} \cdot \boldsymbol{e}$ and hence $\partial_F \mu = 2F\boldsymbol{e} \otimes \boldsymbol{e}$. It follows immediately from (5.53) that

$$N = 2\lambda F\boldsymbol{e} \otimes F\boldsymbol{e}. \tag{5.65}$$

The reaction stress $N$ is a pure tension in the direction of $F\boldsymbol{e}$. □

**Exercise 5.4.1** Determine the reaction stress $N$ for the constraint

$$\mathcal{D} = \{F \in \mathcal{U}(V) \mid \boldsymbol{e} \cdot C\boldsymbol{e} - \boldsymbol{e} \cdot \boldsymbol{e} = 0\},$$

where $\boldsymbol{e}$ is a fixed unit vector in the reference configuration. Interpret this constraint physically.

**Exercise 5.4.2** Show that the reaction stress $N$ is an arbitrary symmetric tensor for the rigidity constraint

$$\mathcal{D} = \{F \in \mathcal{O}(V)\},$$

by taking the tensor equation $\mu(C) = C - \mathit{1} = 0$ as a constraint (equivalent to a multiple condition of six equations).

**Exercise 5.4.3** By using a similar argument employing the Clausius–Duhem inequality, show that for incompressible Navier–Stokes fluids the stress tensor is given by

$$T = -p\mathit{1} + 2\mu D, \qquad \mu \geq 0,$$

with arbitrary pressure $p$ and $\operatorname{tr} D = 0$.

## 5.5 Stability of Equilibrium

For analyzing thermodynamical stability of equilibrium, we consider a supply-free body occupying a region $\mathcal{V}$ with a fixed adiabatic boundary. We have

$$\boldsymbol{v} = \boldsymbol{0}, \quad \boldsymbol{q} \cdot \boldsymbol{n} = 0, \quad \boldsymbol{\Phi} \cdot \boldsymbol{n} = 0 \qquad \text{on } \partial\mathcal{V},$$

and hence the entropy inequality (5.4) and the energy balance (2.68) become

$$\frac{d}{dt}\int_{\mathcal{V}} \rho\eta \, dv \geq 0, \qquad \frac{d}{dt}\int_{\mathcal{V}} \rho(\varepsilon + \tfrac{1}{2}\boldsymbol{v}\cdot\boldsymbol{v})\, dv = 0. \tag{5.66}$$

In other words, the total entropy must increase in time while the total energy remains constant for a body with fixed adiabatic boundary. Statements of this kind are usually called *stability criteria*, as we shall explain in the following example.

We say that an equilibrium state is *stable* if any small disturbance away from it will eventually die out and thus the original state will be restored.

Suppose that the region $\mathcal{V}$ is occupied by a thermoelastic fluid in an equilibrium state at rest with constant mass density $\rho_o$ and internal energy density $\varepsilon_o$. Now let us consider a small disturbance from the equilibrium state at the initial time such that

$$\rho(\boldsymbol{x}, 0) = \hat{\rho}(\boldsymbol{x}), \quad \varepsilon(\boldsymbol{x}, 0) = \hat{\varepsilon}(\boldsymbol{x}), \quad \boldsymbol{v}(\boldsymbol{x}, 0) = 0,$$

and $|\hat{\rho} - \rho_o|$ and $|\hat{\varepsilon} - \varepsilon_o|$ are small quantities. If we assume that the original state is stable then the perturbed state will eventually return to the original state at a later time. Therefore, since the total entropy must increase we conclude that

$$\int_{\mathcal{V}} \rho_o\eta_o \, dv \geq \int_{\mathcal{V}} \hat{\rho}\hat{\eta}\, dv, \tag{5.67}$$

where $\eta_o = \eta(\varepsilon_o, \rho_o)$ and $\hat{\eta} = \eta(\hat{\varepsilon}, \hat{\rho})$ are the final equilibrium entropy and the perturbed initial entropy respectively. Here we have regarded $(\varepsilon, \rho)$ as independent variables instead of $(\rho, \theta)$ for convenience. Expanding $\hat{\eta}$ in Taylor series around the equilibrium state, we obtain from (5.67)

$$\begin{aligned}\int_{\mathcal{V}} \Big\{ & \eta_o(\hat{\rho} - \rho_o) + \frac{\partial \eta}{\partial \rho}\Big|_o \hat{\rho}(\hat{\rho} - \rho_o) + \frac{\partial \eta}{\partial \varepsilon}\Big|_o \hat{\rho}(\hat{\varepsilon} - \varepsilon_o) \\ & + \frac{1}{2}\frac{\partial^2 \eta}{\partial \varepsilon^2}\Big|_o \hat{\rho}(\hat{\varepsilon} - \varepsilon_o)^2 + \frac{1}{2}\frac{\partial^2 \eta}{\partial \rho^2}\Big|_o \hat{\rho}(\hat{\rho} - \rho_o)^2 \\ & + \frac{\partial^2 \eta}{\partial \varepsilon \partial \rho}\Big|_o \hat{\rho}(\hat{\varepsilon} - \varepsilon_o)(\hat{\rho} - \rho_o)\Big\}\, dv + o(3) \leq 0.\end{aligned} \tag{5.68}$$

Since total mass and total energy remain constant, we have

$$\int_{\mathcal{V}} (\hat{\rho} - \rho_o)\, dv = 0, \qquad \int_{\mathcal{V}} (\hat{\rho}\hat{\varepsilon} - \rho_o\varepsilon_o)\, dv = 0,$$

and hence

$$\int_{\mathcal{V}} \hat{\rho}(\hat{\varepsilon} - \varepsilon_o)\, dv = \int_{\mathcal{V}} (\hat{\rho}\hat{\varepsilon} - \rho_o\varepsilon_o)\, dv - \varepsilon_o \int_{\mathcal{V}} (\hat{\rho} - \rho_o)\, dv = 0,$$

$$\int_{\mathcal{V}} \hat{\rho}(\hat{\rho} - \rho_o)\, dv = \int_{\mathcal{V}} (\hat{\rho} - \rho_o)^2 dv = \int_{\mathcal{V}} \frac{\hat{\rho}}{\rho_o}(\hat{\rho} - \rho_o)^2 dv + o(3).$$

Therefore, up to the second-order terms (5.68) becomes

$$\int_{\mathcal{V}} \left\{ \frac{1}{2}\frac{\partial^2\eta}{\partial\varepsilon^2}\bigg|_o \hat{\rho}(\hat{\varepsilon} - \varepsilon_o)^2 + \frac{\partial^2\eta}{\partial\varepsilon\partial\rho}\bigg|_o \hat{\rho}(\hat{\varepsilon} - \varepsilon_o)(\hat{\rho} - \rho_o) + \left(\frac{1}{2}\frac{\partial^2\eta}{\partial\rho^2} + \frac{1}{\rho}\frac{\partial\eta}{\partial\rho}\right)\bigg|_o \hat{\rho}(\hat{\rho} - \rho_o)^2 \right\} dv \leq 0.$$

By the mean value theorem for integrals, it reduces to

$$\left\{ \frac{\partial^2\eta}{\partial\varepsilon^2}\bigg|_o (\varepsilon^* - \varepsilon_o)^2 + 2\frac{\partial^2\eta}{\partial\varepsilon\partial\rho}\bigg|_o (\varepsilon^* - \varepsilon_o)(\rho^* - \rho_o) + \left(\frac{\partial^2\eta}{\partial\rho^2} + \frac{2}{\rho}\frac{\partial\eta}{\partial\rho}\right)\bigg|_o (\rho^* - \rho_o)^2 \right\} \frac{\rho^* V}{2} \leq 0,$$

where $V$ is the volume of the region $\mathcal{V}$ while $\rho^* = \hat{\rho}(\boldsymbol{x}^*)$ and $\varepsilon^* = \hat{\varepsilon}(\boldsymbol{x}^*)$ for some point $\boldsymbol{x}^*$ in $\mathcal{V}$.

Since the non-positiveness of the above quadratic form must hold for any small disturbance and $\rho_o$ and $\varepsilon_o$ are arbitrary, it follows that the matrix

$$\begin{bmatrix} \dfrac{\partial^2\eta}{\partial\varepsilon^2} & \dfrac{\partial^2\eta}{\partial\varepsilon\partial\rho} \\ \dfrac{\partial^2\eta}{\partial\varepsilon\partial\rho} & \dfrac{\partial^2\eta}{\partial\rho^2} + \dfrac{2}{\rho}\dfrac{\partial\eta}{\partial\rho} \end{bmatrix}$$

must be negative semi-definite, or equivalently

$$\frac{\partial^2\eta}{\partial\varepsilon^2} \leq 0, \tag{5.69}$$

$$\frac{\partial^2\eta}{\partial\varepsilon^2}\left(\frac{\partial^2\eta}{\partial\rho^2} + \frac{2}{\rho}\frac{\partial\eta}{\partial\rho}\right) - \left(\frac{\partial^2\eta}{\partial\varepsilon\partial\rho}\right)^2 \geq 0. \tag{5.70}$$

In these relations, partial derivatives are taken with respect to the variables $(\varepsilon, \rho)$.

In order to give more suggestive meanings to the above conditions for stability, we shall reiterate them in terms of the independent variables $(\rho, \theta)$. In making this change of variables we must admit the invertibility of $\varepsilon(\rho, \theta)$ with respect to the temperature, i.e., $\partial\varepsilon/\partial\theta \neq 0$. To avoid confusion, variables held constant in partial differentiations will be indicated.

From the Gibbs relation (5.25), we have

$$\left.\frac{\partial\eta}{\partial\varepsilon}\right|_\rho = \frac{1}{\theta}, \qquad \left.\frac{\partial\eta}{\partial\rho}\right|_\varepsilon = -\frac{p}{\theta\,\rho^2},$$

and by the use of the chain rule for the change of variables (see Exercise 5.5.1), we have

$$\left.\frac{\partial\theta}{\partial\varepsilon}\right|_\rho = \left.\frac{\partial\varepsilon}{\partial\theta}\right|_\rho^{-1}, \qquad \left.\frac{\partial\theta}{\partial\rho}\right|_\varepsilon = -\left.\frac{\partial\varepsilon}{\partial\rho}\right|_\theta \left.\frac{\partial\varepsilon}{\partial\theta}\right|_\rho^{-1}.$$

It follows immediately that

$$\left.\frac{\partial^2\eta}{\partial\varepsilon^2}\right|_\rho = -\frac{1}{\theta^2}\left.\frac{\partial\varepsilon}{\partial\theta}\right|_\rho^{-1}, \qquad \frac{\partial^2\eta}{\partial\varepsilon\partial\rho} = \frac{1}{\theta^2}\left.\frac{\partial\varepsilon}{\partial\rho}\right|_\theta \left.\frac{\partial\varepsilon}{\partial\theta}\right|_\rho^{-1},$$

and

$$\left.\frac{\partial^2\eta}{\partial\rho^2}\right|_\varepsilon + \frac{2}{\rho}\left.\frac{\partial\eta}{\partial\rho}\right|_\varepsilon = \frac{p}{\rho^2\theta^2}\left.\frac{\partial\theta}{\partial\rho}\right|_\varepsilon - \frac{1}{\rho^2\theta}\left.\frac{\partial p}{\partial\rho}\right|_\varepsilon,$$

which, by the relation

$$\left.\frac{\partial p}{\partial\rho}\right|_\varepsilon = \left.\frac{\partial p}{\partial\rho}\right|_\theta + \left.\frac{\partial p}{\partial\theta}\right|_\rho \left.\frac{\partial\theta}{\partial\rho}\right|_\varepsilon,$$

and the integrability condition (5.26), becomes

$$\left.\frac{\partial^2\eta}{\partial\rho^2}\right|_\varepsilon + \frac{2}{\rho}\left.\frac{\partial\eta}{\partial\rho}\right|_\varepsilon = -\frac{1}{\rho^2\theta}\left.\frac{\partial p}{\partial\rho}\right|_\theta - \frac{1}{\theta^2}\left.\frac{\partial\varepsilon}{\partial\rho}\right|_\theta^2 \left.\frac{\partial\varepsilon}{\partial\theta}\right|_\rho^{-1}.$$

Therefore, in terms of the variables $(\rho,\theta)$, the stability condition (5.69) reduces to

$$\frac{\partial\varepsilon}{\partial\theta} > 0, \tag{5.71}$$

since we have already admitted that $\partial\varepsilon/\partial\theta \neq 0$, and the condition (5.70), after simplifications, implies

$$\frac{\partial p}{\partial\rho} \geq 0. \tag{5.72}$$

Obviously, these conditions are restrictions on the constitutive functions $\varepsilon(\rho,\theta)$ and $p(\rho,\theta)$.

We have shown that the stability of equilibrium requires that: (1) the *specific heat* at constant volume (equivalently at constant density), $c_v = \partial\varepsilon/\partial\theta$, must be positive; (2) the isothermal compressibility (see Exercise 5.5.3) must be non-negative. Some other implications of stability conditions are given in the exercises.

**Exercise 5.5.1** Let $u$, $v$, and $w$ be three variables and assume there is a relation between them so that we have the functional relations: $u = u(v, w)$, $v = v(u, w)$, and $w = w(u, v)$. Show that

$$1)\quad \left.\frac{\partial u}{\partial v}\right|_w \left.\frac{\partial v}{\partial u}\right|_w = 1,$$

$$2)\quad \left.\frac{\partial u}{\partial v}\right|_w \left.\frac{\partial v}{\partial w}\right|_u \left.\frac{\partial w}{\partial u}\right|_v = -1.$$

We put the variable as a subscript only to emphasize its constancy held in the partial differentiation. (*Hint*: Write $u = u(v(u, w), w) = \tilde{u}(u, w)$ and compute $\partial\tilde{u}/\partial u$ and $\partial\tilde{u}/\partial w$.)

**Exercise 5.5.2** Let $v = 1/\rho$ be the specific volume, the volume per unit mass. Show that

1) The Gibbs relation can be written as

$$d\psi = -\eta\, d\theta - p\, dv.$$

2) The condition (5.72) implies

$$\left.\frac{\partial^2 \psi}{\partial v^2}\right|_\theta \geq 0.$$

Therefore, the stability of equilibrium requires that the free energy be a concave-upward function of the specific volume.

**Exercise 5.5.3** With pressure and temperature as variables, let $v = v(p, \theta)$ and $\eta = \eta(p, \theta)$. Define

$$c_p = \theta \left.\frac{\partial \eta}{\partial \theta}\right|_p \qquad \text{specific heat at constant pressure,}$$

$$\alpha = \frac{1}{v}\left.\frac{\partial v}{\partial \theta}\right|_p \qquad \text{coefficient of thermal expansion,}$$

$$\kappa_T = -\frac{1}{v}\left.\frac{\partial v}{\partial p}\right|_\theta \qquad \text{isothermal compressibility.}$$

Derive the following expression from the Gibbs relation

$$c_p = c_v - \theta \left.\frac{\partial p}{\partial v}\right|_\theta \left.\frac{\partial v}{\partial \theta}\right|_p^2,$$

in which the condition (5.26) may be used. Furthermore, from (5.72) show that

$$c_p \geq c_v, \qquad \alpha^2 \leq \frac{1}{\theta v} c_p \kappa_T.$$

Therefore, both $c_p$ and $\kappa_T$ are positive, and $\alpha$ must vanish if $\kappa_T$ tends to zero. Note that the specific heat at constant volume can also be defined as

$$c_v = \theta \frac{\partial \eta}{\partial \theta}\bigg|_v .$$

### 5.5.1 Thermodynamic Stability Criteria

To establish a stability criterion for a material system under a different condition, one may try to find a decreasing function of time $\mathcal{A}(t)$ from the balance laws and the entropy inequality in integral forms. Such a function is called the *availability*[4] of the system, since it is the quantity available to the system for its expense in the course toward equilibrium. In the previous example, from $(5.66)_1$ one may define the availability $\mathcal{A}$ of the system as

$$\mathcal{A}(t) = -\int_{\mathcal{V}} \rho\eta \, dv, \qquad \frac{d\mathcal{A}}{dt} \leq 0.$$

As a second example, we consider a supply-free body with a fixed isothermal boundary,

$$\boldsymbol{v} = \mathbf{0}, \quad \theta = \theta_o \quad \text{on } \partial\mathcal{V},$$

and assume that the relation $\boldsymbol{\Phi} = \boldsymbol{q}/\theta$ holds. Then the energy balance (2.68) and the entropy inequality (5.4) lead to

$$\frac{d}{dt}\int_{\mathcal{V}} \rho(\varepsilon + \tfrac{1}{2}\boldsymbol{v}\cdot\boldsymbol{v})\, dv + \int_{\partial\mathcal{V}} \boldsymbol{q}\cdot\boldsymbol{n}\, da = 0,$$

$$\frac{d}{dt}\int_{\mathcal{V}} \rho\eta\, dv + \frac{1}{\theta_o}\int_{\partial\mathcal{V}} \boldsymbol{q}\cdot\boldsymbol{n}\, da \geq 0.$$

Elimination of the terms containing surface integrals from above, gives

$$\frac{d\mathcal{A}}{dt} \leq 0, \quad \mathcal{A}(t) = \int_{\mathcal{V}} \rho(\varepsilon - \theta_o\eta + \tfrac{1}{2}\boldsymbol{v}\cdot\boldsymbol{v})\, dv. \tag{5.73}$$

In this manner we have found a decreasing function of time, the availability $\mathcal{A}(t)$, which characterizes the stability for this system. Note that

$$\int_{\mathcal{V}} \rho(\varepsilon - \theta_o\eta)\, dv$$

[4] Such a function is also known as a *Liapounov function* in the stability theory of dynamic systems, see [33].

is the total free energy if $\theta = \theta_o$ throughout the body. Therefore, it follows that for a body with constant uniform temperature in a fixed region the availability $\mathcal{A}$ reduces to the sum of the free energy and the kinetic energy.

Summarizing the above two situations, we can state the following *thermodynamic stability criteria* of equilibrium:

1) *For a body with fixed adiabatic boundary and constant energy, the entropy tends to a maximum in equilibrium.*

2) *For a body with fixed boundary and constant uniform temperature, the sum of the free energy and the kinetic energy tends to a minimum in equilibrium.*

We have seen in this section that thermodynamic stability criteria, like the entropy principle, impose further restrictions on properties of the constitutive functions, namely, specific heat and compressibility must be positive. On the other hand, such criteria, besides being used in analyzing stability of solutions, are the basic principles for the formulation of equilibrium solutions in terms of minimization (or maximization) problems.

**Exercise 5.5.4** Consider a material region $\mathcal{V}$ subject to a uniform temperature $\theta_o$ and pressure $p_o$ on the boundary $\partial\mathcal{V}$. Show that the linear momentum of the body is constant and the availability of the system can be defined as

$$\mathcal{A} = \int_{\mathcal{V}} \rho(\varepsilon - \theta_o\eta + \frac{p_o}{\rho} + \tfrac{1}{2}\boldsymbol{v}\cdot\boldsymbol{v})\, dv.$$

The relation (2.10) can be used in the derivation.

## 5.6 Phase Equilibrium

We consider a body consisting of two phases with different material properties. The interface will be considered as a moving singular surface so that the phase transition may progress within the body with a continuous and piecewise smooth deformation and temperature (see [41]). Accepting the assumption $(5.9)_1$, the jump condition $(5.8)_2$ of entropy becomes

$$-\rho_\kappa U_\kappa [\![\eta]\!] + \left[\!\!\left[\frac{\boldsymbol{q}_\kappa}{\theta}\right]\!\!\right]\cdot \boldsymbol{n}_\kappa \geq 0.$$

Since the temperature is continuous across the interface, $[\![\theta]\!] = 0$, by eliminating the heat flux from the jump condition of energy $(2.87)_3$ and the use of the kinematic compatibility condition (2.26), the jump condition of entropy takes the form (see (2.88) of Exercise 2.5.3),

$$\frac{1}{\theta} U_\kappa \big( [\![\rho_\kappa(\varepsilon - \theta\eta)]\!] - \langle T_\kappa \boldsymbol{n}_\kappa \rangle \cdot [\![F \boldsymbol{n}_\kappa]\!] \big) \geq 0, \tag{5.74}$$

where $\langle A \rangle = \frac{1}{2}(A^+ + A^-)$ is the mean value of $A$ at the interface. We can rewrite the last term in the above expression as

$$\begin{aligned} \langle T_\kappa \boldsymbol{n}_\kappa \rangle \cdot [\![F \boldsymbol{n}_\kappa]\!] &= [\![T_\kappa \boldsymbol{n}_\kappa \cdot F \boldsymbol{n}_\kappa]\!] - \frac{1}{2} [\![T_\kappa \boldsymbol{n}_\kappa]\!] \cdot F^+ \boldsymbol{n}_\kappa - \frac{1}{2} [\![T_\kappa \boldsymbol{n}_\kappa]\!] \cdot F^- \boldsymbol{n}_\kappa \\ &= [\![T_\kappa \boldsymbol{n}_\kappa \cdot F \boldsymbol{n}_\kappa]\!] - \frac{1}{2} \rho_\kappa U_\kappa^2 [\![F \boldsymbol{n}_\kappa \cdot F \boldsymbol{n}_\kappa]\!], \end{aligned}$$

where in the second passage $(2.88)_2$ has been used. The condition (5.74) now becomes

$$\frac{\rho_\kappa}{\theta} U_\kappa \boldsymbol{n}_\kappa \cdot \Big( [\![P_\kappa]\!] + \frac{1}{2} U_\kappa^2 [\![F^T F]\!] \Big) \boldsymbol{n}_\kappa \geq 0. \tag{5.75}$$

Here, we have introduced Eshelby's energy–momentum tensor ([20]), or simply the *Eshelby tensor*, in the reference configuration by

$$P_\kappa = \psi 1 - \frac{1}{\rho_\kappa} F^T T_\kappa, \tag{5.76}$$

where

$$\psi = \varepsilon - \theta\eta$$

is the free energy density.

The condition (5.75) characterized the process of phase transition, and we can regard the left-hand side as a function of the interface velocity $U_\kappa$, the normal $\boldsymbol{n}_\kappa$, and other properties of the materials adjacent to the interface. Hence,

$$f(U_\kappa; X) \geq 0,$$

where $X$ stands for the other variables involved. Assuming that the function is smooth, then since $f(0, X) = 0$, we have the following condition for $f$ to attain its minimum at $U_\kappa = 0$,

$$\frac{\partial f}{\partial U_\kappa}\Big|_{U_\kappa = 0} = 0,$$

from which (5.75) implies that

$$\boldsymbol{n}_\kappa \cdot [\![P_\kappa]\!] \boldsymbol{n}_\kappa = 0. \tag{5.77}$$

We say that an interface is in equilibrium if there is no surface production of entropy. Therefore, we conclude that the condition of phase equilibrium is characterized by the continuity of the temperature $\theta$ and the Eshelby tensor $P_\kappa$ across the interface.

We can rewrite the Eshelby tensor in terms of the Cauchy stress tensor $T$ and the mass density $\rho$ in the present configuration, from $(2.84)_1$

$$P_\kappa = \psi 1 - \frac{1}{\rho}\, F^T T F^{-T}.$$

Note that the Eshelby tensor is not symmetric, in general, but, of course, only the symmetric part is involved in the condition (5.77) of phase equilibrium.

Moreover, at the singular surface with unit normal $\boldsymbol{n}$ in the present configuration, since from $(2.28)_1$,

$$\boldsymbol{n}_\kappa = \frac{F^T\boldsymbol{n}}{|F^T\boldsymbol{n}|},$$

the condition of phase equilibrium (5.77) becomes

$$\left[\!\left[ \psi - \frac{1}{\rho}\frac{T\boldsymbol{n}\cdot B\boldsymbol{n}}{\boldsymbol{n}\cdot B\boldsymbol{n}} \right]\!\right] = 0, \tag{5.78}$$

where $B = FF^T$ is the left Cauchy–Green tensor.

For fluids, the stress $T$ reduces to a hydrostatic pressure, $T = -p1$. Hence, the phase equilibrium condition (5.77) for fluids becomes

$$\left[\!\left[ \psi + \frac{p}{\rho} \right]\!\right] = 0.$$

Therefore, in phase equilibrium, the *free enthalpy*

$$g = \psi + \frac{p}{\rho} \tag{5.79}$$

is continuous at the fluid–fluid interface. The free enthalpy is also known as the *Gibbs free energy*.

For solids, if the normal $\boldsymbol{n}$ is in the principal direction $\boldsymbol{e}_i$ of the stress tensor $T$,

$$T\boldsymbol{e}_i = -p_i\boldsymbol{e}_i, \qquad i = 1, 2, 3,$$

where $p_i$ is the pressure on the surface perpendicular to the direction $\boldsymbol{e}_i$. Then the condition (5.78) becomes

$$\left[\!\left[ \psi + \frac{p_i}{\rho} \right]\!\right] = 0 \quad \text{at} \quad \boldsymbol{n} = \boldsymbol{e}_i.$$

This result was established by Gibbs [23] as a generalization of the phase equilibrium condition of fluids to non-hydrostatically stressed solids.

# 6. Isotropic Elastic Solids

## 6.1 Constitutive Equations

In the previous chapters, we have discussed the basic equations and constitutive relations for material bodies in general. Now we shall consider some interesting static problems for isotropic elastic bodies.

For elastic materials, the constitutive equation of the stress tensor can be written as

$$T = \mathcal{T}(F),$$

where $F$ is the deformation gradient relative to some reference configuration $\kappa$, which is chosen so that $\det F > 0$.

A deformation in an elastic body can be regarded as an isothermal deformation for a thermoelastic body, for which the temperature is constant and uniform throughout the entire body. From this point of view, the material is also hyperelastic with the free energy $\psi$ as the stored energy function (see (5.28)),

$$\mathcal{T}(F) = \rho \frac{\partial \psi(F)}{\partial F} F^T, \tag{6.1}$$

for compressible elastic materials or

$$T + p\mathit{1} = \mathcal{T}(F) = \rho \frac{\partial \psi(F)}{\partial F} F^T,$$

for incompressible elastic materials.

Moreover, if the material is isotropic, then by (4.26) we have the following representation for the stress tensor,

$$\begin{aligned} T &= t_0 \mathit{1} + t_1 B + t_2 B^2, \\ t_i &= t_i(I_B, II_B, III_B), \end{aligned}$$

or, alternatively, from (4.27),

$$\begin{aligned} T &= s_0 \mathit{1} + s_1 B + s_{-1} B^{-1}, \\ s_i &= s_i(I_B, II_B, III_B), \end{aligned} \tag{6.2}$$

where $I_B$, $II_B$, $III_B$ are the three principal invariants of the left Cauchy–Green tensor $B$.

From (6.1), we can also write

$$T(B) = 2\rho \frac{\partial \psi(B)}{\partial B} B,$$

where the stored energy function $\psi(B) = \psi(I_B, II_B, III_B)$. By performing the tensor gradient of $\psi$ in the above expression and comparing it with (6.2), we can obtain

$$\begin{aligned} s_0 &= 2\rho\Big(III_B \frac{\partial \psi}{\partial III_B} + II_B \frac{\partial \psi}{\partial II_B}\Big), \\ s_1 &= 2\rho \frac{\partial \psi}{\partial I_B}, \\ s_{-1} &= -2\rho III_B \frac{\partial \psi}{\partial II_B}. \end{aligned} \tag{6.3}$$

Similar relations for $t_i$ have been given in (5.34).

For incompressible isotropic elastic materials, we can write

$$\begin{aligned} T &= -p1 + s_1 B + s_{-1} B^{-1}, \\ s_i &= s_i(I_B, II_B), \end{aligned} \tag{6.4}$$

where $B$ is required to be unimodular, i.e., $III_B = 1$. Note that in this case, we have $II_B = \operatorname{tr} B^{-1}$ (see (A.36)). Moreover, we have $\psi = \psi(I_B, II_B)$ and

$$s_1 = 2\rho \frac{\partial \psi}{\partial I_B}, \qquad s_{-1} = -2\rho \frac{\partial \psi}{\partial II_B}. \tag{6.5}$$

The undeterminate pressure $p$ can not be determined from the deformation.

The simplest model for incompressible isotropic materials is to take the linear form for the strain energy function

$$\psi = \alpha(I_B - 3) + \beta(II_B - 3), \tag{6.6}$$

where $\alpha$ and $\beta$ are constants. This defines a *Mooney–Rivlin* material. The stress tensor becomes

$$T = -p1 + s_1 B + s_{-1} B^{-1}, \tag{6.7}$$

where, from (6.5), $s_1 = 2\rho\alpha$ and $s_{-1} = -2\rho\beta$ are constants. This incompressible material model is often adopted as the constitutive equation for rubber-like materials. Experimental data indicate that both $\alpha$ and $\beta$ are positive,[1] and the numerical value of $\beta$ is much smaller than that of $\alpha$, typically, say $\beta \simeq 0.1\alpha$. Therefore, experimental evidence seems to support

$$s_1 > 0, \quad s_{-1} \le 0. \tag{6.8}$$

[1] More discussions on experimental data can be found in Sect. 55 of [71].

They are known as (empirical) *E-inequalities* in elasticity.

If, in addition, the material constant $\beta$ is zero, then the stress tensor is given by

$$T = -p\mathit{1} + s_1 B. \tag{6.9}$$

This defines a *neo-Hookean* material. This material model is also predicted by the kinetic theory of rubber from molecular calculations in the first approximation (see [68]). It provides a reasonable account of the behavior of natural rubber for modest strains.

Before closing this section, we shall add an interesting remark. From the representations (6.2) and (6.4), it is obvious that for compressible or incompressible isotropic elastic materials, the stress tensor and the left Cauchy–Green tensor commute (see also [4]),

$$TB = BT. \tag{6.10}$$

In particular, for a deformation such that the physical components $B_{\langle 13\rangle} = B_{\langle 23\rangle} = 0$, since, according to (6.2) or (6.4), $T_{\langle 13\rangle}$ and $T_{\langle 23\rangle}$ also vanish, the only non-trivial relation of (6.10) is the expression for the $\langle 12\rangle$-component, which reads

$$\frac{T_{\langle 11\rangle} - T_{\langle 22\rangle}}{T_{\langle 12\rangle}} = \frac{B_{\langle 11\rangle} - B_{\langle 22\rangle}}{B_{\langle 12\rangle}}. \tag{6.11}$$

The relations (6.10) and (6.11) between stress and deformation do not depend on any particular constitutive function, and thus they are called *universal relations* of isotropic elastic materials, compressible or incompressible. Relations of this kind are very important for experimental verification of material models, since they reflect a direct consequence from the material symmetry without having to know the constitutive function itself.

## 6.2 Boundary Value Problems in Elasticity

For elastic solid bodies, the equation of motion is given by

$$\operatorname{div} T(F) + \rho\, \boldsymbol{b} = \rho\, \ddot{\boldsymbol{x}}, \qquad \rho = \frac{\rho_\kappa}{|\det F|}, \tag{6.12}$$

or, in terms of referential description,

$$\operatorname{Div} T_\kappa(F) + \rho_\kappa \boldsymbol{b} = \rho_\kappa \ddot{\boldsymbol{x}}, \tag{6.13}$$

where $T_\kappa = JTF^{-T}$ is the Piola Kirchhoff stress tensor and $\rho_\kappa$ is the mass density in the reference configuration $\kappa$. The external body force $\boldsymbol{b}$ is usually given in a specific problem.

A boundary value problem in elastic bodies is a problem of finding solutions, $\boldsymbol{x} = \chi(\boldsymbol{X}, t)$, of (6.12), or (6.13), with certain boundary conditions. The following three types of boundary conditions are often considered.

1) *Traction boundary condition*: The forces acting on the boundary are prescribed,

$$T_\kappa(\boldsymbol{X})\boldsymbol{n}_\kappa(\boldsymbol{X}) = \boldsymbol{f}_\kappa(\boldsymbol{X}), \qquad \boldsymbol{X} \in \partial\mathcal{B}_\kappa, \tag{6.14}$$

where $\boldsymbol{f}_\kappa$ denotes external surface forces exerted on the boundary and $\boldsymbol{n}_\kappa$ denotes the outward unit normal in the reference configuration (see (2.91)).

2) *Place boundary condition*: The position of the boundary is prescribed,

$$\chi(\boldsymbol{X}) = \boldsymbol{x}_0(\boldsymbol{X}), \qquad \boldsymbol{X} \in \partial\mathcal{B}_\kappa, \tag{6.15}$$

where $\boldsymbol{x}_0(\boldsymbol{X})$ is a given function.

3) *Mixed boundary condition*: The traction is prescribed on a part of the boundary, while on the other part of the boundary the position is prescribed.

Boundary value problems for incompressible elastic bodies can be similarly formulated. From (5.62) the constitutive equation for the stress tensor can be written as

$$T = -p\mathit{1} + \mathcal{T}(F), \qquad \det F = 1,$$

and the equation of motion (6.12) becomes

$$-\operatorname{grad} p + \operatorname{div} \mathcal{T}(F) + \rho\,\boldsymbol{b} = \rho\,\ddot{\boldsymbol{x}}, \tag{6.16}$$

or, in referential description,

$$-\operatorname{Grad} p + \operatorname{Div} \mathcal{T}_\kappa(F) + \rho_\kappa\boldsymbol{b} = \rho_\kappa\ddot{\boldsymbol{x}}, \tag{6.17}$$

where $p$ is the undeterminate hydrostatic pressure. The pressure $p$ is not directly related to $F$ and should be determined in order to satisfy both the equation of motion and the boundary conditions.

We shall not discuss the existence and uniqueness of boundary value problems defined above except to make an opportune remark on uniqueness. A thin hemisphere, like a tennis ball cut in half, for example, can be turned inside out, so that there are at least two different solutions to the boundary value problem of a hemisphere with zero traction on the boundary. Therefore, it is clear that an unqualified uniqueness is not to be desired for boundary value problems in general. In the later sections, multiple solutions for traction boundary value problems – biaxial stretching of a square sheet and deformation of a hollow sphere – will be discussed in detail.

We shall consider some solutions, called *controllable* deformations. Such a solution is specified by a certain deformation function satisfying the equation of motion such that the body can be maintained in equilibrium by applying suitable surface traction on the boundary alone. In other words, a controllable deformation is a solution of the equilibrium equation,

$$\operatorname{div} T(F) + \rho\, \boldsymbol{b} = 0, \tag{6.18}$$

for compressible elastic bodies, or

$$-\operatorname{grad} p + \operatorname{div} \mathcal{T}(F) + \rho\, \boldsymbol{b} = 0, \tag{6.19}$$

for incompressible elastic bodies. No additional boundary conditions are prescribed, instead, the boundary tractions are to be determined from (6.14). In the case of incompressible bodies, the pressure field must be suitably chosen so as to satisfy the equilibrium equation.

In general, a controllable deformation for a certain elastic material may not be controllable for a different elastic material, since the equilibrium equation depends on the constitutive equation. If a deformation function is controllable for a certain type of elastic materials, it will be called a *universal solution* of such materials. It has been shown by Ericksen ([16] or Sect. 91 of [71]) that homogeneous deformations are the only class of universal solutions for compressible isotropic elastic bodies. However, being allowed to choose a suitable pressure field in order to satisfy the equilibrium equation gives an additional freedom for possible solutions in the case of incompressible bodies. And indeed, there are several other well-known classes of universal solutions for incompressible elastic bodies. The search for universal solutions is known as "Ericksen's problem" in the literature.[2] We shall consider some of these solutions in the succeeding sections.

## 6.3 Homogeneous Stretch

Consider a homogeneous deformation for which the deformation gradient $F$ is a constant tensor,

$$\boldsymbol{x} = F(\boldsymbol{X} - \boldsymbol{X}_o) + \boldsymbol{x}_o,$$

where $\boldsymbol{X}_o$ and $\boldsymbol{x}_o$ denote the origins in the reference and the deformed configurations, respectively. Since $F$ is a constant tensor the stress $T(F)$ is also constant. Therefore, in the absence of external body force $\boldsymbol{b}$, the equilibrium equation (6.18) is identically satisfied and the boundary condition (6.14) can be fulfilled by suitably applied surface forces on the body. For incompressible materials, the equilibrium equation (6.19) can also be satisfied by a constant pressure field $p$. Therefore, a homogeneous deformation is a controllable deformation for any elastic body with no body force. Moreover, it is a universal solution of elastic materials, since it does not depend on any particular form of constitutive equations.

[2] Many such solutions are discussed in [71]. See also [1, 4, 76].

### 6.3.1 Uniaxial Stretch

We consider an isotropic elastic body subject to a homogeneous stretch of the form

$$x = \lambda_1 X, \quad y = \lambda_2 Y, \quad z = \lambda_3 Z, \tag{6.20}$$

where $\lambda_1$, $\lambda_2$, $\lambda_3$ are constants. Then the component of $F$ is given by

$$[F_{\langle i\alpha\rangle}] = \operatorname{diag}[\lambda_1, \lambda_2, \lambda_3], \qquad \det F = \lambda_1\lambda_2\lambda_3,$$

and if we write the stress tensor as

$$T = s_0 1 + \mathcal{T}, \qquad T = -p1 + \mathcal{T},$$

for compressible and incompressible bodies, respectively, then

$$[\mathcal{T}_{\langle ij\rangle}] = s_1 \begin{bmatrix} \lambda_1^2 & 0 & 0 \\ 0 & \lambda_2^2 & 0 \\ 0 & 0 & \lambda_3^2 \end{bmatrix} + s_{-1} \begin{bmatrix} \lambda_1^{-2} & 0 & 0 \\ 0 & \lambda_2^{-2} & 0 \\ 0 & 0 & \lambda_3^{-2} \end{bmatrix},$$

where the coefficients $s_i = s_i(\lambda_1^2, \lambda_2^2, \lambda_3^2)$ and for incompressible bodies, $\lambda_1\lambda_2\lambda_3 = 1$. In the following we shall consider the cases of uniaxial and biaxial stretches.

For uniaxial stretch in the $x$-direction, the lateral stresses must vanish. Therefore, for compressible bodies,

$$\begin{aligned} T_{\langle yy\rangle} &= s_0 + s_1\lambda_2^2 + s_{-1}\lambda_2^{-2} = 0, \\ T_{\langle zz\rangle} &= s_0 + s_1\lambda_3^2 + s_{-1}\lambda_3^{-2} = 0. \end{aligned} \tag{6.21}$$

Taking the difference between $T_{\langle xx\rangle}$ and $T_{\langle yy\rangle}$, we then obtain the axial stress,

$$T_{\langle xx\rangle} = (\lambda_1^2 - \lambda_2^2)\left(s_1 - s_{-1}\frac{1}{\lambda_1^2\lambda_2^2}\right). \tag{6.22}$$

Moreover, from $T_{\langle yy\rangle} - T_{\langle zz\rangle}$, we also have

$$(\lambda_2^2 - \lambda_3^2)\left(s_1 - s_{-1}\frac{1}{\lambda_2^2\lambda_3^2}\right) = 0,$$

which implies a symmetric solution $\lambda_2 = \lambda_3$, and perhaps an asymmetric solution $\lambda_2 \neq \lambda_3$. The asymmetric solution would be impossible if the E-inequalities, $s_1 > 0$ and $s_{-1} \leq 0$, are valid.

If $s_i = s_i(\lambda_1^2, \lambda_2^2, \lambda_3^2)$ are known, the equations (6.21) may be solved for $\lambda_2$ and $\lambda_3$ in terms of $\lambda_1$. Then, (6.22) determines the axial stress $T_{\langle xx\rangle}$ uniquely in terms of the axial extension $\lambda_1$. On the other hand, since the stress $T_{\langle xx\rangle}$ and the extensions $\lambda_1$, $\lambda_2$ are measurable quantities, the relation (6.22) may be used for the experimental determination of $s_1$ and $s_{-1}$, while $s_0$ can be determined directly from (6.21).

For incompressible bodies, since $\lambda_1\lambda_2\lambda_3 = 1$ and if, in addition, we assume $\lambda_2 = \lambda_3$, then we have

$$\lambda_2 = \lambda_3 = \frac{1}{\sqrt{\lambda_1}}.$$

The axial stress $T_{\langle xx\rangle}$ becomes

$$T_{\langle xx\rangle} = \Big(\lambda_1^2 - \frac{1}{\lambda_1}\Big)\Big(s_1 - s_{-1}\frac{1}{\lambda_1}\Big).$$

For Mooney–Rivlin material, it asserts a linear relation between $\lambda_1^{-1}$ and $T_{\langle xx\rangle}/(\lambda_1^2 - \lambda_1^{-1})$. Such a relation is verified experimentally for rubber within moderate strains. The hydrostatic pressure $p$, which helps to maintain the lateral surface free of stress, is given by

$$p = s_1\frac{1}{\lambda_1} + s_{-1}\lambda_1.$$

### 6.3.2 Biaxial Stretch

We consider an incompressible body subject to homogeneous stretches in the $x$- and $y$-directions only, hence

$$T_{\langle zz\rangle} = 0, \quad \text{and} \quad \lambda_3 = \frac{1}{\lambda_1\lambda_2},$$

which implies that the undeterminate pressure has the value,

$$p = s_1\frac{1}{\lambda_1^2\lambda_2^2} + s_{-1}\lambda_1^2\lambda_2^2.$$

The stresses $T_{\langle xx\rangle}$ and $T_{\langle yy\rangle}$ are given by

$$T_{\langle xx\rangle} = \Big(\lambda_1^2 - \frac{1}{\lambda_1^2\lambda_2^2}\Big)(s_1 - s_{-1}\lambda_2^2),$$
$$T_{\langle yy\rangle} = \Big(\lambda_2^2 - \frac{1}{\lambda_1^2\lambda_2^2}\Big)(s_1 - s_{-1}\lambda_1^2),$$

where $s_i = s_i(\lambda_1^2, \lambda_2^2)$. Since cross-sectional areas change after deformation, it is more convenient to measure Piola–Kirchhoff stresses as forces per unit reference area. From (2.84), $T_\kappa = JTF^{-T}$, the Piola–Kirchhoff stresses are given by

$$T_{\kappa\langle xx\rangle} = \frac{T_{\langle xx\rangle}}{\lambda_1} = \Big(\lambda_1 - \frac{1}{\lambda_1^3\lambda_2^2}\Big)(s_1 - s_{-1}\lambda_2^2),$$
$$T_{\kappa\langle yy\rangle} = \frac{T_{\langle yy\rangle}}{\lambda_2} = \Big(\lambda_2 - \frac{1}{\lambda_1^2\lambda_2^3}\Big)(s_1 - s_{-1}\lambda_1^2).$$

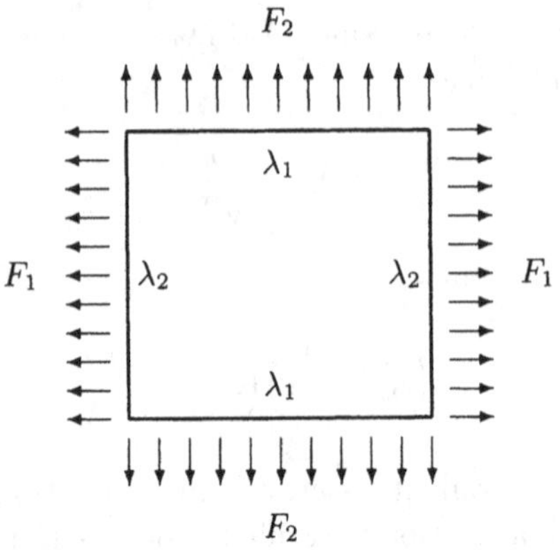


**Fig. 6.1.** Biaxial stretch

If we consider the body as a square sheet of uniform thickness, subjected to biaxial stretching, then the equilibrium of the sheet must be maintained by the distributed forces along the four lateral surfaces (Fig. 6.1). The total forces on the lateral surfaces in the $x$- and $y$-directions are given by $F_1 = A T_{\kappa\langle xx\rangle}$ and $F_2 = A T_{\kappa\langle yy\rangle}$, or

$$\begin{aligned} F_1 &= A\big(s_1 - s_{-1}\lambda_2^2\big)\Big(\lambda_1 - \frac{1}{\lambda_1^3\lambda_2^2}\Big), \\ F_2 &= A\big(s_1 - s_{-1}\lambda_1^2\big)\Big(\lambda_2 - \frac{1}{\lambda_1^2\lambda_2^3}\Big), \end{aligned} \tag{6.23}$$

where $A$ is the reference cross-sectional area.

In biaxial stretching experiments, for some prescribed values of extensions $\lambda_1$ and $\lambda_2$, the required axial forces $F_1$ and $F_2$ can be measured. Therefore, from the above equations, the material parameters $s_1$ and $s_{-1}$ can be determined as functions of $(\lambda_1, \lambda_2)$. One may also determine the stored energy function $\psi(I_B, II_B)$ from the relation (6.5), by suitably controlling the extensions $\lambda_1$ and $\lambda_2$ so as to maintain either constant $I_B$ or constant $II_B$ while measuring the axial forces.

## 6.4 Symmetric Loading of a Square Sheet

Now, consider the body as a square sheet of rubber, loaded on four lateral surfaces with equal forces, $F_1 = F_2$. Intuitively, one would expect the rubber sheet to remain square at least if the forces are not too large. However, not only shall we see that it may also become a rectangular one, but also the square one is unstable as long as the loading is large enough.

Let us consider rubber as a Mooney–Rivlin material, with constants $s_1 > 0$ and $s_{-1} < 0$, and denote $h = -s_{-1}/s_1$. Then from (6.23) the condition $F_1 = F_2$ can be rewritten as

$$(\lambda_1 - \lambda_2)\Big((1 + \lambda_1^3\lambda_2^3)(1 - h\lambda_1\lambda_2) + h(\lambda_1 + \lambda_2)^2\Big) = 0, \qquad (6.24)$$

which immediately gives the symmetric solution, $\lambda_1 = \lambda_2$. Since $\lambda_1$, $\lambda_2$, and $h$ are positive quantities, no other solution exists if $h\lambda_1\lambda_2 < 1$, which incidentally rules out the possibility for an asymmetric solution in the case of neo-Hookean materials ($h = 0$).

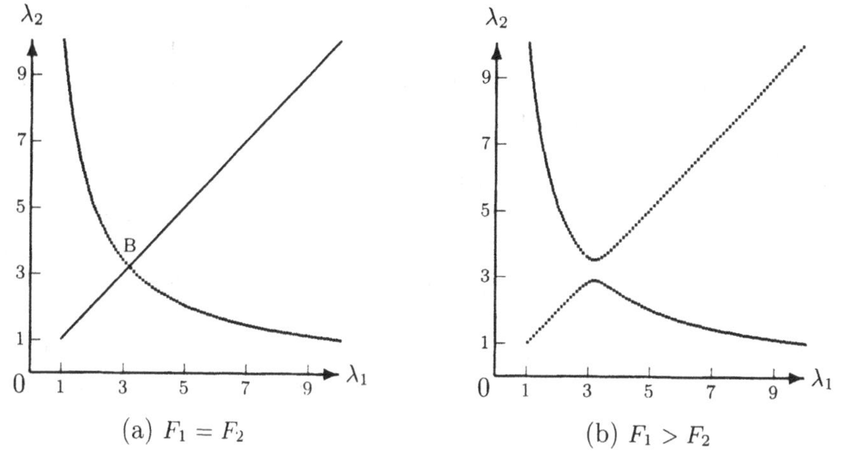


**Fig. 6.2.** Biaxial stretching of a square sheet

The asymmetric solution may exist and can be found from the equation

$$(1 + \lambda_1^3\lambda_2^3)(1 - h\lambda_1\lambda_2) + h(\lambda_1 + \lambda_2)^2 = 0, \quad \text{if} \quad h\lambda_1\lambda_2 > 1. \qquad (6.25)$$

For such a solution, in general, $\lambda_1$ and $\lambda_2$ are different, and the square sheet becomes rectangular after stretching. In Fig. 6.2(a) we have plotted $\lambda_1$ against $\lambda_2$ for a typical value $h = 0.1$ (say). The straight line represents the symmetric solution $\lambda_1 = \lambda_2$, and the curve corresponds to the asymmetric solution $\lambda_1 \neq \lambda_2$. The point, the bifurcation point B, where the two intercept each other has the value $\lambda_B = \lambda_1 = \lambda_2 = 3.1685$.

In the linear theory of elasticity, solutions of traction boundary value problems, (6.18) and (6.14), are unique up to a rigid deformation, and hence the stress fields are unique. However, for large enough forces, there are two possible solutions with different stress fields, corresponding to the one with $\lambda_1 = \lambda_2$ and the other with $\lambda_1 \neq \lambda_2$, to the traction boundary value problem of a square sheet with biaxial symmetric loading. Therefore, it is clear that, unlike the linear theory, one should not expect unqualified uniqueness of solutions to the boundary value problems for theories of finite deformation, in general.

When a square sheet is slowly stretched with equal forces $F_1 = F_2$, it will remain square when $\lambda_1 < \lambda_B$. As the stretch $\lambda_1$ reaches the value $\lambda_B$, one may ask whether the sheet will remain square or become rectangular, the two possible solutions. In other words, which one is a stable solution?

In order to answer this question, we shall consider an imperfect problem for which the square sheet is stretched with unequal forces $F_1 = (1+\varepsilon)F_2$, namely, $F_1$ is slightly larger than $F_2$. From (6.23) we obtain the following equation,

$$(\lambda_1^4\lambda_2^3 - \lambda_2)(1 + h\lambda_2^2) = (1+\varepsilon)(\lambda_1^3\lambda_2^4 - \lambda_1)(1 + h\lambda_1^2).$$

This is a relation between $\lambda_2$ and $\lambda_1$ for a given value of $\varepsilon$. The equation can be solved numerically and the results are plotted in Fig. 6.2(b) for $\varepsilon = 0.005$ and $h = 0.1$. It is interesting to note that there are also two solution curves, but they do not intercept each other. They approach the two intercepting curves for $F_1 = F_2$ when $\varepsilon$ approaches zero. Therefore, in stretching a square sheet by equal forces, it is clear that beyond the bifurcation point B, any tiny imbalance of forces will cause the sheet to change into a rectangular shape with the long side along the direction of the slightly larger force, i.e., $\lambda_1 > \lambda_2$ for $F_1 > F_2$. This is what would happen most probably because it is extremely difficult to keep the forces "exactly" equal by any practical means. Therefore, we conclude that beyond the bifurcation point, the symmetric solution is unstable.

### 6.4.1 Stability of a Square Sheet

The stability of a square sheet under symmetric biaxial stretching may also be analyzed via a thermodynamic criterion following the procedure discussed in Sect. 5.5.

Consider a body in a region $\mathcal{V}$ at a uniform constant temperature and free of external supplies, then we have from the energy equation (2.68),

$$\frac{d}{dt}\int_{\mathcal{V}} \rho(\varepsilon + \frac{1}{2}\dot{\boldsymbol{x}}\cdot\dot{\boldsymbol{x}})\,dv + \int_{\partial\mathcal{V}} (\boldsymbol{q} - T\dot{\boldsymbol{x}})\cdot\boldsymbol{n}\,da = 0, \tag{6.26}$$

and the entropy inequality (5.4),

$$\frac{d}{dt}\int_{\mathcal{V}} \rho\eta\,dv + \frac{1}{\theta}\int_{\partial\mathcal{V}} \boldsymbol{q}\cdot\boldsymbol{n}\,da \geq 0. \tag{6.27}$$

By eliminating the heat flux, we obtain

$$\frac{d}{dt}\int_{\mathcal{V}} \rho(\psi + \frac{1}{2}\dot{\boldsymbol{x}}\cdot\dot{\boldsymbol{x}})\,dv - \int_{\partial\mathcal{V}} \dot{\boldsymbol{x}}\cdot T\boldsymbol{n}\,da \leq 0, \tag{6.28}$$

where $\psi = \varepsilon - \theta\eta$ is the free energy. If the region occupied by the body in the reference state is denoted by $\mathcal{V}_\kappa$ then the above condition can be written in the reference state as

$$\frac{d}{dt}\int_{\mathcal{V}_\kappa} \rho(\psi + \frac{1}{2}\dot{\boldsymbol{x}}\cdot\dot{\boldsymbol{x}})J\,dv - \int_{\partial\mathcal{V}_\kappa} \dot{\boldsymbol{x}}\cdot T_\kappa \boldsymbol{n}_\kappa\,da \leq 0. \tag{6.29}$$

Let the region $\mathcal{V}_\kappa$ occupied by the square sheet in the reference state be given by $0 \leq X \leq 1$, $0 \leq Y \leq 1$, and $-D \leq Z \leq D$. We have the following boundary conditions:

$$T_\kappa \boldsymbol{n}_\kappa\Big|_{X=1} = T_\kappa\langle xx\rangle\,\boldsymbol{e}_x, \qquad T_\kappa \boldsymbol{n}_\kappa\Big|_{Y=1} = T_\kappa\langle yy\rangle\,\boldsymbol{e}_y, \qquad T_\kappa \boldsymbol{n}_\kappa\Big|_{Z=\pm D} = 0,$$

and from (6.20),

$$\dot{\boldsymbol{x}}\Big|_{X=0} = 0, \qquad \dot{\boldsymbol{x}}\Big|_{X=1} = \dot{\lambda}_1\boldsymbol{e}_x,$$
$$\dot{\boldsymbol{x}}\Big|_{Y=0} = 0, \qquad \dot{\boldsymbol{x}}\Big|_{Y=1} = \dot{\lambda}_2\boldsymbol{e}_y,$$

for which the biaxial stretching is regarded as a continuous process, while $F_1 = 2DT_\kappa\langle xx\rangle$ and $F_2 = 2DT_\kappa\langle yy\rangle$ are the total forces prescribed at the four lateral surfaces (see Fig. 6.1). Therefore, it follows that

$$\int_{\partial\mathcal{V}_\kappa} T_\kappa\dot{\boldsymbol{x}}\cdot\boldsymbol{n}_\kappa\,da = \int_{A_1} T_\kappa\langle xx\rangle\dot{\lambda}_1\,da + \int_{A_2} T_\kappa\langle yy\rangle\dot{\lambda}_2\,da,$$

where $A_1$ and $A_2$ are the lateral surfaces of the region at $X=1$ and $Y=1$, respectively. Therefore, assuming the process is quasi-static (with negligible acceleration), then from (6.29) we have

$$\begin{aligned}
&\frac{d}{dt}\int_{\mathcal{V}_\kappa}\rho\,\psi\,dv - \int_{A_1} T_\kappa\langle xx\rangle\dot{\lambda}_1\,da - \int_{A_2} T_\kappa\langle yy\rangle\dot{\lambda}_2\,da\\
&\quad= \frac{d}{dt}\int_{\mathcal{V}_\kappa}\rho\,\psi\,dv - \int_{A_1}\frac{F_1}{2D}\dot{\lambda}_1\,da - \int_{A_2}\frac{F_2}{2D}\dot{\lambda}_2\,da\\
&\quad= \frac{d}{dt}\Big(\int_{\mathcal{V}_\kappa}\rho\,\psi\,dv - \int_{A_1}\frac{1}{2D}F_1\lambda_1\,da - \int_{A_2}\frac{1}{2D}F_2\lambda_2\,da\Big) \leq 0,
\end{aligned}$$

which, upon integration gives

$$\frac{d}{dt}\Big(\rho\psi\,2D - F_1\lambda_1 - F_2\lambda_2\Big) \leq 0,$$

since the deformation is homogeneous. Therefore, we can define the availability function of the square sheet as

$$\mathcal{A} = \rho\psi - \tilde{F}_1\lambda_1 - \tilde{F}_2\lambda_2, \tag{6.30}$$

where $\tilde{F}_1 = F_1/2D$ e $\tilde{F}_2 = F_2/2D$, so that the last relation becomes

$$\frac{d\mathcal{A}}{dt} \leq 0. \tag{6.31}$$

This allows us to establish a stability criterion for the deformation of the square sheet. The availability $\mathcal{A}(t)$ is a decreasing function of time. Therefore, if we assume the deformed state characterized by the stretches $(\bar{\lambda}_1, \bar{\lambda}_2)$ is a stable equilibrium state, then any small perturbation from this state will eventually return to this state as time tends to infinity. Suppose that such a perturbation is represented by a process $(\lambda_1(t), \lambda_2(t))$, it follows that

$$\lim_{t\to\infty} \lambda_1(t) = \bar{\lambda}_1, \qquad \lim_{t\to\infty} \lambda_2(t) = \bar{\lambda}_2,$$

and hence by the condition (6.31) the availability $\mathcal{A}(t) = \mathcal{A}(\lambda_1(t). \lambda_2(t))$ will attain its minimum at $(\bar{\lambda}_1, \bar{\lambda}_2)$. This criterion is equivalent to the following conditions:

$$\left.\frac{\partial \mathcal{A}}{\partial \lambda_1}\right|_E = 0, \qquad \left.\frac{\partial \mathcal{A}}{\partial \lambda_2}\right|_E = 0, \tag{6.32}$$

and the matrix

$$\begin{bmatrix} \left.\frac{\partial^2 \mathcal{A}}{\partial \lambda_1^2}\right|_E & \left.\frac{\partial^2 \mathcal{A}}{\partial \lambda_1 \partial \lambda_2}\right|_E \\ \left.\frac{\partial^2 \mathcal{A}}{\partial \lambda_1 \partial \lambda_2}\right|_E & \left.\frac{\partial^2 \mathcal{A}}{\partial \lambda_2^2}\right|_E \end{bmatrix}$$

is positive semi-definite or, equivalently,

$$\left.\frac{\partial^2 \mathcal{A}}{\partial \lambda_1^2}\right|_E \geq 0, \qquad \left.\frac{\partial^2 \mathcal{A}}{\partial \lambda_1^2}\right|_E \left.\frac{\partial^2 \mathcal{A}}{\partial \lambda_2^2}\right|_E - \left.\frac{\partial^2 \mathcal{A}}{\partial \lambda_1 \partial \lambda_2}\right|_E^2 \geq 0, \tag{6.33}$$

where $E$ denotes the evaluation at the equilibrium state $(\bar{\lambda}_1, \bar{\lambda}_2)$.

Now we have the free energy $\psi = \psi(I_B, II_B)$, where from (6.20)

$$I_B = \lambda_1^2 + \lambda_2^2 + \frac{1}{\lambda_1^2 \lambda_2^2}, \qquad II_B = \frac{1}{\lambda_1^2} + \frac{1}{\lambda_2^2} + \lambda_1^2 \lambda_2^2.$$

By (6.5) and (6.30), the conditions (6.32) become

$$\begin{aligned} \tilde{F}_1 &= (s_1 - s_{-1}\lambda_2^2)\Big(\lambda_1 - \frac{1}{\lambda_1^3 \lambda_2^2}\Big), \\ \tilde{F}_2 &= (s_1 - s_{-1}\lambda_1^2)\Big(\lambda_2 - \frac{1}{\lambda_1^2 \lambda_2^3}\Big), \end{aligned} \tag{6.34}$$

where the equations are evaluated at the equilibrium state and the overhead bars are suppressed for simplicity. Note that they are identical to the relations

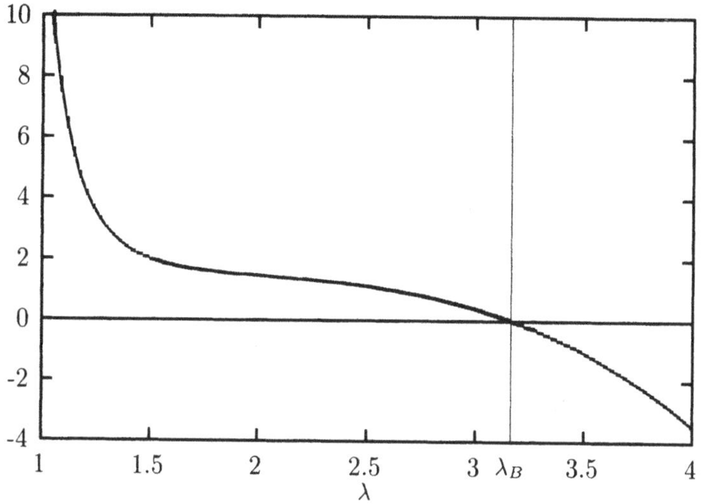


**Fig. 6.3.** $f(\lambda, \lambda)$ symmetric solution

(6.23). Meanwhile the condition (6.33) leads to

$$1 + \frac{3}{\lambda_1^4 \lambda_2^2} + h\Big(\frac{1}{\lambda_1^4} + \lambda_2^2\Big) \geq 0, \tag{6.35}$$

which is identically satisfied since $h = -s_{-1}/s_1 > 0$, and

$$\Big(1 + \frac{3}{\lambda_1^4 \lambda_2^2} + h\big(\frac{3}{\lambda_1^4} + \lambda_2^2\big)\Big)\Big(1 + \frac{3}{\lambda_2^4 \lambda_1^2} + h\big(\frac{3}{\lambda_2^4} + \lambda_1^2\big)\Big) \\ - \Big(\frac{2}{\lambda_1^3 \lambda_2^3} + 2h\lambda_1\lambda_2\Big)^2 \geq 0. \tag{6.36}$$

Let the left-hand side of the relation (6.36) be denoted by $f(\lambda_1, \lambda_2)$, then we have

$$f(\lambda_1, \lambda_2) \geq 0, \tag{6.37}$$

which is the condition for an equilibrium state $(\lambda_1, \lambda_2)$ to be stable.

For symmetric loading, putting $\tilde{F}_1 = \tilde{F}_2$, from the equations (6.34) we recover the relation (6.24), which gives the symmetric solution $\lambda_1 = \lambda_2$ and the asymmetric solution $\lambda_1 \neq \lambda_2$ given by (6.25).

We have plotted the function $f(\lambda, \lambda)$ against $\lambda$ for the symmetric solution in Fig. 6.3 for $h = 0.1$. This shows that the function $f(\lambda, \lambda)$ is positive for $\lambda < \lambda_B = 3.1685$, which corresponds to the bifurcation point in Fig. 6.2(a), and therefore, according to the condition (6.37), the symmetric solution is stable. However, beyond the value $\lambda_B$, the function $f(\lambda, \lambda)$ becomes negative and hence the square sheet is no longer stable.

For an asymmetric solution $\lambda_1 \neq \lambda_2$, from the condition (6.25) one can solve for $\lambda_2$ in terms of $\lambda_1$ so that $\lambda_2 = g(\lambda_1)$ and hence $f(\lambda_1, g(\lambda_1))$ becomes

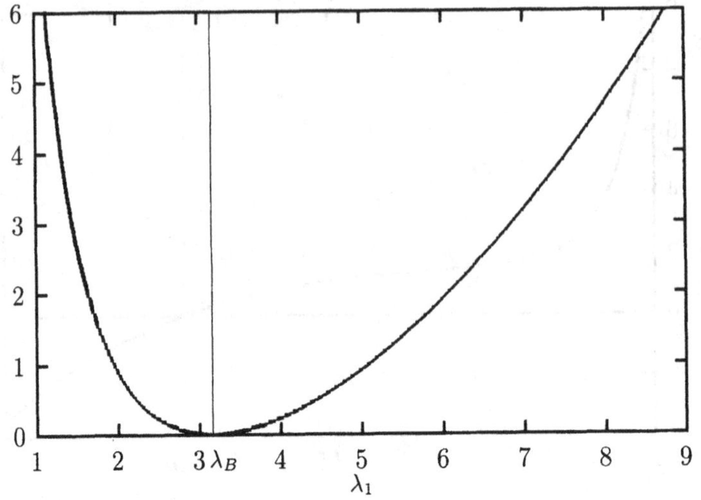


**Fig. 6.4.** $f(\lambda_1, \lambda_2)$ – asymmetric solution

a function of $\lambda_1$ only. Doing this numerically, we can easily verify the condition (6.37) by plotting the function $f(\lambda_1, \lambda_2)$ against $\lambda_1$, as shown in Fig. 6.4, from which we conclude that the asymmetric solution is always stable. Hence, a square sheet turns into a rectangular one when symmetric loading gradually increases beyond the bifurcation point.[3]

## 6.5 Simple Shear

We consider an isotropic elastic body subject to a deformation of simple shear given by (see Example 1.2.1, Fig. 1.3)

$$x = X + \kappa Y, \quad y = Y, \quad z = Z,$$

where the amount of shear $\kappa$ is a constant, we have the left Cauchy–Green tensor

$$[B_{\langle ij\rangle}] = \begin{bmatrix} 1+\kappa^2 & \kappa & 0 \\ \kappa & 1 & 0 \\ 0 & 0 & 1 \end{bmatrix}, \tag{6.38}$$

and its inverse

$$[B_{\langle ij\rangle}]^{-1} = \begin{bmatrix} 1 & -\kappa & 0 \\ -\kappa & 1+\kappa^2 & 0 \\ 0 & 0 & 1 \end{bmatrix},$$

[3] See also the stability analysis in [32, 56] and more discussions in [18].

so that its principal invariants are given by

$$I_B = 3 + \kappa^2, \quad II_B = 3 + \kappa^2, \quad III_B = 1.$$

Since the simple shear is a homogeneous deformation, it satisfies the equilibrium equation with no body force. The stress tensor is a constant tensor given by

$$\begin{aligned}[T_{\langle ij\rangle}] =& (s_0 + s_1 + s_{-1}) \begin{bmatrix} 1 & 0 & 0 \\ 0 & 1 & 0 \\ 0 & 0 & 1 \end{bmatrix} + (s_1 - s_{-1})\kappa \begin{bmatrix} 0 & 1 & 0 \\ 1 & 0 & 0 \\ 0 & 0 & 0 \end{bmatrix} \\ &+ s_1\kappa^2 \begin{bmatrix} 1 & 0 & 0 \\ 0 & 0 & 0 \\ 0 & 0 & 0 \end{bmatrix} + s_{-1}\kappa^2 \begin{bmatrix} 0 & 0 & 0 \\ 0 & 1 & 0 \\ 0 & 0 & 0 \end{bmatrix},\end{aligned} \tag{6.39}$$

where $s_i = s_i(3+\kappa^2, 3+\kappa^2, 1) = s_i(\kappa^2)$. Note that simple shear is a volume-preserving deformation and the above results are also valid for an incompressible elastic body, provided that the material parameter $s_0$ is replaced by $-p$, the undeterminate pressure.

The shear stress $T_{\langle xy\rangle}$ on the surface, $Y = Y_0$, has the value

$$T_{\langle xy\rangle} = \hat{\mu}\kappa, \tag{6.40}$$

where

$$\hat{\mu}(\kappa^2) = s_1(\kappa^2) - s_{-1}(\kappa^2)$$

is called the *shear modulus* of the material. For a small $\kappa$, then

$$\hat{\mu}(\kappa^2) = \mu + o(\kappa^2),$$

where $\mu = \hat{\mu}(0)$ is the classical shear modulus. Therefore, any discrepancy from the classical result for the shear stress is at least of third-order in the amount of shear $\kappa$. Moreover, unlike elastic fluids, for which $\hat{\mu}(\kappa^2) = 0$ for any $\kappa$, elastic solids with this trivial property will be ignored in our discussion. Therefore, by (6.40) the shear stress $T_{\langle xy\rangle}$ does not vanish if $\kappa \neq 0$.

For simple shear, from (6.11) and (6.38) we have the following universal relation,

$$T_{\langle xx\rangle} - T_{\langle yy\rangle} = \kappa T_{\langle xy\rangle}, \tag{6.41}$$

which can be checked immediately from (6.39). If we denote the normal stress on the slanted surface of the block, corresponding to the plane in the reference state $X = X_0$, by $N$ (see Fig. 6.5), the universal relation (6.41) can be rewritten as (see Exercise 6.5.1)

$$T_{\langle yy\rangle} - N = \frac{\kappa}{1+\kappa^2} T_{\langle xy\rangle}. \tag{6.42}$$

From this expression, it is clear that in order to effect a simple shear on a rectangular block, besides shear stresses, normal stresses must also be applied on the surfaces of the block, since from the above relation the two normal stresses can not vanish simultaneously or even be equal to each other unless there is no shear at all. Moreover, the normal stress difference is a second-order effect in the amount of shear according to (6.40) and (6.42). Therefore, the normal stress difference is more significant than the discrepancy in the shear stress as an indication for the departure from the classical theory. The existence of a normal stress difference is usually known as the *Poynting effect* or simply as the *normal stress effect.*

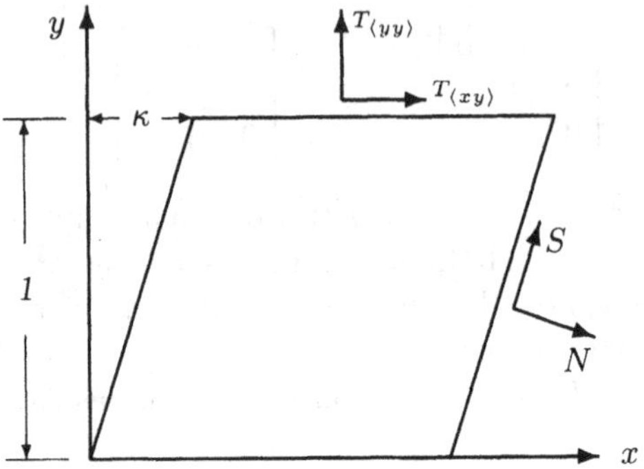


**Fig. 6.5.** Simple shear

For compressible bodies, there is another way of looking at the impossibility of having a simple shear deformation by shear stress alone. From (6.39) the mean hydrostatic pressure is given by

$$-\frac{1}{3}\operatorname{tr}T = -(s_0 + s_1 + s_{-1}) - \frac{1}{3}(s_1 + s_{-1})\kappa^2.$$

If we assume that the reference configuration is a natural state, namely, no stress when $\kappa = 0$, then from (6.39)

$$s_0(0) + s_1(0) + s_{-1}(0) = 0,$$

and hence the mean hydrostatic pressure is of second-order in the amount of shear. Therefore a non-vanishing hydrostatic pressure (or tension according to the sign) is needed to effect a simple shear in a compressible isotropic elastic body. This is called the *Kelvin effect.*

**Exercise 6.5.1** Let the normal stress and shear stress on the slanted surface, corresponding to the surface in the reference state $X = X_0$, of the block be denoted by $N$ and $S$, respectively (see Fig. 6.5), show that

$$N = T_{\langle yy\rangle} - \frac{\kappa}{1+\kappa^2}T_{\langle xy\rangle},$$

$$S = \frac{1}{1+\kappa^2}T_{\langle xy\rangle}.$$

For infinitesimal $\kappa$, if the second-order terms in $\kappa$ are neglected, it follows that $N \simeq T_{\langle yy\rangle}$ and $S \simeq T_{\langle xy\rangle}$.

**Exercise 6.5.2** Consider a torsion and extension of a cylinder given by the deformation function (1.16),

$$r = \sqrt{a}R, \quad \theta = \Theta + \tau Z, \quad z = \frac{1}{a}Z.$$

1) Show that it is a controllable universal solution for incompressible isotropic materials.
2) Determine the stress fields of a hollow cylinder free of stress at the outer surface for Mooney–Rivlin materials.
3) Discuss the presence of the normal stress effect.

## 6.6 Pure Shear of a Square Block

We have noticed that by applying shear stresses alone on the surface of a rectangular block, the body will tend to contract or expand if normal stresses were not supplied properly. To examine such changes quantitatively, we consider a deformation that consists of a homogeneous stretch followed by a simple shear,

$$x = \lambda_1 X + \kappa\lambda_2 Y, \quad y = \lambda_2 Y, \quad z = \lambda_3 Z.$$

Since this is a homogeneous deformation, it is a controllable universal solution for an elastic body. The deformation gradient relative to the Cartesian coordinate system is given by

$$[F_{\langle i\alpha\rangle}] = \begin{bmatrix} \lambda_1 & \kappa\lambda_2 & 0 \\ 0 & \lambda_2 & 0 \\ 0 & 0 & \lambda_3 \end{bmatrix}. \tag{6.43}$$

The left Cauchy–Green tensor is given by

$$[B_{\langle ij\rangle}] = \begin{bmatrix} \lambda_1^2 + \kappa^2\lambda_2^2 & \kappa\lambda_2^2 & 0 \\ \kappa\lambda_2^2 & \lambda_2^2 & 0 \\ 0 & 0 & \lambda_3^2 \end{bmatrix}, \tag{6.44}$$

and its inverse

$$[(B^{-1})_{\langle ij\rangle}] = \begin{bmatrix} \dfrac{1}{\lambda_1^2} & -\dfrac{\kappa}{\lambda_1^2} & 0 \\ -\dfrac{\kappa}{\lambda_1^2} & \dfrac{1}{\lambda_2^2} + \dfrac{\kappa^2}{\lambda_1^2} & 0 \\ 0 & 0 & \dfrac{1}{\lambda_3^2} \end{bmatrix}.$$

For isotropic elastic body, the stress tensor $T_{\langle ij\rangle}$ can be calculated from (6.2) for a compressible body or from (6.4) for an incompressible body. In particular, the shear stress on the surface, $Y = Y_0$, is given by

$$T_{\langle xy\rangle} = \kappa\Big(s_1\lambda_2^2 - s_{-1}\frac{1}{\lambda_1^2}\Big), \tag{6.45}$$

where $s_i = s_i(I_B, II_B, III_B)$. This holds for either compressible or incompressible ($III_B = 1$) bodies. Moreover, from (6.44), the universal relation (6.11) takes the following form,

$$T_{\langle xx\rangle} - T_{\langle yy\rangle} = \frac{\lambda_1^2 - \lambda_2^2 + \kappa^2\lambda_2^2}{\kappa\lambda_2^2} T_{\langle xy\rangle}. \tag{6.46}$$

Unlike the case of simple shear discussed in the previous section, for a fixed $\kappa$, it is now possible to determine the three constants $\lambda_1$, $\lambda_2$, and $\lambda_3$ in such a way that three normal stresses vanish on the surface of the block. In the case of an incompressible body, the three conditions for vanishing normal stresses can be used to determine $\lambda_1$, $\lambda_2$ and the pressure $p$, while the condition of incompressibility, $\lambda_1\lambda_2\lambda_3 = 1$, determines $\lambda_3$.

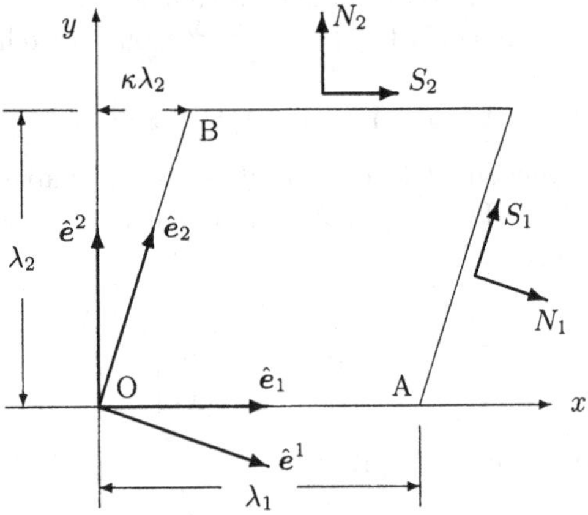


**Fig. 6.6.** Pure shear

We consider a square block, with sides of unit length, and introduce a new basis in the deformed configuration (see Fig. 6.6) defined by

$$\hat{\mathbf{e}}_1 = \mathbf{e}_x, \quad \hat{\mathbf{e}}_2 = \kappa\mathbf{e}_x + \mathbf{e}_y, \quad \hat{\mathbf{e}}_3 = \mathbf{e}_z,$$

and its dual basis

$$\hat{\mathbf{e}}^1 = \mathbf{e}_x - \kappa\mathbf{e}_y, \quad \hat{\mathbf{e}}^2 = \mathbf{e}_y, \quad \hat{\mathbf{e}}^3 = \mathbf{e}_z.$$

Note that $|\hat{e}^1| = |\hat{e}_2| = \sqrt{1+\kappa^2}$ and the base vectors $\hat{e}^1$ and $\hat{e}^2$ are normal to the deformed surfaces of $X = X_0$ and $Y = Y_0$, respectively. Therefore, normal stresses $N_1$ and $N_2$ on these surfaces are given by

$$\begin{aligned} N_1 &= \frac{\hat{e}^1}{|\hat{e}^1|} \cdot T\frac{\hat{e}^1}{|\hat{e}^1|} = \frac{\widehat{T}^{11}}{1+\kappa^2}, \\ N_2 &= \hat{e}^2 \cdot T\hat{e}^2 = \widehat{T}^{22}, \qquad N_3 = \hat{e}^3 \cdot T\hat{e}^3 = \widehat{T}^{33}, \end{aligned} \tag{6.47}$$

where we have denoted the contravariant components of $T$ relative to the new basis by $\widehat{T}^{ij} = \hat{e}^i \cdot T\hat{e}^j$ according to (A.5). The components $\widehat{T}^{ij}$ can be obtained from the Cartesian components $T_{\langle mn\rangle}$ by the following transformation rule (see (A.14)),

$$[\widehat{T}^{ij}] = [M_m{}^i]^{-T}[T_{\langle mn\rangle}][M_n{}^j]^{-1}, \tag{6.48}$$

where $[M_k{}^i]$ is the transformation matrix of change of basis from $\{e_i\}$ to $\{\hat{e}_k\}$, from (A.11) it is given by $M_k{}^i = \hat{e}_k \cdot e^i$ or

$$[M_k{}^i] = \begin{bmatrix} 1 & 0 & 0 \\ \kappa & 1 & 0 \\ 0 & 0 & 1 \end{bmatrix}.$$

We obtain

$$[\widehat{T}^{ij}] = \begin{bmatrix} T_{\langle xx\rangle} - 2\kappa T_{\langle xy\rangle} + \kappa^2 T_{\langle yy\rangle} & T_{\langle xy\rangle} - \kappa T_{\langle yy\rangle} & 0 \\ T_{\langle xy\rangle} - \kappa T_{\langle yy\rangle} & T_{\langle yy\rangle} & 0 \\ 0 & 0 & T_{\langle zz\rangle} \end{bmatrix}. \tag{6.49}$$

From (6.47) the vanishing of normal stresses $N_1 = N_2 = N_3 = 0$ would require $T_{\langle yy\rangle} = T_{\langle zz\rangle} = 0$ and $T_{\langle xx\rangle} - 2\kappa T_{\langle xy\rangle} = 0$. The last condition combined with the relation (6.46) leads to[4]

$$\lambda_1^2 = (1+\kappa^2)\lambda_2^2. \tag{6.50}$$

It is interesting to point out that this relation is also a universal relation for elastic bodies and it admits a very simple geometric interpretation, namely, $\overline{OA} = \overline{OB}$, as shown in Fig. 6.6. Furthermore, from (6.49) the stress tensor relative to the product basis $\{\hat{e}_i \otimes \hat{e}_j\}$ takes the following simple form,

$$T = \tau(\hat{e}_1 \otimes \hat{e}_2 + \hat{e}_2 \otimes \hat{e}_1),$$

where from (6.45) and (6.50), the shear stress $\tau$ is given by

$$\tau = T_{\langle xy\rangle} = \kappa\Big(s_1 \frac{\lambda_1^2}{1+\kappa^2} - s_{-1}\frac{1}{\lambda_1^2}\Big). \tag{6.51}$$

[4] See also the derivation in [61] and more discussion on pure shear in [48].

Besides the vanishing of normal stresses, the shear stresses on the surfaces, $X = X_0$ and $Y = Y_0$, are equal, i.e., $S_1 = S_2 = \tau$. Such a state of stress is called a *pure shear.* Thus we have seen that to effect a state of pure shear on a square block, it is only necessary to apply equal shear stresses on the four surfaces. The amount of shear $\kappa$ and the stretches $\lambda_1$ and $\lambda_2$ are adjusted in such a way that the length of the four sides remains the same. A square block becoming a rhombic block is also what one would expect intuitively in a pure shear.

To determine the stretches for a given amount of $\kappa$, explicit constitutive expressions would be needed. As an example, for a neo-Hookean material given by (6.9), we have the contravariant components of the stress

$$\widehat{T}^{ij} = -p\,\hat{g}^{ij} + s_1 \widehat{B}^{ij},$$

where the metric tensor $\hat{g}^{ij} = \hat{\mathbf{e}}^i \cdot \hat{\mathbf{e}}^j$ is given by

$$[\hat{g}^{ij}] = \begin{bmatrix} 1+\kappa^2 & -\kappa & 0 \\ -\kappa & 1 & 0 \\ 0 & 0 & 1 \end{bmatrix},$$

and the contravariant components of $B$ can be calculated similar to (6.48),

$$[\widehat{B}^{ij}] = \begin{bmatrix} \lambda_1^2 & 0 & 0 \\ 0 & \lambda_2^2 & 0 \\ 0 & 0 & \lambda_3^2 \end{bmatrix}.$$

Therefore, by putting $\widehat{T}^{11} = \widehat{T}^{22} = \widehat{T}^{33} = 0$ and by the incompressibility condition $\lambda_1 \lambda_2 \lambda_3 = 1$, we obtain

$$\begin{aligned} p &= s_1 (1+\kappa^2)^{-1/3}, \\ \lambda_1 &= (1+\kappa^2)^{1/3}, \\ \lambda_2 &= \lambda_3 = (1+\kappa^2)^{-1/6}. \end{aligned}$$

Moreover, from (6.51) the shear stress $\tau$ has the value,

$$\tau = s_1 \kappa (1+\kappa^2)^{-1/3}.$$

Note that the value of $\lambda_1$ is greater than 1, while that of $\lambda_2$ is less than 1. Therefore the block lengthens in the $x$-direction and shortens in the $y$- and $z$-directions. This is a quantitative description of the Poynting effect mentioned in the case of simple shear.

**Exercise 6.6.1** The relation (6.10) can be written in component form as

$$T^{ij} g_{jk} B^{kl} = B^{ij} g_{jk} T^{kl}.$$

Show that the universal relation (6.50) follows directly from this equation relative to the basis $\{\hat{\mathbf{e}}_i\}$ and the conditions $\widehat{T}^{11} = \widehat{T}^{22} = 0$.

**Exercise 6.6.2** Show that the right stretch tensor $U$ and the rotation tensor $R$ of the deformation gradient $F$ in (6.43) with the condition (6.50) are given by

$$[U_{\langle\beta\alpha\rangle}] = \begin{bmatrix} \lambda_1\cos\theta & \lambda_1\sin\theta & 0 \\ \lambda_1\sin\theta & \lambda_1\cos\theta & 0 \\ 0 & 0 & \lambda_3 \end{bmatrix}, \quad [R_{\langle i\beta\rangle}] = \begin{bmatrix} \cos\theta & \sin\theta & 0 \\ -\sin\theta & \cos\theta & 0 \\ 0 & 0 & 1 \end{bmatrix},$$

where $\theta$ is half the angle between the base vectors $\hat{\mathbf{e}}^2$ and $\hat{\mathbf{e}}_2$ (see Fig. 6.6) or

$$\tan 2\theta = \kappa.$$

Verify that the vectors $(1,1,0)$ and $(-1,1,0)$ are eigenvectors of $U$ and hence, this deformation is a biaxial stretch along two diagonals of the undeformed square (such a deformation is often called a "pure shear" in the literature) followed by a rotation that brings the base back to the horizontal position. Note the difference between this and the results for the simple shear in Example 1.2.1.

## 6.7 Finite Deformation of Spherical Shells

We shall consider some problems of spherical shells. In terms of spherical coordinates $(R,\Theta,\Phi)$ and $(r,\theta,\phi)$ in the reference and the deformed configurations, given a deformation of the form,

$$r = r(R), \quad \theta = \Theta, \quad \phi = c\Phi, \tag{6.52}$$

where $c$ is a constant, the matrix of the deformation gradient is

$$[F^i{}_\alpha] = \begin{bmatrix} r' & 0 & 0 \\ 0 & 1 & 0 \\ 0 & 0 & c \end{bmatrix},$$

which in terms of physical components becomes

$$[F_{\langle i\alpha\rangle}] = \begin{bmatrix} r' & 0 & 0 \\ 0 & \dfrac{r}{R} & 0 \\ 0 & 0 & c\dfrac{r\sin\theta}{R\sin\Theta} \end{bmatrix} = \begin{bmatrix} r' & 0 & 0 \\ 0 & \dfrac{r}{R} & 0 \\ 0 & 0 & c\dfrac{r}{R} \end{bmatrix}.$$

The matrix of the left Cauchy–Green tensor is then

$$[B_{\langle ij\rangle}] = \operatorname{diag}\left[r'^2, \frac{r^2}{R^2}, c^2\frac{r^2}{R^2}\right],$$

and its inverse,

$$[B^{-1}_{\langle ij\rangle}] = \operatorname{diag}\left[\frac{1}{r'^2}, \frac{R^2}{r^2}, \frac{R^2}{c^2r^2}\right],$$

are functions of $r$ only.

For incompressible isotropic materials, $\det F = 1$, and shells for which $T_{\langle\theta\theta\rangle} = T_{\langle\phi\phi\rangle}$, we have from the above expressions and (6.4) that

$$c\frac{r^2}{R^2}r' = 1, \qquad B_{\langle\theta\theta\rangle} = B_{\langle\phi\phi\rangle}.$$

Hence we obtain

$$r(R) = (A + cR^3)^{1/3}, \qquad c = \pm 1. \tag{6.53}$$

where $A$ is an integration constant.

It is easy to verify that the deformation given by (6.52) and (6.53) is a controllable universal solution for incompressible isotropic materials. The equilibrium equation (6.19) in the spherical coordinate system $(r, \theta, \phi)$ now reads (see (A.83))

$$\begin{aligned}&\frac{\partial T_{\langle rr\rangle}}{\partial r} = -\frac{2}{r}(T_{\langle rr\rangle} - T_{\langle\theta\theta\rangle}),\\ &\frac{\partial p}{\partial \theta} = 0, \qquad \frac{\partial p}{\partial \phi} = 0.\end{aligned} \tag{6.54}$$

From this and (6.4) we obtain

$$\begin{aligned}T_{\langle rr\rangle} = f(r) &= 2\int\left\{s_1\left(\frac{r^2}{R^2} - \frac{R^4}{r^4}\right) + s_{-1}\left(\frac{R^2}{r^2} - \frac{r^4}{R^4}\right)\right\}\frac{dr}{r}\\ &= 2\int\left(\frac{r^2}{R^2} - \frac{R^4}{r^4}\right)\left(s_1 - s_{-1}\frac{r^2}{R^2}\right)\frac{dr}{r}.\end{aligned} \tag{6.55}$$

Hence we have

$$p = s_1\frac{R^4}{r^4} + s_{-1}\frac{r^4}{R^4} - f(r),$$

and

$$T_{\langle\theta\theta\rangle} = T_{\langle\phi\phi\rangle} = s_1\left(\frac{r^2}{R^2} - \frac{R^4}{r^4}\right) + s_{-1}\left(\frac{R^2}{r^2} - \frac{r^4}{R^4}\right) + f(r). \tag{6.56}$$

All the other stress components are zero. Moreover, the material parameters $s_1$ and $s_{-1}$ are functions of $r$,

$$s_1(r) = s_1(I_B, II_B), \qquad s_{-1}(r) = s_{-1}(I_B, II_B),$$

where

$$I_B = \frac{R^4}{r^4} + 2\frac{r^2}{R^2}, \qquad II_B = \frac{r^4}{R^4} + 2\frac{R^2}{r^2}. \tag{6.57}$$

For Mooney-Rivlin materials, with constant $s_1$ and $s_{-1}$, (6.55) can be integrated. Let $\xi = r/R$, then by (6.53)

$$d\xi = (\xi - c\xi^4)\frac{dr}{r}, \tag{6.58}$$

and hence, $f(r)$ becomes

$$\begin{aligned} f(r) &= 2\int(\xi^2 - \xi^{-4})(s_1 - s_{-1}\xi^2)(\xi - c\xi^4)^{-1}d\xi \\ &= -2\int(1 + c\xi^3)(s_1 - s_{-1}\xi^2)\xi^{-5}d\xi \\ &= s_1(\frac{1}{2}\xi^{-4} + 2c\xi^{-1}) - s_{-1}(\xi^{-2} - 2c\xi) + K. \end{aligned}$$

where $K$ is an integration constant. Therefore, the stress fields for Mooney-Rivlin materials are given by

$$\begin{aligned} T_{\langle rr\rangle} &= s_1\left(\frac{1}{2}\frac{R^4}{r^4} + 2c\frac{R}{r}\right) - s_{-1}\left(\frac{R^2}{r^2} - 2c\frac{r}{R}\right) + K, \\ T_{\langle\theta\theta\rangle} &= T_{\langle\phi\phi\rangle} = s_1\left(\frac{r^2}{R^2} - \frac{1}{2}\frac{R^4}{r^4} + 2c\frac{R}{r}\right) - s_{-1}\left(\frac{r^4}{R^4} - 2c\frac{r}{R}\right) + K. \end{aligned} \tag{6.59}$$

### 6.7.1 Eversion of a Spherical Shell

For the case $c = -1$, we have

$$r = (A - R^3)^{1/3}, \quad \theta = \Theta, \quad \phi = -\Phi.$$

This deformation is an eversion that turns the shell inside out. Since for $R_0 < R_1$ we have $A - R_0^3 > A - R_1^3$, and hence $r_0 = r(R_0) > r_1 = r(R_1)$, in other words, the inner surface becomes the outer surface after deformation and vice versa. Moreover, for this to be physically possible, it is necessary that $r_1$ be positive, or the constant $A$ must satisfy

$$A > R_1^3. \tag{6.60}$$

To effect such a deformation for a complete spherical shell, it would be necessary to make a cut somewhere first and then join together again after the eversion.

If we require that the shell be free of traction at the inner and the outer surfaces in the everted state, $T_{\langle rr\rangle}(r_1) = T_{\langle rr\rangle}(r_0) = 0$, then (6.55) implies

$$\int_{r_0}^{r_1}\left(\frac{r^2}{R^2} - \frac{R^4}{r^4}\right)\left(s_1 - s_{-1}\frac{r^2}{R^2}\right)\frac{dr}{r} = 0, \tag{6.61}$$

where the material parameters $s_1$ and $s_{-1}$ are functions of $R = (A - r^3)^{1/3}$ and $r$.

The constant $A$ is to be determined as a solution of the above equation for a traction-free everted state. We can show that such a solution exists for any incompressible isotropic materials if the E-inequalities (6.8) hold and the shell is not too thick. Indeed, by the mean value theorem for definite integrals, we have

$$\left(\frac{\bar{r}^2}{R(\bar{r})^2} - \frac{R(\bar{r})^4}{\bar{r}^4}\right)\left(s_1(\bar{r}) - s_{-1}(\bar{r})\frac{\bar{r}^2}{R(\bar{r})^2}\right)\frac{r_1 - r_0}{\bar{r}} = 0$$

from (6.61), for some value $\bar{r}$ such that $r_0 \geq \bar{r} \geq r_1$. Since $s_1 > 0$ and $s_{-1} \leq 0$, it implies

$$\frac{\bar{r}^2}{R(\bar{r})^2} - \frac{R(\bar{r})^4}{\bar{r}^4} = 0.$$

The only real positive root of this equation is $R(\bar{r}) = \bar{r}$, or $(A - \bar{r}^3)^{1/3} = \bar{r}$, yielding

$$A = 2\bar{r}^3,$$

and hence $A$ must satisfy

$$2r_0^3 \geq A \geq 2r_1^3,$$

or equivalently,

$$2R_0^3 \leq A \leq 2R_1^3.$$

Since, from (6.60), the constant $A$ must also be greater than $R_1^3$, we conclude that the shell can always be everted if $R_1^3 < 2R_0^3$. In other words, if the thickness $(R_1 - R_0)$ is less than $(\sqrt[3]{2} - 1)$ times (or about 26% of) the inner radius $R_0$, it is possible to turn the shell inside out freely for any incompressible isotropic elastic materials. It has been proved by Ericksen [17] that the eversion is possible for Mooney–Rivlin materials.

We have seen here another example of non-uniqueness of a traction boundary value problem, that for a spherical shell with null traction, there are at least two possible solutions, namely, the undeformed state with zero stresses and the everted equilibrium state, with non-vanishing stresses in the interior.

### 6.7.2 Inflation of a Spherical Shell

Consider the case $c = 1$,

$$r = (A + R^3)^{1/3}, \quad \theta = \Theta, \quad \phi = \Phi. \tag{6.62}$$

For a shell of radius $R_0 \leq R \leq R_1$, the inner and the outer surfaces must be maintained by suitable pressures, i.e.,

$$T_{\langle rr \rangle}(r_0) = -p_0, \qquad T_{\langle rr \rangle}(r_1) = -p_1, \tag{6.63}$$

where $r_0 = r(R_0)$ and $r_1 = r(R_1)$ are the inner and the outer radii of the shell in the deformed configuration. Let $[\![p]\!] = p_0 - p_1$ denote the pressure difference between the inner and the outer surfaces, then

$$[\![p]\!] = T_{\langle rr \rangle}(r_1) - T_{\langle rr \rangle}(r_0). \tag{6.64}$$

We define the thickness parameter $a$ and the expansion ratio $\lambda$ by

$$a = \frac{R_1}{R_0} \quad \text{and} \quad \lambda = \frac{r_1}{R_1}, \tag{6.65}$$

respectively, with $a > 1$ and $\lambda > 1$. We have

$$r_1 = a\lambda R_0, \qquad r_0 = \left(1 + a^3(\lambda^3 - 1)\right)^{1/3} R_0, \tag{6.66}$$

where the second relation is obtained by the elimination of the constant $A$ from $(6.62)_1$. Therefore, for a given thickness parameter $a$, the pressure difference $[\![p]\!]$ is a function of the expansion ratio $\lambda$. For Mooney–Rivlin materials, this function can be written out explicitly from (6.59). Such pressure–radius relations are shown in Fig. 6.7, for a typical value $s_{-1}/s_1 = -0.1$ and several values of $a$.

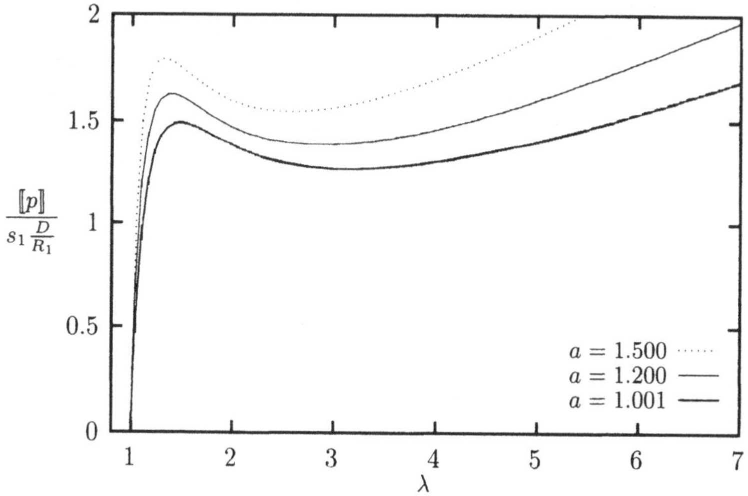


**Fig. 6.7.** Pressure–radius relation

We notice that the pressure–radius curves for a thick shell, $a = 1.5$, a thickness, $D = R_1 - R_0$, half the value of the inner radius, and for a thin shell, $a = 1.001$, a thickness one thousandth of that of the inner radius, are characteristically similar. After a steep rise, the curve reaches a local maximum and then decreases to a local minimum before it slowly rises again. Such a non-monotone function may usually lead to certain unstable behavior, which we shall consider in the next section.

For thin shells, the expression (6.64) for the pressure difference can be approximated by the Taylor expansion

$$[\![p]\!] = T_{\langle rr\rangle}(r_1) - T_{\langle rr\rangle}(r_0) = \frac{\partial T_{\langle rr\rangle}}{\partial r}\bigg|_{r_0} d + o(d^2),$$

where $d = r_1 - r_0$. By using $(6.54)_1$ through (6.56), and by neglecting the higher-order terms, this last equation takes the form

$$\begin{aligned} [\![p]\!] &= \frac{2}{r} d\,(T_{\langle\theta\theta\rangle} - T_{\langle rr\rangle}) \\ &= \frac{2}{r} d\Big(\frac{r^2}{R^2} - \frac{R^4}{r^4}\Big)\Big(s_1 - s_{-1}\frac{r^2}{R^2}\Big), \end{aligned}$$

which can also be written as

$$[\![p]\!] = 2\frac{D}{R}\Big(\frac{R}{r} - \frac{R^7}{r^7}\Big)\Big(s_1 - s_{-1}\frac{r^2}{R^2}\Big). \tag{6.67}$$

In the above expression, since the shell is thin, we have dropped the subindex $_0$ in both $r$ and $R$ for simplicity, and employed the relation $r^2 d \simeq R^2 D$ for incompressibility. Note that the derivation of the pressure-radius relation (6.67) for a thin spherical shell is valid for any incompressible isotropic material. This relation has been used as the constitutive equation for spherical rubber balloons (for a different derivation see [47, 55]) and its curve for Mooney–Rivlin materials is almost identical to that shown in Fig. 6.7 for $a = 1.001$.

Since $T_{\langle\theta\theta\rangle} = T_{\langle\phi\phi\rangle}$ the tangential stress in the shell is a pure tension. Let the surface tension (per unit length in the inflated state) be denoted by $\sigma$, then from the classical formula,

$$[\![p]\!] = 2\frac{\sigma}{r},$$

by comparison with the relation (6.67), the surface tension for a spherical rubber balloon can be expressed as

$$\sigma = D\Big(1 - \frac{R^6}{r^6}\Big)\Big(s_1 - s_{-1}\frac{r^2}{R^2}\Big).$$

Note that even though the pressure curve is non-monotone, one can easily verify that the surface tension is a monotone increasing function of radius if the E-inequalities (6.8) hold.

**Exercise 6.7.1** Consider the deformation of a cylinder given by

$$r = r(R), \quad \theta = \Theta, \quad z = \frac{1}{c}Z,$$

where $c$ is a constant, for incompressible isotropic elastic materials.

1) Show that incompressibility implies that $r(R) = (cR^2 + A)^{1/2}$, where $A$ is a constant.
2) Show that it is a controllable universal solution and determine the stress fields.
3) Discuss the inflation and the possibility of eversion for hollow cylinders.

## 6.8 Stability of Spherical Shells

The pressure–radius relation for the inflation of a spherical shell, as shown in Fig. 6.7, is non-monotone, a feature that may lead to unstable behavior. We shall investigate the stability of thin spherical shells under two different boundary conditions.

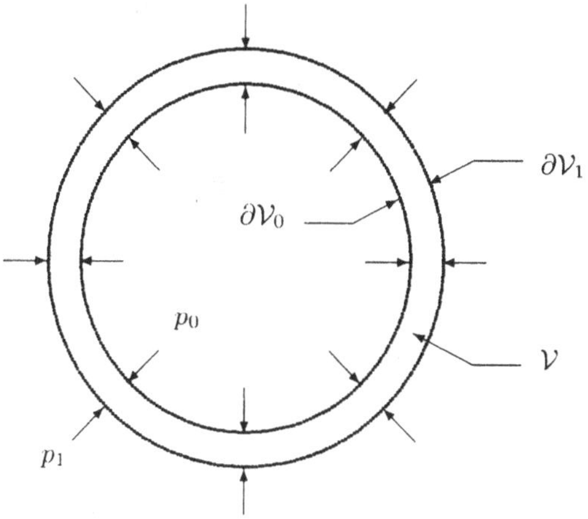


**Fig. 6.8.** Spherical shell

Let the spherical shell lie in a region $\mathcal{V}$ between $\partial\mathcal{V}_0$ and $\partial\mathcal{V}_1$, where $\mathcal{V}_0$ and $\mathcal{V}_1$ are spherical balls with radii $r_0$ and $r_1$, respectively (Fig. 6.8). Assume that the shell is subject to uniform temperature $\theta$ and free from external supplies. Then with the same consideration leading to (6.28), we have

$$\frac{d}{dt}\int_{\mathcal{V}} \rho(\psi + \frac{1}{2}\boldsymbol{v}\cdot\boldsymbol{v})dv - \int_{\partial\mathcal{V}} \boldsymbol{v}\cdot T\boldsymbol{n}\,da \leq 0, \tag{6.68}$$

where $\psi = \varepsilon - \theta\eta$ is the free energy. Moreover, from the boundary conditions

(6.63), we have

$$\begin{aligned}\int_{\partial\mathcal{V}} \boldsymbol{v}\cdot T\boldsymbol{n}\,da &= p_0 \int_{\partial\mathcal{V}_0} \boldsymbol{v}\cdot\boldsymbol{n}\,da - p_1 \int_{\partial\mathcal{V}_1} \boldsymbol{v}\cdot\boldsymbol{n}\,da \\ &= [\![p]\!]\,\frac{d}{dt}\int_{\mathcal{V}_0} dv - p_1\,\frac{d}{dt}\int_{\mathcal{V}_1-\mathcal{V}_0} dv = [\![p]\!]\frac{dV_0}{dt},\end{aligned} \tag{6.69}$$

In this derivation, we have used the relation (2.10) and the incompressibility condition that the volume of $\mathcal{V} = \mathcal{V}_1 - \mathcal{V}_0$ does not change. We have also denoted $V_0$ as the volume of the spherical region $\mathcal{V}_0$.

Combining (6.68) and (6.69), we obtain for quasi-static problems (with negligible acceleration)

$$\frac{d}{dt}\int_{\mathcal{V}} \rho\psi\,dv - [\![p]\!]\frac{dV_0}{dt} \leq 0. \tag{6.70}$$

Since the pressure–radius relations are essentially the same irrespective of thickness, for simplicity in the following analysis we shall consider only thin shells of Mooney–Rivlin materials. The free energy can be expressed as

$$\rho\psi = \frac{s_1}{2}(I_B - 3) - \frac{s_{-1}}{2}(II_B - 3), \tag{6.71}$$

where $I_B$ and $II_B$ are given by (6.57).

### 6.8.1 Stability under Constant Pressures

We shall analyze the stability of a spherical shell under prescribed constant internal and external pressures. In order to be able to maintain constant internal pressure, we have tacitly assumed that a suitable device is provided, such as a tube interconnecting the interior to a constant pressure chamber. However, as long as the interior can be maintained at constant pressure, such a device is irrelevant to the problem.

The relation (6.70) can now be written as

$$\frac{d}{dt}\left\{4\pi\int_{r_0}^{r_1} \rho\psi\, r^2 dr - \frac{4}{3}\pi r_0^3 [\![p]\!]\right\} \leq 0.$$

Therefore, we can define the availability $\mathcal{A}$, a monotone decreasing function of time, for a thin shell as

$$\mathcal{A} = 4\pi r^2 d\,\rho\psi - \frac{4}{3}\pi r^3\,[\![p]\!].$$

Since $\psi = \psi(r)$ and $r^2 d \simeq R^2 D$, we have

$$\mathcal{A}(r) = 4\pi R^2 D\,\rho\psi(r) - \frac{4}{3}\pi r^3\,[\![p]\!].$$

According to the stability criterion, if $r = \bar{r}$ corresponds to a stable equilibrium state, then $\mathcal{A}(r)$ must attain its minimum at $r = \bar{r}$. The necessary and sufficient conditions for this to happen are

$$\left.\frac{d\mathcal{A}}{dr}\right|_{r=\bar{r}} = 0, \qquad \left.\frac{d^2\mathcal{A}}{dr^2}\right|_{r=\bar{r}} \geq 0.$$

From the expressions (6.71) and (6.57) for the free energy, we obtain after simple differentiations,

$$\frac{d\mathcal{A}}{dr} = 4\pi r^2 \Big(F(r) - [\![p]\!]\Big),$$
$$\frac{d^2\mathcal{A}}{dr^2} = 8\pi r \Big(F(r) - [\![p]\!]\Big) + 4\pi r^2 \frac{dF(r)}{dr},$$

where

$$F(r) = 2\frac{D}{R}\left\{ s_1 \left(\frac{R}{r} - \frac{R^7}{r^7}\right) - s_{-1}\left(\frac{r}{R} - \frac{R^5}{r^5}\right)\right\}. \tag{6.72}$$

Therefore, we have the following stability conditions:

$$[\![p]\!] = F(\bar{r}), \qquad \frac{dF(\bar{r})}{dr} \geq 0. \tag{6.73}$$

The first condition is merely the pressure–radius relation (6.67), while the other condition implies that the pressure–radius curve $F(r)$ must have a positive slope at a stable equilibrium state.

Therefore, the branch with negative slope in the pressure–radius curve shown in Fig. 6.7 (with $a = 1.001$ for thin shells) corresponds to unstable equilibrium states under prescribed constant internal and external pressures. A similar analysis for thick spherical shells also leads to the same conclusion for stability [42].

### 6.8.2 Stability for an Enclosed Spherical Shell

For an enclosed spherical shell, if we assume that the air inside the shell is an ideal gas, then the internal pressure is inversely proportional to the interior volume under isothermal conditions (Boyle's law). Therefore, for the present case we suppose that the external pressure $p_1$ is constant while the internal pressure $p_0$ is given by

$$p_0 = \frac{k}{V_0}, \tag{6.74}$$

where $k$ is a constant.

We can rewrite the condition (6.70) as

$$\frac{d}{dt}\left\{4\pi \int_{r_0}^{r_1} \rho\psi\, r^2 dr - k \log V_0 + p_1 V_0\right\} \leq 0.$$

Therefore, for a thin shell, we can define the availability $\mathcal{A}$ as

$$\mathcal{A} = 4\pi r^2 d\,\rho\psi(r) - k\log V_0 + p_1 V_0.$$

Since $V_0 = \frac{4}{3}\pi r^3$, we obtain

$$\frac{d\mathcal{A}}{dr} = 4\pi r^2\Big(F(r) - \frac{k}{V_0} + p_1\Big),$$

where $F(r)$ is given by (6.72). The equilibrium condition becomes

$$\frac{k}{V_0} - p_1 = p_0 - p_1 = F(\bar{r}), \tag{6.75}$$

by (6.74). This is again the pressure–radius relation (6.67).

Moreover, the second derivative of $\mathcal{A}$ is

$$\frac{d^2\mathcal{A}}{dr^2} = 8\pi r\Big(F(r) - p_0 + p_1\Big) + 4\pi r^2\Big(\frac{dF(r)}{dr} + \frac{k}{V_0^2}\frac{dV_0}{dr}\Big).$$

Hence, the stability condition requires that

$$\begin{aligned}\frac{d^2\mathcal{A}}{dr^2}\bigg|_{r=\bar{r}} &= 4\pi r^2\Big(\frac{dF(r)}{dr} + \frac{3}{r}p_0\Big)\bigg|_{r=\bar{r}} \\ &= 12\pi r\Big(\frac{r}{3}\frac{dF(r)}{dr} + F(r) + p_1\Big)\bigg|_{r=\bar{r}} \geq 0,\end{aligned} \tag{6.76}$$

where (6.74) and (6.75) have been used. By the use of (6.72), after a simple calculation, we obtain the expression

$$\frac{r}{3}\frac{dF(r)}{dr} + F(r) = \frac{4}{3}\frac{D}{R}\Big\{s_1\Big(\frac{R}{r} + 2\frac{R^7}{r^7}\Big) - s_{-1}\Big(2\frac{r}{R} + \frac{R^5}{r^5}\Big)\Big\},$$

which is always positive if the E-inequalities (6.8) holds. Therefore, for this case the stability condition (6.76) is identically satisfied at any equilibrium state. In other words, every equilibrium state is stable.

It is worthwhile to emphasize that the stability conditions depend on the particular situation a body happens to encounter. For different situations the stability conditions are, in general, different as we have seen in the above examples.[5]

**Exercise 6.8.1** Proceed with a similar analysis to determine the equilibrium and the stability conditions for
1) thick spherical shells subject to prescribed constant pressures,
2) enclosed spherical shells of finite thickness.

[5] A different stability analysis for these examples can be found in Sect. 8.3, [55]. See also [56].

# 7. Thermodynamics with Lagrange Multipliers

## 7.1 Supply-Free Bodies

The entropy principle based on the Clausius–Duhem inequality has been widely adopted in the development of modern rational thermodynamics, after the fundamental memoir of Coleman and Noll [11]. The main assumptions,

$$\boldsymbol{\Phi} = \frac{1}{\theta}\,\boldsymbol{q}, \qquad s = \frac{1}{\theta}\,r, \tag{7.1}$$

while at least tacit in all classical theories of continuum mechanics, are not particularly well motivated for materials in general. In fact, the relation $(7.1)_1$ is known to be inconsistent with the result from the kinetic theory of ideal gases, and is also found to be inappropriate to account for thermodynamics of diffusion. Other assumptions and formulations of the second law of thermodynamics have been proposed elsewhere.[1] In [50] Müller proposed to abandon the assumption $(7.1)_1$ by treating the entropy flux $\boldsymbol{\Phi}$ and the energy flux $\boldsymbol{q}$ as independent constitutive quantities and hence leaving the entropy inequality in its general form,

$$\rho\dot{\eta} + \operatorname{div}\boldsymbol{\Phi} - \rho s \geq 0. \tag{7.2}$$

But, in addition, he also proposed that if the body is free of external supplies, the entropy supply must also vanish, i.e.,

$$s = 0 \quad \text{if} \quad r = 0 \quad \text{and} \quad \boldsymbol{b} = 0,$$

which is certainly much weaker than the assumption $(7.1)_2$. He argued that since constitutive properties of a material should not depend on external supplies, in exploiting constitutive restrictions it suffices to consider only supply-free bodies. For supply-free bodies, the entropy inequality becomes

$$\rho\dot{\eta} + \operatorname{div}\boldsymbol{\Phi} \geq 0. \tag{7.3}$$

It has been shown that the entropy principle imposes severe restrictions on constitutive functions and the exploitation for such restrictions based on

[1] For more general discussions and historical notes on the second law of thermodynamics, see [69].

the Clausius–Duhem inequality are relatively easy. On the other hand, the exploitation of the entropy principle based on the general entropy inequality (7.3) for supply-free bodies, first considered by Müller ([52, 53]) was much more difficult and later its procedure has been improved greatly by the use of Lagrange multipliers in a more systematic manner proposed by Liu [37].

In this chapter we shall illustrate the procedure for exploiting the entropy principle by the use of Lagrange multipliers for viscous heat-conducting fluids. For convenience, we shall use component notations in a Cartesian coordinate system for mathematical manipulations, which are often simpler.

## 7.2 Viscous Heat-Conducting Fluid

The determination of the basic fields $\{\rho(\boldsymbol{x},t), v_i(\boldsymbol{x},t), \theta(\boldsymbol{x},t)\}$ are based on the balance laws, which, for a supply-free body, can be written as

$$\begin{aligned} &\dot\rho + \rho v_{k,k} = 0, \\ &\rho\dot v_i - T_{ik,k} = 0, \\ &\rho\dot\varepsilon + q_{k,k} - T_{ik}v_{i,k} = 0. \end{aligned} \tag{7.4}$$

We consider the viscous heat-conducting fluid, for which the constitutive quantities, the symmetric stress tensor $T_{ik}$, the heat flux $q_k$, and the internal energy $\varepsilon$ are given by constitutive equations of the form

$$\mathcal{C} = \widehat{\mathcal{C}}(\rho, v_i, \theta, \theta_{,i}, v_{i,k}).$$

The principle of material objectivity requires that the constitutive function of an objective constitutive quantity must be independent of the observer. As we have shown earlier (see for example, Sect. 3.2.2), this requirement implies that the constitutive functions are independent of the velocity $v_i$, and depend on the velocity gradient only through its symmetric part $D_{ik}$. Therefore, they can be written as

$$\begin{aligned} T_{ik} &= T_{ik}(\rho, \theta, \theta_{,n}, D_{mn}), \\ q_k &= q_k(\rho, \theta, \theta_{,n}, D_{mn}), \\ \varepsilon &= \varepsilon(\rho, \theta, \theta_{,n}, D_{mn}). \end{aligned} \tag{7.5}$$

Moreover, the constitutive functions are isotropic functions.

The field equations are obtained by introducing constitutive equations (7.5) into the balance equations (7.4). Any solution $\{\rho(\boldsymbol{x},t), v_i(\boldsymbol{x},t), \theta(\boldsymbol{x},t)\}$ of the field equations is called a *thermodynamic process* in the supply-free body.

Since our objective is to derive restrictions imposed on the constitutive equations that are independent of external supplies, we need only consider supply-free bodies without loss of generality. Accordingly, the entropy principle requires that the entropy inequality

$$\rho\dot{\eta} + \Phi_{k,k} \geq 0 \tag{7.6}$$

must hold for every thermodynamic process in supply-free bodies. The specific entropy density $\eta$ and the entropy flux $\Phi_i$ are also constitutive quantities,

$$\begin{aligned} \eta &= \eta(\rho, \theta, \theta_{,n}, D_{mn}), \\ \Phi_k &= \Phi_k(\rho, \theta, \theta_{,n}, D_{mn}). \end{aligned} \tag{7.7}$$

This constraint on thermodynamic processes can be stated in a different way, namely, the fields that satisfy the entropy inequality are constrained by the requirement that they must be solutions of the field equations. We can take care of this requirement by the use of Lagrange multipliers much like that in the classical problems of finding the extrema with constraints:

**Method of Lagrange multipliers.** *There exist Lagrange multipliers $\Lambda^\rho$, $\Lambda^{v_i}$ and $\Lambda^\varepsilon$ such that the inequality*

$$\begin{aligned} \rho\dot{\eta} + \Phi_{k,k} - \Lambda^\rho(\dot{\rho} + \rho D_{kk}) - \Lambda^{v_i}(\rho\dot{v}_i - T_{ik,k}) \\ - \Lambda^\varepsilon(\rho\dot{\varepsilon} + q_{k,k} - T_{ik}D_{ik}) \geq 0 \end{aligned} \tag{7.8}$$

*is valid under no constraints, in other words, it must hold for any fields $\{\rho(x_m,t), v_i(x_m,t), \varepsilon(x_m,t)\}$. Moreover, the Lagrange multipliers are functions of $(\rho, \theta, \theta_{,m}, D_{mn})$.*

The validity of this statement will be justified in the next section.

To exploit the consequence of the inequality (7.8), we first observe that since constitutive functions are independent of the velocity $v_i$, the inequality contains an explicit term linear in $\dot{v}_i$. Its coefficient, the Lagrange multiplier $\Lambda^{v_i}$, must vanish because the inequality must hold for any field $v_i(x_m,t)$, and, in particular, for any value of $\dot{v}_i$ at any particular position and time. Therefore, we have

$$\Lambda^{v_i} = 0, \tag{7.9}$$

and the inequality after the constitutive equations are introduced becomes

$$\begin{aligned} &\rho\Big(\frac{\partial\eta}{\partial\rho} - \Lambda^\varepsilon\frac{\partial\varepsilon}{\partial\rho} - \frac{1}{\rho}\Lambda^\rho\Big)\dot{\rho} + \rho\Big(\frac{\partial\eta}{\partial\theta} - \Lambda^\varepsilon\frac{\partial\varepsilon}{\partial\theta}\Big)\dot{\theta} \\ &+ \rho\Big(\frac{\partial\eta}{\partial\theta_{,k}} - \Lambda^\varepsilon\frac{\partial\varepsilon}{\partial\theta_{,k}}\Big)(\theta_{,k})^{\cdot} + \rho\Big(\frac{\partial\eta}{\partial D_{kl}} - \Lambda^\varepsilon\frac{\partial\varepsilon}{\partial D_{kl}}\Big)\dot{D}_{kl} \\ &+ \Big(\frac{\partial\Phi_k}{\partial\theta_{,l}} - \Lambda^\varepsilon\frac{\partial q_k}{\partial\theta_{,l}}\Big)\theta_{,lk} + \Big(\frac{\partial\Phi_k}{\partial D_{jl}} - \Lambda^\varepsilon\frac{\partial q_k}{\partial D_{jl}}\Big)D_{jl,k} \\ &+ \Big(\frac{\partial\Phi_k}{\partial\rho} - \Lambda^\varepsilon\frac{\partial q_k}{\partial\rho}\Big)\rho_{,k} + \Big(\frac{\partial\Phi_k}{\partial\theta} - \Lambda^\varepsilon\frac{\partial q_k}{\partial\theta}\Big)\theta_{,k} - \big(\Lambda^\rho\rho\,\delta_{ik} - \Lambda^\varepsilon T_{ik}\big)D_{ik} \geq 0. \end{aligned}$$

Note that the above inequality is linear in $\dot{\rho}$, $\dot{\theta}$, $(\theta_{,k})^{\cdot}$, $\dot{D}_{kl}$, and $\rho_{,k}$, $\theta_{,lk}$, $D_{jl,k}$ and by the same argument, it must hold for any values of these quantities. Therefore, the coefficients of these derivatives must vanish, which leads to the following results:

$$\begin{aligned}
&\frac{\partial \eta}{\partial \rho} = \Lambda^{\varepsilon}\frac{\partial \varepsilon}{\partial \rho} + \frac{1}{\rho}\Lambda^{\rho}, \qquad && \frac{\partial \eta}{\partial \theta_{,k}} = \Lambda^{\varepsilon}\frac{\partial \varepsilon}{\partial \theta_{,k}},\\
&\frac{\partial \eta}{\partial \theta} = \Lambda^{\varepsilon}\frac{\partial \varepsilon}{\partial \theta}, && \frac{\partial \eta}{\partial D_{kl}} = \Lambda^{\varepsilon}\frac{\partial \varepsilon}{\partial D_{kl}},
\end{aligned} \tag{7.10}$$

and

$$\begin{aligned}
&\frac{\partial \Phi_k}{\partial \rho} - \Lambda^{\varepsilon}\frac{\partial q_k}{\partial \rho} = 0,\\
&\Big(\frac{\partial \Phi_k}{\partial \theta_{,l}} + \frac{\partial \Phi_l}{\partial \theta_{,k}}\Big) - \Lambda^{\varepsilon}\Big(\frac{\partial q_k}{\partial \theta_{,l}} + \frac{\partial q_l}{\partial \theta_{,k}}\Big) = 0,\\
&\Big(\frac{\partial \Phi_k}{\partial D_{jl}} + \frac{\partial \Phi_l}{\partial D_{jk}}\Big) - \Lambda^{\varepsilon}\Big(\frac{\partial q_k}{\partial D_{jl}} + \frac{\partial q_l}{\partial D_{jk}}\Big) = 0.
\end{aligned} \tag{7.11}$$

In obtaining $(7.11)_{2,3}$, we have taken into account the symmetry of the second derivatives, therefore only the symmetric parts of their coefficients need to vanish. More specifically, the term involving $D_{jl,k}$ can be written as

$$\Big(\frac{\partial \Phi_k}{\partial D_{jl}} - \Lambda^{\varepsilon}\frac{\partial q_k}{\partial D_{jl}}\Big)D_{jl,k} = \frac{1}{2}\Big\{\Big(\frac{\partial \Phi_k}{\partial D_{jl}} + \frac{\partial \Phi_l}{\partial D_{jk}}\Big) - \Lambda^{\varepsilon}\Big(\frac{\partial q_k}{\partial D_{jl}} + \frac{\partial q_l}{\partial D_{jk}}\Big)\Big\}\, v_{j,lk},$$

since $D_{jl} = D_{lj}$. Due to the symmetry of $D_{jl}$, the relation $(7.11)_3$ can further be reduced to

$$\frac{\partial \Phi_k}{\partial D_{jl}} - \Lambda^{\varepsilon}\frac{\partial q_k}{\partial D_{jl}} = 0. \tag{7.12}$$

Indeed, if we denote the left-hand side of (7.12) by $H^k_{jl}$ then $H^k_{jl} = H^k_{lj}$ and the relation $(7.11)_3$ becomes

$$H^k_{jl} + H^l_{jk} = 0,$$

which implies that

$$H^k_{jl} = -H^l_{jk} = -H^l_{kj} = H^j_{kl} = H^j_{lk} = -H^k_{lj} = -H^k_{jl},$$

and hence $H^k_{jl} = 0$.

### 7.2.1 General Results

We can now obtain the consequence of the relations (7.10) through (7.12). First of all, we claim that the relations $(7.11)_2$ and (7.12), for the entropy flux and the heat flux, together with the fact that they are isotropic functions, imply the following parallel relation,

$$\Phi_k = \Lambda^\varepsilon q_k, \tag{7.13}$$

and the Lagrange multiplier $\Lambda^\varepsilon$ does not depend on $\theta_{,i}$ and $D_{ik}$. The proof of this claim is not trivial, in general, and will be given in Sect. 7.4.

Since now we have

$$\Lambda^\varepsilon = \Lambda^\varepsilon(\rho, \theta),$$

from (7.13), the relation $(7.11)_1$ implies that

$$\frac{\partial \Lambda^\varepsilon}{\partial \rho} q_k = 0.$$

Of course, $q_k \neq 0$, in general, therefore $\Lambda^\varepsilon$ must be independent of the density, hence

$$\Lambda^\varepsilon = \Lambda^\varepsilon(\theta). \tag{7.14}$$

Furthermore, from $(7.10)_2$ and $(7.10)_3$, the integrability condition for $\eta$ with respect to $\theta$ and $\theta_{,k}$ gives

$$\frac{\partial \Lambda^\varepsilon}{\partial \theta} \frac{\partial \varepsilon}{\partial \theta_{,k}} = 0,$$

which shows that $\varepsilon$ is independent of $\theta_{,k}$ since we will not consider the possibility of $\Lambda^\varepsilon$ being a constant, in general. Similarly, the integrability condition with respect to $\theta$ and $D_{kl}$ from $(7.10)_2$ and $(7.10)_4$ leads to the independence of $\varepsilon$ from $D_{kl}$. Therefore, we conclude that

$$\varepsilon = \varepsilon(\rho, \theta), \qquad \eta = \eta(\rho, \theta),$$

and the first two relations of (7.10) can then be put together to give the total differential of $\eta$,

$$d\eta = \Lambda^\varepsilon d\varepsilon + \frac{\Lambda^\rho}{\rho} d\rho, \tag{7.15}$$

which also shows that

$$\Lambda^\rho = \Lambda^\rho(\rho, \theta).$$

After elimination of those linear terms, the inequality (7.8) becomes

$$\left( \frac{\partial \Phi_k}{\partial \theta} - \Lambda^\varepsilon \frac{\partial q_k}{\partial \theta} \right) \theta_{,k} - \left( \Lambda^\rho \rho\, \delta_{ik} - \Lambda^\varepsilon T_{ik} \right) D_{ik} \geq 0. \tag{7.16}$$

The left-hand side, denoted by $\sigma$, is the entropy production density, which is a non-negative quantity for any values of $(\rho, \theta, \theta_{,k}, D_{ik})$. We call a state with no entropy production an *equilibrium state*. Since $\sigma$ takes its minimum value at the equilibrium state, namely zero, at $\theta_{,k} = 0$ and $D_{ik} = 0$. One of the necessary conditions for the minimum is

$$\left.\frac{\partial \sigma}{\partial D_{ik}}\right|_0 = -(\rho \Lambda^\rho + p \Lambda^\varepsilon)\delta_{ik} = 0,$$

where the index 0 indicates the evaluation at the equilibrium state and the equilibrium pressure $p$ is defined as

$$T_{ik}|_0 = -p(\rho, \theta)\, \delta_{ik}.$$

Hence, we have

$$\Lambda^\rho = -\Lambda^\varepsilon \frac{p}{\rho}. \tag{7.17}$$

Insertion of this into (7.15) gives

$$d\eta = \Lambda^\varepsilon \Big( d\varepsilon - \frac{p}{\rho^2} d\rho \Big). \tag{7.18}$$

Therefore, by comparison with the Gibbs relation (5.25) for elastic fluids in thermostatics, we can identify the Lagrange multiplier $\Lambda^\varepsilon$ as the reciprocal of the absolute temperature $\theta$,

$$\Lambda^\varepsilon(\theta) = \frac{1}{\theta}. \tag{7.19}$$

An immediate consequence follows from (7.13),

$$\Phi_k = \frac{1}{\theta}\, q_k, \tag{7.20}$$

which assures that the entropy flux is the heat flux divided by the absolute temperature for the fluid under consideration. This is what is usually taken as an assumption for Clausius–Duhem inequality.

With (7.19) and (7.20) the remaining inequality (7.16) becomes

$$\sigma = -\frac{1}{\theta^2}\, q_k \theta_{,k} + \frac{1}{\theta}\Big( T_{ik} + p\, \delta_{ik} \Big) D_{ik} \geq 0. \tag{7.21}$$

### 7.2.2 Navier–Stokes–Fourier Fluids

Further consequences from the remaining inequality (7.21) will only concern the linear terms of the constitutive functions. Therefore, up to the first-order terms in the gradient $\theta_{,m}$ and $D_{mn}$, we may write

$$\begin{aligned} T_{ik} &= -p\, \delta_{ik} + \lambda\, D_{nn} \delta_{ik} + 2\mu\, D_{ik} + o(2), \\ q_k &= -\kappa\, \theta_{,k} + o(2), \end{aligned} \tag{7.22}$$

where the viscosities $\lambda$ and $\mu$, as well as the thermal conductivity $\kappa$, are functions of $(\rho, \theta)$. The inequality (7.21) with the linear constitutive relations now takes the form

$$\sigma = \frac{1}{\theta^2}\kappa\,\theta_{,k}\,\theta_{,k} + \frac{1}{\theta}\Big((\lambda + \tfrac{2}{3}\mu)\,D_{kk}^2 + 2\mu\,\hat{D}_{ik}\hat{D}_{ik}\Big) \geq 0,$$

where $\hat{D}_{ik} = D_{ik} - \frac{1}{3}D_{nn}\delta_{ik}$ is the traceless part (or deviatoric part, see (4.36)) of $D_{ik}$.

Since the trace $D_{kk}$ and the traceless part $\hat{D}_{ik}$ are mutually independent, the left-hand side of the inequality is quadratic in three independent variables $\{\theta_{,k}, D_{kk}, \hat{D}_{ik}\}$, and hence it follows that the thermal conductivity, and the bulk and the shear viscosities, are positive (see (4.37)),

$$\kappa \geq 0 \qquad 3\lambda + 2\mu \geq 0, \qquad \mu \geq 0. \tag{7.23}$$

These inequalities ensure that the heat flux points in the direction to the colder region and that the fluid particles tend to flow in the direction of a shear force.

**Remark.** Even though we have shown that for the viscous heat-conducting fluids the general entropy inequality does reduce to the Clausius–Duhem inequality, we have seen in this example that the method of Lagrange multipliers enables us to avoid the unnecessary assumptions $\boldsymbol{\Phi} = \boldsymbol{q}/\theta$ – questionable in general – and the wishful argument that the external supplies can be arbitrarily adjusted in the exploitation procedure based on the Clausius–Duhem inequality (see the remark on p. 134).

## 7.3 Method of Lagrange Multipliers

We consider a system of linear algebraic equations with an inequality constraint. Suppose that we have a linear system of $m$ equations and $n$ unknowns

$$A_{ab}X_b + B_a = 0, \tag{7.24}$$

and a linear inequality

$$\alpha_b X_b + \beta \geq 0, \tag{7.25}$$

where $a = 1, \cdots, m$, $b = 1, \cdots, n$, and the summation convention is used. As indicated, $\boldsymbol{A}$ is an $m \times n$ matrix, and $\boldsymbol{B} \in I\!R^m$, $\boldsymbol{\alpha} \in I\!R^n$, are vectors. If $\boldsymbol{A}$ and $\boldsymbol{B}$ are given, of course, in general, the solution $\boldsymbol{X} \in I\!R^n$ of (7.24) may not satisfy the inequality (7.25).

We have seen from the example in the previous section that after introducing the constitutive relations into the system of balance laws and the entropy inequality, we can put them in the form of the above relations in spite of the fact that they are partial differential relations.

### 7.3.1 An Algebraic Problem

The entropy principle requires that the constitutive functions be imposed in such a way that the entropy inequality is satisfied for any solutions of the balance equations. For the algebraic relations (7.24) and (7.25), we can also pose the following problem:

> *What conditions must be imposed on the matrix $\boldsymbol{A}$ and the vector $\boldsymbol{B}$ so that every solution of the linear system* (7.24) *will always satisfy the inequality constraint* (7.25)*?*

This is similar to the problem of exploiting the entropy principle, but is much simpler of course, because it is purely an algebraic one. The answer is stated in the following lemma [37].

**Lemma 7.3.1** *Let* $S = \{\boldsymbol{X} \in I\!R^n \mid A_{ab}X_b + B_a = 0\}$, *the solution set of* (7.24), *be neither empty nor the whole space* $I\!R^n$ *and* $\boldsymbol{\alpha}$ *is not a zero vector* (otherwise the cases are trivial). *Then, the following two statements are equivalent:*

$$1)\ \ \alpha_b X_b + \beta \geq 0 \qquad \forall\, \boldsymbol{X} \in S; \tag{7.26}$$

2) *There exists a vector* $\boldsymbol{\Lambda} \in I\!R^m$ *such that*

$$\alpha_b X_b + \beta - \Lambda_a(A_{ab}X_b + B_a) \geq 0 \qquad \forall\, \boldsymbol{X} \in I\!R^n. \tag{7.27}$$

*Proof.* It is obvious that the condition (2) implies the condition (1). To prove the converse, let us observe that the condition (7.27) can be written as

$$(\alpha_b - \Lambda_a A_{ab})X_b + (\beta - \Lambda_a B_a) \geq 0, \qquad \forall\, \boldsymbol{X} \in I\!R^n. \tag{7.28}$$

We claim that the coefficient of $\boldsymbol{X}$ must vanish, and hence

$$\begin{aligned} \alpha_b - \Lambda_a A_{ab} &= 0, \\ \beta - \Lambda_a B_a &\geq 0. \end{aligned} \tag{7.29}$$

Indeed, if this does not hold, then for some index $b$, suppose that $\alpha_b - \Lambda_a A_{ab} \neq 0$. Since $\boldsymbol{X} \in I\!R^n$ is arbitrary, we can choose $X_i = 0$ for any $i \neq b$ and $X_b$ in such a way that the inequality (7.28) is violated. Therefore (7.28) and (7.29) are equivalent. To complete the proof of the lemma, we need only show that (1) implies the existence of $\boldsymbol{\Lambda}$ so that (7.29) hold. The proof can better be given in geometric terms.

Let $H = \{\boldsymbol{X} \in I\!R^n \mid \alpha_b X_b + \beta \geq 0\}$ denote the half-space defined by the inequality (7.25). Then the statement (1) means that the hyperplane $S$ must lie inside the half-space $H$, or simply $S \subset H$. Let $H_o = \{\boldsymbol{X} \in I\!R^n \mid \alpha_b X_b = 0\}$ be the $(n-1)$-dimensional hyperplane parallel to $H$ and passing through the origin, and $S_o = \{\boldsymbol{X} \in I\!R^n \mid A_{ab}X_b = 0\}$ be the hyperplane parallel to $S$ and passing through the origin. Note that both $S_o$ and $H_o$ are subspaces, and

$\dim S_o$ is smaller than $\dim H_o$. We claim that if the hyperplane $S$ is contained in the half-space $H$ then the subspace $S_o$ is also contained in the subspace $H_o$. Since if this is not true, then there is some vector $\boldsymbol{Z}$ in $S_o$ but not in $H_o$. Then for any constant $c$, if $\boldsymbol{Y} \in S$, $\boldsymbol{Y} + c\boldsymbol{Z}$ is also in $S$. But since $\boldsymbol{Z} \notin H_o$, which means that $\alpha_b Z_b \neq 0$. then it is possible to choose a value of $c$ such that $\alpha_b(Y_b + cZ_b) + \beta$ becomes negative. This means that $\boldsymbol{Y} + c\boldsymbol{Z}$ is not in $H$, which contradicts the assumption $S \subset H$.

If we denote the orthogonal complements of $S_o$ and $H_o$ by $S_o^\perp$ and $H_o^\perp$, respectively, then it follows from $S_o \subset H_o$ that $S_o^\perp \supset H_o^\perp$. Now, since $\boldsymbol{\alpha}$ is in $H_o^\perp$, therefore $\boldsymbol{\alpha}$ is also a vector in $S_o^\perp$. Since we know from linear algebra that $\dim S_o^\perp = \operatorname{rank} \boldsymbol{A}$, and the row vectors of the matrix $A_{ab}$ with $a = 1, \cdots, m$ are in $S_o^\perp$, this set of vectors span the space $S_o^\perp$. Therefore, there exists a $\boldsymbol{\Lambda}$ such that $\boldsymbol{\alpha}$ can be expressed as

$$\alpha_b - \Lambda_a A_{ab} = 0,$$

and the first part of (7.29) is proved. The second part follows easily, because for any $\boldsymbol{X} \in S$, we now have

$$\alpha_b X_b + \beta = \Lambda_a A_{ab} X_b + \beta = -\Lambda_a B_a + \beta,$$

which implies $\beta - \Lambda_a B_a \geq 0$ by the assumption that $S$ is in $H$. This completes the proof of the lemma. □

If the $m$ row-vectors of the matrix $A_{ab}$ are linearly independent, then they form a basis of $S_o^\perp$. In this case, the components $\Lambda_a$ are uniquely determined by $\boldsymbol{\alpha}$. On the other hand, if the $i$-th row-vector $A_{ib}$ is zero, then $\Lambda_i$ is arbitrary, which implies from $(7.29)_2$ that $B_i$ must vanish. In this case, we can simply eliminate the $i$-th equation from the system (7.24), because it is merely $B_i = 0$, which contains no $\boldsymbol{X}$ at all.

We call $\Lambda_a$ Lagrange multipliers, because the replacement of the statement (1) by the statement (2) of the lemma is similar to the use of Lagrange multipliers in classical problems of maximizing or minimizing a function with constraints.

### 7.3.2 Local Solvability

Now let us take viscous heat-conducting fluids considered in the previous section as an example. Incorporated with the constitutive equations (7.5), the system of balance laws (7.4) takes the following quasi-linear form,

$$A_{ab}(y) X_b + B_a(y) = 0, \tag{7.30}$$

and so does the entropy inequality (7.6),

$$\alpha_b(y) X_b + \beta(y) \geq 0, \tag{7.31}$$

where

$$(X_b) = (\dot{\rho}, \dot{v}_i, \dot{\theta}, (\theta_{,i})^{\cdot}, \dot{D}_{ij}, \rho_{,i}, \theta_{,ij}, D_{ij,k}),$$
$$(y_c) = (\rho, v_i, \theta, \theta_{,i}, D_{ij}), \tag{7.32}$$

for $a = 1, \cdots, 5$, $b = 1, \cdots, 41$, and $c = 1, \cdots, 14$. In counting the numbers of corresponding components, symmetry of indices is taken into account. The matrix $A_{ab}$ as well as $B_a$, $\alpha_b$, $\beta$ can all be written out easily. Their explicit expressions are left as an exercise.

The entropy principle requires that the inequality (7.31) be satisfied for any solution $\{\rho(\boldsymbol{x}, t), v_i(\boldsymbol{x}, t), \varepsilon(\boldsymbol{x}, t)\}$ of the system of partial differential equations (7.30). This requirement will reduce to a purely algebraic problem stated above if the following assumption can be justified.

**Local solvability**. *The system (7.30) is locally solvable at $(\boldsymbol{x}^o, t^o)$ if for any values of $(y_c, X_b)$ satisfying (7.30) algebraically, there is a solution $\{\rho(\boldsymbol{x}, t), v_i(\boldsymbol{x}, t), \varepsilon(\boldsymbol{x}, t)\}$ of the system of differential equations (7.30) in a neighborhood of $(\boldsymbol{x}^o, t^o)$ at which the solution is consistent with the given values.*

The local solvability assumption is the key point in employing the method of Lagrange multipliers for the exploitation of the entropy principle. In most "normal" situations, like the one we consider here, this assumption can be justified. The justification is usually based on the existence of solutions for initial-value problems.

In the present example, the balance laws (7.4) or (7.30) can be written as a system of partial differential equations in the form:

$$\frac{\partial \rho}{\partial t} = f_\rho(\rho, \rho_{,k}, v_k, v_{k,l}),$$
$$\frac{\partial v_i}{\partial t} = f_{v_i}(\rho, \rho_{,k}, \theta, \theta_{,k}, \theta_{,kl}, v_k, v_{k,l}, v_{k,lm}), \tag{7.33}$$
$$\frac{\partial \varepsilon}{\partial t} = f_\varepsilon(\rho, \rho_{,k}, \theta, \theta_{,k}, \theta_{,kl}, v_k, v_{k,l}, v_{k,lm}),$$

where the explicit expressions of the functions $f_\rho$, $f_{v_i}$ and $f_\varepsilon$ on the right-hand side are left as an exercise. By the Cauchy–Kowalewsky theorem (see, for example, [12]) for initial value problems, there exists a unique analytic solution $\{\rho(\boldsymbol{x}, t), v_i(\boldsymbol{x}, t), \varepsilon(\boldsymbol{x}, t)\}$ of the system (7.33) in a neighborhood of a point, say (0,0), for any analytic initial values $\{\rho(\boldsymbol{x}, 0), v_i(\boldsymbol{x}, 0), \varepsilon(\boldsymbol{x}, 0)\}$. Since the initial values are arbitrary, by Taylor series expansion such as

$$\rho(\boldsymbol{x}, 0) = \rho(0,0) + \rho_{,k}(0,0)x_k + \frac{1}{2}\rho_{,kl}(0,0)x_k x_l + \cdots,$$

$(\rho, \theta, v_k)$ and their gradients of any order, $(\rho_{,k}, \theta_{,k}, \theta_{,kl}, v_{k,l}, v_{k,lm}, v_{k,lmn}, \cdots)$ can be assigned any values at (0,0).

In order to show that the system (7.33) is locally solvable at (0,0), we note that for the existence of a solution, we can give any values to all the elements of $(y_c, X_b)$ except those five with time derivative, namely $(\dot{\rho}, \dot{v}_i, \dot{\theta}, (\theta_{,k})^{\cdot}, \dot{D}_{ij})$, or equivalently, $(\frac{\partial \rho}{\partial t}, \frac{\partial v_i}{\partial t}, \frac{\partial \theta}{\partial t}, \frac{\partial \theta_{,i}}{\partial t}, \frac{\partial v_{i,j}}{\partial t})$. The first three of them can be calculated so that $(y_c, X_b)$ satisfies (7.33). Therefore, to complete our argument we only have to show that the values of $\frac{\partial v_{i,j}}{\partial t}$ and $\frac{\partial \theta_{,i}}{\partial t}$ can also be given arbitrarily at (0,0). To see this, we take the derivatives of the equation $(7.33)_{2,3}$ to obtain

$$\frac{\partial v_{i,j}}{\partial t} = \frac{\partial f_{v_i}}{\partial \rho_k} \rho_{,kj} + \frac{\partial f_{v_i}}{\partial \theta_{,kl}} \theta_{,klj} + \frac{\partial f_{v_i}}{\partial v_{k,lm}} v_{k,lmj} + \cdots,$$

$$\frac{\partial \theta_{,i}}{\partial t} = \left( \frac{\partial f_\varepsilon}{\partial \rho_k} \rho_{,kj} + \frac{\partial f_\varepsilon}{\partial \theta_{,kl}} \theta_{,klj} + \frac{\partial f_\varepsilon}{\partial v_{k,lm}} v_{k,lmj} + \cdots \right) \left( \rho \frac{\partial \varepsilon}{\partial \theta} \right)^{-1} + \cdots.$$

This equation can be satisfied for any given value of $\frac{\partial v_{i,j}}{\partial t}$ and $\frac{\partial \theta_{,i}}{\partial t}$ by an appropriate choice of values of the higher gradients of $\{\rho_{,kj}, \theta_{klj}, v_{k,lmj}\}$, which does not belong to the chosen values of $(y_c, X_b)$ but for the existence of a solution it is completely arbitrary. This proves our claim that indeed for any given values of $(y_c, X_b)$ satisfying the system (7.33) algebraically at a point, there is a solution of the system in the neighborhood of that point. Therefore, our problem of exploiting the entropy principle in this case reduces to the algebraic problem posed at the beginning of this section, and the use of Lagrange multipliers in the previous section is justified.

To justify the local solvability assumption for the use of Lagrange multipliers in the exploitation of the entropy principle usually requires a careful analysis of the system of field equations, and sometimes additional conditions between the derivatives of the variables must be considered in order to prove the local solvability.[2] Of course, the exploitation of the entropy principle can also be done by direct elimination of quantities imposed by the equations of balance, albeit very tediously (see, for example, [53]). The use of Lagrange multipliers greatly simplifies this task and makes the procedure for the exploitation of the entropy principle systematic and straightforward. Thermodynamics with Lagrange multipliers has since become the routine procedure in formulating constitutive theories.

**Exercise 7.3.1** Write out explicitly the matrix $A_{ab}$ as well as $B_a$, $\alpha_b$, and $\beta$ in the relations (7.30).

**Exercise 7.3.2** Give the explicit expressions of the right-hand side of the field equations (7.33).

---

[2] For more details, the reader is referred to the original paper [37].

**Exercise 7.3.3** Derive thermodynamic restrictions on constitutive equations (5.15), $\mathcal{C} = \mathcal{F}(F, \theta, \boldsymbol{g})$, for thermoelastic materials based on the general entropy inequality (7.6) by the use of Lagrange multipliers. (Formulate the problem in material description and think about the problem caused by the use of deformation gradient as a variable – the compatibility condition $\dot{F} = \operatorname{Grad} \boldsymbol{v}$.)

## 7.4 Relation Between Entropy Flux and Heat Flux

First of all, it is easy to prove the parallel relation (7.13) between the entropy flux and the heat flux from the condition $(7.11)_2$ and (7.12) if only linear constitutive equations are concerned. Indeed, since $q_k$ and $\Phi_k$ are vector isotropic functions of $(\rho, \theta, \theta_{,m}, D_{mn})$, their linear representations in these variables are given by

$$q_k = k\,\theta_{,k}, \qquad \Phi_k = \phi\,\theta_{,k},$$

where $k$ and $\phi$ are independent of $\theta_{,m}$ and $D_{mn}$. Then, the condition $(7.11)_2$ implies that

$$\phi = \Lambda^{\varepsilon} k,$$

which immediately leads to the relation,

$$\Phi_k = \Lambda^{\varepsilon}\, q_k,$$

and $\Lambda^{\varepsilon}$ is independent of $\theta_{,k}$ and $D_{kj}$ by introducing the above relation into $(7.11)_2$ and (7.12).

This is indeed almost trivial, nevertheless, we remark that although general representations for isotropic functions are well known, a direct proof of the parallel relation based on such representations (see (4.51)) would be too complicated if not entirely impractical (such a proof has been given in [36]).

### 7.4.1 Theorem of Parallel Isotropic Vector Functions

To prove the parallel relation between the entropy flux and the heat flux, from conditions similar to the equations (7.11), is a typical problem in exploiting thermodynamic restrictions by the use of Lagrange multipliers with the general entropy inequality, see for example: for viscous fluids [52], for thermoelastic solids [53], for fluids and thermoelastic solids in electromagnetic field [30, 45], for rigid heat conductors [3] and for mixtures of fluids [54]. This problem is usually trivial if linear constitutive relations are assumed, as we have seen above. Otherwise, it can be quite difficult, even though such a relation may still be expected. In the following, we shall treat this problem without recourse to general representations for isotropic functions considered in Chap. 4. The proof will be based on the following lemma:

**Lemma 7.4.1** *Let $\boldsymbol{h}(A, \boldsymbol{v})$ be an isotropic vector function. Then the function $\boldsymbol{h}$ satisfies the following relation,*

$$(\delta_{ij} h_k - \delta_{ik} h_j) = \Big(\frac{\partial h_i}{\partial v_j} v_k - \frac{\partial h_i}{\partial v_k} v_j\Big) + \Big(\frac{\partial h_i}{\partial A_{jl}} A_{kl} + \frac{\partial h_i}{\partial A_{lj}} A_{lk} - \frac{\partial h_i}{\partial A_{kl}} A_{jl} - \frac{\partial h_i}{\partial A_{lk}} A_{lj}\Big).$$

*Proof.* If we define

$$\mathcal{F}(Q) = \boldsymbol{h}(QAQ^T, Q\boldsymbol{v}) - Q\,\boldsymbol{h}(A, \boldsymbol{v}) \qquad \forall Q \in \mathcal{L}(V),$$

then by taking the gradient with respect to $Q$ we obtain

$$\partial_Q \mathcal{F}(1)[W] = \partial_A \boldsymbol{h}[WA + AW^T] + \partial_{\boldsymbol{v}} \boldsymbol{h}[W\boldsymbol{v}] - W\boldsymbol{h}.$$

Since $\boldsymbol{h}(A, \boldsymbol{v})$ is an isotropic function, we have $\mathcal{F}(Q) = 0$ for any $Q \in \mathcal{O}(V)$, and hence from Lemma 4.2.4, proved in Chap. 4, it follows that

$$\partial_Q \mathcal{F}(1)[W] = 0,$$

for any $W \in Skw(V)$. Therefore we have, in the terms of components,

$$\Big(\frac{\partial h_i}{\partial A_{jl}} A_{kl} + \frac{\partial h_i}{\partial A_{lj}} A_{lk} + \frac{\partial h_i}{\partial v_j} v_k - \delta_{ij} h_k\Big) W_{jk} = 0,$$

which proves the lemma. □

One can easily extend the above lemma to the case for an isotropic vector function of an arbitrary number of vector and tensor variables.

Now we shall state the problem in the following theorem, but first let us introduce the following abbreviations:

$$k_i = \Phi_i - \Lambda q_i, \quad H^v_{ij} = \frac{\partial \Phi_i}{\partial v_j} - \Lambda \frac{\partial q_i}{\partial v_j}, \quad H^A_{ijk} = \frac{\partial \Phi_i}{\partial A_{jk}} - \Lambda \frac{\partial q_i}{\partial A_{jk}}. \tag{7.34}$$

We shall consider constitutive classes depending on an arbitrary number of vector and tensor variables. Dependence of scalar variables is irrelevant in isotropic functions and hence will not be mentioned explicitly.

**Theorem 7.4.2.** *Let $\boldsymbol{\Phi}$ and $\boldsymbol{q}$ be isotropic vector functions, and $\Lambda$ be an isotropic scalar function, of an arbitrary number of vector and tensor variables. Assume that*

i) *for $N$ vector variables $\boldsymbol{v}^a$, $a = 1, \cdots, N$,*

$$\Big(\frac{\partial \Phi_i}{\partial v^a_j} + \frac{\partial \Phi_j}{\partial v^a_i}\Big) - \Lambda\Big(\frac{\partial q_i}{\partial v^a_j} + \frac{\partial q_j}{\partial v^a_i}\Big) = 0, \tag{7.35}$$

ii) *for every other vector variable* $\boldsymbol{u}$,

$$\frac{\partial \Phi_i}{\partial u_j} - \Lambda \frac{\partial q_i}{\partial u_j} = 0, \tag{7.36}$$

iii) *for every tensor variable* $A$,

$$\frac{\partial \Phi_i}{\partial A_{jk}} - \Lambda \frac{\partial q_i}{\partial A_{jk}} = 0. \tag{7.37}$$

*Then for* $N = 1$, *the function* $\Lambda$ *is independent of all vector and tensor variables, and*

$$\boldsymbol{\Phi} = \Lambda \boldsymbol{q}.$$

*The above conclusion also holds for* $N = 2$ *if* $\boldsymbol{q}$ *and* $\boldsymbol{v}^1 \times \boldsymbol{v}^2$ *are functionally independent.*[3]

*Proof.* With the abbreviations (7.34), the assumption (i) implies that ${H^{v^a}}_{ij}$ is skew-symmetric,

$${H^{v^a}}_{ij} + {H^{v^a}}_{ji} = 0, \tag{7.38}$$

and the assumptions (ii) and (iii) give

$${H^u}_{ij} = 0, \qquad {H^A}_{ikj} = 0.$$

Applying Lemma 7.4.1 to the vectors $\boldsymbol{\Phi}$ and $\boldsymbol{q}$, we have

$$\begin{aligned}
(\delta_{ij}\Phi_k - \delta_{ik}\Phi_j) &= \sum \Big( \frac{\partial \Phi_i}{\partial v_j} v_k - \frac{\partial \Phi_i}{\partial v_k} v_j \Big) \\
&\quad + \sum \Big( \frac{\partial \Phi_i}{\partial A_{jl}} A_{kl} + \frac{\partial \Phi_i}{\partial A_{lj}} A_{lk} - \frac{\partial \Phi_i}{\partial A_{kl}} A_{jl} - \frac{\partial \Phi_i}{\partial A_{lk}} A_{lj} \Big), \\
(\delta_{ij} q_k - \delta_{ik} q_j) &= \sum \Big( \frac{\partial q_i}{\partial v_j} v_k - \frac{\partial q_i}{\partial v_k} v_j \Big) \\
&\quad + \sum \Big( \frac{\partial q_i}{\partial A_{jl}} A_{kl} + \frac{\partial q_i}{\partial A_{lj}} A_{lk} - \frac{\partial q_i}{\partial A_{kl}} A_{jl} - \frac{\partial q_i}{\partial A_{lk}} A_{lj} \Big),
\end{aligned}$$

where the summations are taken over all vector variables (including the vectors $\boldsymbol{v}^a$ and all other vectors $\boldsymbol{u}$) and all tensor variables, respectively.

Multiplying the second equation with $\Lambda$ and subtracting it from the first one, we obtain, by the use of the abbreviations (7.34) and the assumptions (ii) and (iii),

$$(\delta_{ij} k_k - \delta_{ik} k_j) = \sum_{a=1}^{N} ({H^a}_{ij}\, v_k^a - {H^a}_{ik}\, v_j^a), \tag{7.39}$$

[3] For the case $N > 2$ and some other more general conditions see [43].

in which for simplicity, we have written ${H^a}_{ij}$ for ${H^{v^a}}_{ij}$. This implies that

$$k_1 = \sum_{a=1}^{N} {H^a}_{12}\, v_2^a, \qquad k_2 = \sum_{a=1}^{N} {H^a}_{23}\, v_3^a, \qquad k_3 = \sum_{a=1}^{N} {H^a}_{31}\, v_1^a, \tag{7.40}$$

and the following system of six equations:

$$\begin{aligned}
&\sum_{a=1}^{N}({H^a}_{12}\, v_2^a + {H^a}_{31}\, v_3^a) = 0, \qquad &&\sum_{a=1}^{N}({H^a}_{12}\, v_3^a + {H^a}_{31}\, v_2^a) = 0,\\
&\sum_{a=1}^{N}({H^a}_{23}\, v_3^a + {H^a}_{12}\, v_1^a) = 0, \qquad &&\sum_{a=1}^{N}({H^a}_{23}\, v_1^a + {H^a}_{12}\, v_3^a) = 0,\\
&\sum_{a=1}^{N}({H^a}_{31}\, v_1^a + {H^a}_{23}\, v_2^a) = 0, \qquad &&\sum_{a=1}^{N}({H^a}_{31}\, v_2^a + {H^a}_{23}\, v_1^a) = 0,
\end{aligned} \tag{7.41}$$

by the use of the relation (7.38). The three equations on the right of (7.41) reduce to

$$\sum_{a=1}^{N} {H^a}_{12}\, v_3^a = 0, \qquad \sum_{a=1}^{N} {H^a}_{23}\, v_1^a = 0, \qquad \sum_{a=1}^{N} {H^a}_{31}\, v_2^a = 0. \tag{7.42}$$

We shall now proceed to prove the theorem for $N = 1$ and for $N = 2$ separately.

For $N = 1$, since $\boldsymbol{v}^1 \neq 0$ in general, from (7.42) we obtain

$${H^1}_{12} = 0, \qquad {H^1}_{23} = 0, \qquad {H^1}_{31} = 0.$$

Therefore, from (7.40) it follows that $\boldsymbol{k}$ must vanish, in other words, the relation $\boldsymbol{\Phi} = \Lambda\, \boldsymbol{q}$ holds.

By the substitution of the relation $\boldsymbol{\Phi} = \Lambda\, \boldsymbol{q}$ into the conditions (7.35), (7.36), and (7.37), it follows immediately that the partial derivatives of $\Lambda$ with respect to all the vector and the tensor variables must vanish, since, in general, the vector function $\boldsymbol{q}$ need not vanish, and the theorem is proved for $N = 1$.

For $N = 2$, from (7.41) and (7.42) we have the following linear system of six equations for six variables $(H^1_{23}, H^1_{31}, H^1_{12}, H^2_{23}, H^2_{31}, H^2_{12})$,

$$\begin{aligned}
&{H^1}_{12}\, v_2^1 + {H^1}_{31}\, v_3^1 + {H^2}_{12}\, v_2^2 + {H^2}_{31}\, v_3^2 = 0,\\
&{H^1}_{23}\, v_3^1 + {H^1}_{12}\, v_1^1 + {H^2}_{23}\, v_3^2 + {H^2}_{12}\, v_1^2 = 0,\\
&{H^1}_{31}\, v_1^1 + {H^1}_{23}\, v_2^1 + {H^2}_{31}\, v_1^2 + {H^2}_{23}\, v_2^2 = 0,\\
&{H^1}_{12}\, v_3^1 + {H^2}_{12}\, v_3^2 = 0,\\
&{H^1}_{23}\, v_1^1 + {H^2}_{23}\, v_1^2 = 0,\\
&{H^1}_{31}\, v_2^1 + {H^2}_{31}\, v_2^2 = 0.
\end{aligned} \tag{7.43}$$

The coefficient matrix of this system is of rank equal to 5, and the system admits a one-parameter solution given by

$$\frac{H^1{}_{23}}{v_1^2} = \frac{H^1{}_{31}}{v_2^2} = \frac{H^1{}_{12}}{v_3^2} = -\frac{H^2{}_{23}}{v_1^1} = -\frac{H^2{}_{31}}{v_2^1} = -\frac{H^2{}_{12}}{v_3^1} = \gamma,$$

which imply from (7.40) that

$$\boldsymbol{k} = \boldsymbol{\Phi} - \Lambda\,\boldsymbol{q} = \gamma\,(\boldsymbol{v}^1 \times \boldsymbol{v}^2), \tag{7.44}$$

where $\gamma$ is a scalar function of the vector and the tensor variables.

By the relation (7.44), the assumptions (ii) and (iii) lead to

$$\frac{\partial \Lambda}{\partial X}\,q_i + \frac{\partial \gamma}{\partial X}\,(\boldsymbol{v}^1 \times \boldsymbol{v}^2)_i = 0,$$

where $X$ stands for the components of any vector variable $\boldsymbol{u}$ and any tensor variable $A$. Since $\boldsymbol{q}$ and $\boldsymbol{v}^1 \times \boldsymbol{v}^2$ are functionally independent by assumption, the above relations are possible only if both $\Lambda$ and $\gamma$ are independent of $\boldsymbol{u}$ and $A$. Therefore, $\Lambda$ and $\gamma$ are functions of $\boldsymbol{v}^1$ and $\boldsymbol{v}^2$ only,

$$\Lambda = \Lambda(\boldsymbol{v}^1, \boldsymbol{v}^2), \qquad \gamma = \gamma(\boldsymbol{v}^1, \boldsymbol{v}^2).$$

Consequently, we also have $\boldsymbol{k} = \boldsymbol{k}(\boldsymbol{v}^1, \boldsymbol{v}^2)$.

On the other hand, since $\boldsymbol{k} = \boldsymbol{\Phi} - \Lambda\,\boldsymbol{q}$ is an isotropic vector function, it can be represented by,

$$\boldsymbol{k} = k_1\,\boldsymbol{v}^1 + k_2\,\boldsymbol{v}^2,$$

or we have

$$\gamma\,\boldsymbol{v}^1 \times \boldsymbol{v}^2 = k_1\,\boldsymbol{v}^1 + k_2\,\boldsymbol{v}^2,$$

where $k_1$ and $k_2$ are isotropic scalar functions of $(\boldsymbol{v}^1, \boldsymbol{v}^2)$. Taking the inner product of this relation with $\boldsymbol{v}^1 \times \boldsymbol{v}^2$ we obtain

$$\gamma\,(\boldsymbol{v}^1 \times \boldsymbol{v}^2) \cdot (\boldsymbol{v}^1 \times \boldsymbol{v}^2) = 0,$$

which implies that $\gamma$ must vanish. Therefore, $\boldsymbol{k} = 0$ and the relation $\boldsymbol{\Phi} = \Lambda\,\boldsymbol{q}$ holds.

Finally, by the substitution of $\boldsymbol{\Phi} = \Lambda\,\boldsymbol{q}$ into (7.35), it follows that the partial derivatives of $\Lambda$ with respect to $\boldsymbol{v}^1$ and $\boldsymbol{v}^2$ must vanish. Therefore, $\Lambda$ is independent of any vector and tensor variables. This completes the proof for $N = 2$. □

The problem of proving the relation (7.13), $\boldsymbol{\Phi} = \Lambda^{\varepsilon}\boldsymbol{q}$, for the viscous heat-conducting fluids considered in this chapter is a special case of the above theorem (N=1) with one vector variable $\operatorname{grad}\theta$ and one symmetric tensor variable $D$.

# 8. Rational Extended Thermodynamics

## 8.1 Introduction

In the constitutive theories of materials we have considered so far, there is one essential feature, namely, the basic equations are based upon the principle of balance of mass, momentum, and energy, while the diversity of materials is characterized solely by the functional complexity of constitutive equations.

A different approach, known as *Extended Thermodynamics*, has been proposed by Liu and Müller in [46], in which, in addition to the densities of mass, momentum, and energy, the momentum flux and the energy flux are also taken as basic field quantities. For these extended field quantities, balance equations of momentum flux and energy flux are postulated and the theory is formulated in the framework of rational thermodynamics laid down in the previous chapters.

Unlike the constitutive equations of ordinary thermodynamics, in which constitutive functions may depend on the history and space gradients of the basic field variables, the constitutive functions of extended thermodynamics are assumed to be instantaneous and local, i.e., they are functions of basic field variables only. As a consequence, the resulting field equations form a system of first-order quasi-linear partial differential equations – in contrast to the ordinary theory, for example, the system is of second-order in space and first-order in time in the case of Navier–Stokes equations (see (4.40)). Owing to the simple constitutive equations in the extended theories, the complexity of materials is no longer characterized by constitutive equations alone, rather it is also characterized by the choice of different basic fields. In other words, the complexity of materials rests upon the formulation of the system of balance equations.

Extended thermodynamics has been an area of active research in the last two decades. It has been found that it not only has more systematic mathematical structures – symmetric hyperbolic systems – it also has yielded very explicit constitutive relations, in complete agreement with the results of the kinetic theory of gases, in spite of being a phenomenological theory based on the framework of rational thermodynamics considered in this book. In this chapter, we shall give an outlook of rational extended thermodynamics. Most of the main results and references of recent developments can be found in the book of Müller and Ruggeri [57].

## 8.2 Formal Structure of System of Balance Equations

We consider a system of balance equations relative to an inertial frame

$$\frac{\partial \boldsymbol{u}}{\partial t} + \operatorname{div} \boldsymbol{H}(\boldsymbol{u}) = \boldsymbol{g}(\boldsymbol{u}) \tag{8.1}$$

for the basic field $\boldsymbol{u}(\boldsymbol{x},t)$ in $I\!R^N$. The flux $\boldsymbol{H}(\boldsymbol{x},t)$ and the production density $\boldsymbol{g}(\boldsymbol{x},t)$ are assumed to be functions of the basic field $\boldsymbol{u}(\boldsymbol{x},t)$. Following the idea of rational thermodynamics – the entropy principle – the constitutive function of $\boldsymbol{H}(\boldsymbol{u})$ and the production $\boldsymbol{g}(\boldsymbol{u})$ have to be determined in such a way that the system (8.1) is consistent with the entropy condition: (also see [22])

$$\frac{\partial h(\boldsymbol{u})}{\partial t} + \operatorname{div} \boldsymbol{\Pi}(\boldsymbol{u}) = \Sigma(\boldsymbol{u}) \geq 0, \tag{8.2}$$

where $h, \Sigma \in I\!R$ and $\boldsymbol{\Pi} \in I\!R^3$. Note that all the constitutive quantities are assumed to be functions of the basic variables $\boldsymbol{u}$ only.

This requirement leads to the existence of Lagrange multiplier $\boldsymbol{\Lambda}$ in $I\!R^N$ (see Sect. 7.3 for justification), such that

$$\left\{\frac{\partial h(\boldsymbol{u})}{\partial t} + \operatorname{div} \boldsymbol{\Pi}(\boldsymbol{u}) - \Sigma(\boldsymbol{u})\right\} - \boldsymbol{\Lambda} \cdot \left\{\frac{\partial \boldsymbol{u}}{\partial t} + \operatorname{div} \boldsymbol{H}(\boldsymbol{u}) - \boldsymbol{g}(\boldsymbol{u})\right\} = 0$$

holds for any $\boldsymbol{u}(\boldsymbol{x},t) \in I\!R^N$. In particular, it must hold for arbitrary values of $\dfrac{\partial \boldsymbol{u}}{\partial t}$ and $\dfrac{\partial \boldsymbol{u}}{\partial x_i}$. Since the above relation is linear in both of them, it follows immediately that

$$\frac{\partial h}{\partial \boldsymbol{u}} = \boldsymbol{\Lambda}, \qquad \frac{\partial \Pi_i}{\partial \boldsymbol{u}} = \boldsymbol{\Lambda} \cdot \frac{\partial \boldsymbol{H}_i}{\partial \boldsymbol{u}}, \qquad \Sigma = \boldsymbol{\Lambda} \cdot \boldsymbol{g}. \tag{8.3}$$

We can rewrite the relations $(8.3)_{1,2}$ in the differential form:

$$dh = \boldsymbol{\Lambda} \cdot d\boldsymbol{u}, \qquad d\Pi_i = \boldsymbol{\Lambda} \cdot d\boldsymbol{H}_i, \tag{8.4}$$

which we shall refer to as the *entropy-entropy flux integrability relations*, since the Lagrange multiplier $\boldsymbol{\Lambda}$ appears as the common integration factor of the right-hand sides for $h$ and $\Pi_i$, $i = 1,2,3$.

Furthermore, if we assume the invertibility of $\boldsymbol{\Lambda}(\boldsymbol{u})$, we can write the system (8.1) in terms of the variable $\boldsymbol{\Lambda}$ as

$$\frac{\partial \boldsymbol{u}}{\partial \boldsymbol{\Lambda}} \frac{\partial \boldsymbol{\Lambda}}{\partial t} + \sum_{i=1}^{3} \frac{\partial \boldsymbol{H}_i}{\partial \boldsymbol{\Lambda}} \frac{\partial \boldsymbol{\Lambda}}{\partial x_i} = \boldsymbol{g}. \tag{8.5}$$

Now, if we introduce $\widehat{h}(\boldsymbol{\Lambda})$ and $\widehat{\boldsymbol{\Pi}}(\boldsymbol{\Lambda})$ defined as

$$\widehat{h} = \boldsymbol{\Lambda} \cdot \boldsymbol{u} - h, \qquad \widehat{\Pi}_i = \boldsymbol{\Lambda} \cdot \boldsymbol{H}_i - \Pi_i, \tag{8.6}$$

which by the relation (8.4) lead to

$$d\widehat{h} = \boldsymbol{u} \cdot d\boldsymbol{\Lambda}, \qquad d\widehat{\Pi}_i = \boldsymbol{H}_i \cdot d\boldsymbol{\Lambda},$$

or equivalently,

$$\boldsymbol{u} = \frac{\partial \widehat{h}}{\partial \boldsymbol{\Lambda}}, \qquad \boldsymbol{H}_i = \frac{\partial \widehat{\Pi}_i}{\partial \boldsymbol{\Lambda}}. \tag{8.7}$$

Therefore, the system (8.5) becomes

$$\frac{\partial^2 \widehat{h}}{\partial \boldsymbol{\Lambda} \partial \boldsymbol{\Lambda}} \frac{\partial \boldsymbol{\Lambda}}{\partial t} + \sum_{i=1}^{3} \frac{\partial^2 \widehat{\Pi}_i}{\partial \boldsymbol{\Lambda} \partial \boldsymbol{\Lambda}} \frac{\partial \boldsymbol{\Lambda}}{\partial x_i} = \boldsymbol{g}. \tag{8.8}$$

In other words, the differential operator of the system is determined solely by a pair of a scalar and a vector function. For this reason, the functions $(\widehat{h}, \widehat{\boldsymbol{\Pi}})$ have been called the *generator functions* of the system [65]. Note that the coefficient matrices of this system are all symmetric $N \times N$ matrices.

**Remark.** From (8.3), it follows that

$$\frac{\partial \Pi_i}{\partial \boldsymbol{u}} = \frac{\partial h}{\partial \boldsymbol{u}} \cdot \frac{\partial \boldsymbol{H}_i}{\partial \boldsymbol{u}}. \tag{8.9}$$

In the literature of mathematical theory of hyperbolic conservation laws (see [13]), a pair of scalar and vector functions $(h(\boldsymbol{u}), \boldsymbol{\Pi}(\boldsymbol{u}))$ is called an *entropy–entropy flux pair* for the system of conservation laws,

$$\frac{\partial \boldsymbol{u}}{\partial t} + \operatorname{div} \boldsymbol{H}(\boldsymbol{u}) = 0, \tag{8.10}$$

if it satisfies (8.9).

For a given system of the form (8.10), if an entropy–entropy flux pair exists, it is usually not unique. Even though such a pair may not bear any physical significance, in general, it certainly reflects the notion of an entropy balance equation in continuum thermodynamics. Indeed, in extended thermodynamics, besides a system of balance equations, an additional entropy inequality is also postulated. The system is then required to be consistent with the entropy inequality, and hence the entropy and the entropy flux form a pair in the above sense. Consequently, extended thermodynamics constitutes a class of physical models for mathematical analysis of hyperbolic conservation laws. □

### 8.2.1 Symmetric Hyperbolic System

Consider a quasi-linear system of variable $\boldsymbol{u} \in I\!R^N$,

$$A(\boldsymbol{u}) \frac{\partial \boldsymbol{u}}{\partial t} + \sum_{i=1}^{3} B_i(\boldsymbol{u}) \frac{\partial \boldsymbol{u}}{\partial x_i} = \boldsymbol{g}(\boldsymbol{u}), \tag{8.11}$$

where $A$ and $B_i$ are $N \times N$ matrices.

**Definition.** The system (8.11) is called *symmetric* if $A$ and $B_i$ are all symmetric matrices.

**Definition.** The system (8.11) is called *hyperbolic* if $\det A \neq 0$ and for any $\boldsymbol{\nu} \in I\!R^3$, $\|\boldsymbol{\nu}\| = 1$, the eigenvalue problem

$$\Big(\sum_{i=1}^{3} B_i \nu_i - \lambda A\Big)\boldsymbol{v} = 0$$

has real eigenvalues $\lambda$ and admits $N$ linearly independent eigenvectors $\boldsymbol{v}$.

In the theory of wave propagation, $\boldsymbol{\nu}$ and $\lambda$ have the physical meaning of unit normal and normal speed of the wavefront, respectively. Moreover, from the spectral theorem (see p. 258), it follows that if the system is symmetric and $\det A \neq 0$, then it is also hyperbolic.

Now, returning to our system (8.8), we have the following theorem (see [22]):

**Theorem 8.2.1.** *Suppose that $h(\boldsymbol{u})$ is a concave function of $\boldsymbol{u}$, then the system (8.1), consistent with the entropy condition (8.2), can be reduced to a symmetric hyperbolic system.*

*Proof.* Note that the concavity of $h(\boldsymbol{u})$ is equivalent to the condition that the Hessian matrix

$$\frac{\partial^2 h}{\partial \boldsymbol{u} \partial \boldsymbol{u}} \quad \text{is negative definite.}$$

Therefore, from (8.3) the Jacobian matrix $\dfrac{\partial \boldsymbol{\Lambda}}{\partial \boldsymbol{u}}$ is non-singular, which ensures the invertibility of $\boldsymbol{\Lambda}(\boldsymbol{u})$. By the previous arguments leading to the system (8.8), which is symmetric, it is sufficient to show that it is also hyperbolic.

Since $h(\boldsymbol{u})$ is a concave function, for any variation $\delta\boldsymbol{u} \neq 0$, we have

$$\delta\boldsymbol{u} \cdot \frac{\partial^2 h}{\partial \boldsymbol{u} \partial \boldsymbol{u}} \delta\boldsymbol{u} < 0. \tag{8.12}$$

By the use of (8.3) and (8.7), we can write

$$\delta\boldsymbol{u} \cdot \frac{\partial^2 h}{\partial \boldsymbol{u} \partial \boldsymbol{u}} \delta\boldsymbol{u} = \delta\boldsymbol{u} \cdot \delta\Big(\frac{\partial h}{\partial \boldsymbol{u}}\Big) = \delta\boldsymbol{u} \cdot \delta\boldsymbol{\Lambda} = \delta\Big(\frac{\partial \widehat{h}}{\partial \boldsymbol{\Lambda}}\Big) \cdot \delta\boldsymbol{\Lambda} = \delta\boldsymbol{\Lambda} \cdot \frac{\partial^2 \widehat{h}}{\partial \boldsymbol{\Lambda} \partial \boldsymbol{\Lambda}} \delta\boldsymbol{\Lambda},$$

which implies that

$$A(\boldsymbol{\Lambda}) = \frac{\partial^2 \widehat{h}}{\partial \boldsymbol{\Lambda} \partial \boldsymbol{\Lambda}} \quad \text{is also negative definite.}$$

Therefore, $\det A(\boldsymbol{\Lambda}) \neq 0$ and the system (8.8) is hyperbolic. □

Some important properties of symmetric hyperbolic systems include well-posedness of Cauchy initial value problems and finite speeds of wave propagation [21, 63]. Such properties are almost essential for material behaviors that we might regard the symmetric hyperbolicity as a desirable property for material models.

In the following, we shall see that the Euler equations of an elastic fluid is an example of a symmetric hyperbolic system with concave entropy function.

**Example 8.2.1** Elastic fluids:
From (4.42) we have the following system of balance equations:

$$\begin{aligned}
&\frac{\partial \rho}{\partial t} + \frac{\partial}{\partial x_j}(\rho v_j) = 0, \\
&\frac{\partial}{\partial t}(\rho v_i) + \frac{\partial}{\partial x_j}(\rho v_i v_j + p\,\delta_{ij}) = 0, \\
&\frac{\partial}{\partial t}(\rho\varepsilon + \frac{1}{2}\rho v^2) + \frac{\partial}{\partial x_j}\Big(\rho\varepsilon\, v_j + \frac{1}{2}\rho v^2 v_j + p\, v_j\Big) = 0.
\end{aligned} \tag{8.13}$$

For an elastic fluid, the internal energy density $\varepsilon(\rho,\theta)$ and the pressure $p(\rho,\theta)$ also satisfy the Gibbs relation (5.25),

$$d\eta = \frac{1}{\theta}\Big(d\varepsilon - \frac{p}{\rho^2}d\rho\Big), \tag{8.14}$$

where $\theta$ is the absolute temperature and $\eta$ is the entropy density.

In order to show that the system (8.13) can be reduced to a symmetric hyperbolic system, it suffices to show the existence of an entropy condition (8.2) and the concavity of $h$ according to the above theorem.

In this case, we have

$$\begin{aligned}
\boldsymbol{u} &= \Big(\rho,\ \rho v_i,\ \rho(\varepsilon + \frac{v^2}{2})\Big) \in I\!R^5, \\
\boldsymbol{H}_j &= \Big(\rho v_j,\ \rho v_i v_j + p\delta_{ij},\ \rho(\varepsilon + \frac{v^2}{2} + \frac{p}{\rho})v_j\Big),
\end{aligned} \tag{8.15}$$

and the Lagrange multipliers are defined as

$$\boldsymbol{\Lambda} = \Big(-\frac{1}{\theta}(g - \frac{v^2}{2}),\ -\frac{1}{\theta}v_i,\ \frac{1}{\theta}\Big), \tag{8.16}$$

where $g$ is the free enthalpy defined in (5.79),

$$g = \varepsilon - \theta\eta + \frac{p}{\rho}.$$

Taking the inner product of the system (8.13) with $\boldsymbol{\Lambda}$, we obtain

$$\boldsymbol{\Lambda}\cdot\Big(\frac{\partial \boldsymbol{u}}{\partial t} + \frac{\partial \boldsymbol{H}_j}{\partial x_j}\Big) = 0, \tag{8.17}$$

which, after simplification using the Gibbs relation (8.14), becomes

$$\frac{\partial}{\partial t}(\rho\eta) + \frac{\partial}{\partial x_j}(\rho\eta v_j) = 0. \tag{8.18}$$

This is the entropy condition (8.2) for an elastic fluid by setting

$$h = \rho\eta, \qquad \Pi_j = \rho\eta v_j, \qquad \Sigma = 0.$$

Finally, in order to show the concavity of $h(\boldsymbol{u})$, let us consider the quadratic form,

$$\begin{aligned} \delta\boldsymbol{u}\cdot\frac{\partial^2 h}{\partial\boldsymbol{u}\partial\boldsymbol{u}}\delta\boldsymbol{u} &= \delta\boldsymbol{u}\cdot\delta\boldsymbol{\Lambda} \\ &= -\frac{\rho}{\theta}\left\{(\delta v_k)(\delta v_k) + \frac{1}{\theta}\frac{\partial\varepsilon}{\partial\theta}(\delta\theta)^2 + \frac{1}{\rho^2}\frac{\partial p}{\partial\rho}(\delta\rho)^2\right\}. \end{aligned} \tag{8.19}$$

In the above calculation the relations (8.14), (8.15) and (8.16) are used.

Since the density and the temperature are positive quantities, the function $h(\boldsymbol{u})$ is concave if and only if

$$\frac{\partial\varepsilon}{\partial\theta} > 0, \qquad \frac{\partial p}{\partial\rho} > 0,$$

i.e., the specific heat and the compressibility are positive, which are the conditions (5.71) and (5.72) of thermodynamic stability. In other words, the concavity condition is equivalent to the thermodynamic stability conditions in this case. □

Theories of mixtures and granular materials can also be formulated with hyperbolic field equations [54, 77, 78]. But field equations of ordinary thermodynamic theories are generally not even hyperbolic. Indeed, the Navier–Stokes equation (4.40) is parabolic. However, we shall see later that it is closely related to the extended theory with hyperbolic field equations, consisting of additional balance equations for the stress and the heat flux.

**Exercise 8.2.1** Derive the entropy condition (8.18) from (8.17).

**Exercise 8.2.2** Verify the calculations in (8.19).

### 8.2.2 Galilean Invariance

It is known that relative to inertial frames, dynamic laws, such as balance equations of mass, momentum, and energy, are Galilean invariant (see p. 42, Prop. 2.3.2). Likewise, in formulating a general system of balance equations relative to inertial frames, Galilean invariance must be regarded as a fundamental requirement.

Recall that a change of frame from $(\boldsymbol{x}, t)$ to $(\boldsymbol{x}^*, t^*)$ is a Galilean transformation if

$$\begin{aligned} \boldsymbol{x}^* &= Q(\boldsymbol{x} - \boldsymbol{x}_\circ) + \boldsymbol{V}t + \boldsymbol{c}_\circ, \\ t^* &= t + a, \end{aligned} \tag{8.20}$$

where $\boldsymbol{V}$ is a constant vector and $Q$ is a constant orthogonal tensor.

The requirement of Galilean invariance imposes rather a specific dependence of the flux and the production on the velocity field as we shall see. To begin with, let us express the basic field $\boldsymbol{u}$ in terms of the velocity $\boldsymbol{v}$ and the variables $\boldsymbol{w}$,

$$\boldsymbol{u} = \boldsymbol{F}(\boldsymbol{v}, \boldsymbol{w}) \in \mathbb{R}^N, \qquad \boldsymbol{w} \in \mathbb{R}^{N-3}. \tag{8.21}$$

The variables $\boldsymbol{w}$ are assumed to be objective quantities (with respective to an arbitrary change of frame, i.e., Euclidean transformations, see Sect. 1.7) of various tensorial order. We also split the flux $\boldsymbol{H}_k$ into the convective and non-convective parts,

$$\boldsymbol{H}_k(\boldsymbol{u}) = \boldsymbol{F}(\boldsymbol{v}, \boldsymbol{w})v_k + \boldsymbol{G}_k(\boldsymbol{v}, \boldsymbol{w}). \tag{8.22}$$

Let the velocity-independent part be denoted by

$$\tilde{\boldsymbol{F}}(\boldsymbol{w}) = \boldsymbol{F}(0, \boldsymbol{w}), \qquad \tilde{\boldsymbol{G}}_k(\boldsymbol{w}) = \boldsymbol{G}_k(0, \boldsymbol{w}), \qquad \tilde{\boldsymbol{g}}(\boldsymbol{w}) = \boldsymbol{g}(0, \boldsymbol{w}), \tag{8.23}$$

called the internal parts of the respective quantities. The internal quantities are assumed to be objective.

The balance equation (8.1) can now be written as

$$\frac{\partial}{\partial t}\boldsymbol{F}(\boldsymbol{v}, \boldsymbol{w}) + \frac{\partial}{\partial x_k}\Big(\boldsymbol{F}(\boldsymbol{v}, \boldsymbol{w})v_k + \boldsymbol{G}_k(\boldsymbol{v}, \boldsymbol{w})\Big) = \boldsymbol{g}(\boldsymbol{v}, \boldsymbol{w}). \tag{8.24}$$

**Proposition 8.2.2** *If the system of balance equations (8.24) is Galilean invariant, then there exists an $N \times N$ matrix $X(\boldsymbol{v})$ such that ([64])*

$$\begin{aligned} \boldsymbol{F}(\boldsymbol{v}, \boldsymbol{w}) &= X(\boldsymbol{v})\tilde{\boldsymbol{F}}(\boldsymbol{w}), \\ \boldsymbol{G}_k(\boldsymbol{v}, \boldsymbol{w}) &= X(\boldsymbol{v})\tilde{\boldsymbol{G}}_k(\boldsymbol{w}), \\ \boldsymbol{g}(\boldsymbol{v}, \boldsymbol{w}) &= X(\boldsymbol{v})\tilde{\boldsymbol{g}}(\boldsymbol{w}). \end{aligned} \tag{8.25}$$

*Moreover, $X(\boldsymbol{v})$ has the following properties: For any $\boldsymbol{v}^1, \boldsymbol{v}^2 \in \mathbb{R}^3$*

$$X(\boldsymbol{v}^1 + \boldsymbol{v}^2) = X(\boldsymbol{v}^1)X(\boldsymbol{v}^2), \qquad X(0) = \mathit{1}. \tag{8.26}$$

*Proof.* Consider a Galilean transformation (8.20) with $Q$ being the identity tensor, then we have

$$\boldsymbol{v}^* = \boldsymbol{v} + \boldsymbol{V}, \qquad \boldsymbol{w}^* = \boldsymbol{w}, \tag{8.27}$$

and

$$\frac{\partial}{\partial x_i^*} = \frac{\partial}{\partial x_i}, \qquad \frac{\partial}{\partial t^*} = \frac{\partial}{\partial t} - V_i \frac{\partial}{\partial x_i}. \tag{8.28}$$

Since the system (8.24) is Galilean invariant, it takes the same form in the frame $(\boldsymbol{x}^*, t^*)$,

$$\frac{\partial}{\partial t^*}\boldsymbol{F}(\boldsymbol{v}^*, \boldsymbol{w}^*) + \frac{\partial}{\partial x_k^*}\Big(\boldsymbol{F}(\boldsymbol{v}^*, \boldsymbol{w}^*)v_k^* + \boldsymbol{G}_k(\boldsymbol{v}^*, \boldsymbol{w}^*)\Big) = \boldsymbol{g}(\boldsymbol{v}^*, \boldsymbol{w}^*),$$

which from (8.27) and (8.28) reduces to

$$\frac{\partial}{\partial t}\boldsymbol{F}(\boldsymbol{v}+\boldsymbol{V}, \boldsymbol{w}) + \frac{\partial}{\partial x_k}\Big(\boldsymbol{F}(\boldsymbol{v}+\boldsymbol{V}, \boldsymbol{w})v_k + \boldsymbol{G}_k(\boldsymbol{v}+\boldsymbol{V}, \boldsymbol{w})\Big) = \boldsymbol{g}(\boldsymbol{v}+\boldsymbol{V}, \boldsymbol{w}). \tag{8.29}$$

This system must be equivalent to the system (8.24) and hence by comparison, it requires that one system must be a linear combination of the other. Consequently, there exists a non-singular $N \times N$ matrix $X(\boldsymbol{V})$ such that

$$\begin{aligned}
\boldsymbol{F}(\boldsymbol{v}+\boldsymbol{V}, \boldsymbol{w}) &= X(\boldsymbol{V})\boldsymbol{F}(\boldsymbol{v}, \boldsymbol{w}),\\
\boldsymbol{G}_k(\boldsymbol{v}+\boldsymbol{V}, \boldsymbol{w}) &= X(\boldsymbol{V})\boldsymbol{G}_k(\boldsymbol{v}, \boldsymbol{w}),\\
\boldsymbol{g}(\boldsymbol{v}+\boldsymbol{V}, \boldsymbol{w}) &= X(\boldsymbol{V})\boldsymbol{g}(\boldsymbol{v}, \boldsymbol{w}),
\end{aligned}$$

holds for any $\boldsymbol{v}$ and $\boldsymbol{V}$.

In particular, for $\boldsymbol{v} = 0$ we have for any vector $\boldsymbol{V}$

$$\begin{aligned}
\boldsymbol{F}(\boldsymbol{V}, \boldsymbol{w}) &= X(\boldsymbol{V})\boldsymbol{F}(0, \boldsymbol{w}),\\
\boldsymbol{G}_k(\boldsymbol{V}, \boldsymbol{w}) &= X(\boldsymbol{V})\boldsymbol{G}_k(0, \boldsymbol{w}),\\
\boldsymbol{g}(\boldsymbol{V}, \boldsymbol{w}) &= X(\boldsymbol{V})\boldsymbol{g}(0, \boldsymbol{w}).
\end{aligned}$$

Therefore, the relations (8.25) hold and from which the properties (8.26) follow immediately. □

The properties (8.26) define $X(\boldsymbol{v})$ as an exponential operator. Indeed, by taking the derivative of $(8.26)_1$ with respect to $\boldsymbol{v}^1$ and evaluating at $\boldsymbol{v}^1 = 0$, we obtain

$$\frac{\partial X}{\partial v_k} = A_k X, \qquad X(0) = \mathit{1}, \tag{8.30}$$

where $A_k$ for $k = 1, 2, 3$ are constant $N \times N$ matrices defined by

$$A_k = \left.\frac{\partial X}{\partial v_k}\right|_{\boldsymbol{v}=0}. \tag{8.31}$$

The solution of (8.30) is given by an exponential operator (see p. 71)

$$X(\boldsymbol{v}) = \exp(v_k A_k) = \mathit{1} + v_k A_k + \frac{1}{2} v_k v_l A_k A_l + \cdots. \tag{8.32}$$

We shall give explicit forms of these matrices for an elastic fluid in the following example.

**Example 8.2.2** For an elastic fluid, let $\boldsymbol{w} = (\rho, \varepsilon) \in I\!R^2$, then from $(8.15)_1$ we can write

$$\boldsymbol{F}(\boldsymbol{v}, \boldsymbol{w}) = \Big(\rho,\ \rho v_i,\ \rho(\varepsilon + \frac{v^2}{2})\Big),$$
$$\tilde{\boldsymbol{F}}(\boldsymbol{w}) = (\rho, 0, 0, 0, \rho\varepsilon),$$

and from $(8.15)_2$ we have the non-convective flux

$$\boldsymbol{G}_k(\boldsymbol{v}, \boldsymbol{w}) = (0,\ p\delta_{jk},\ pv_k),$$
$$\tilde{\boldsymbol{G}}_k(\boldsymbol{w}) = (0,\ p\delta_{jk},\ 0).$$

Hence, by comparison with the relation (8.25) we obtain

$$X(\boldsymbol{v}) = \begin{bmatrix} 1 & 0 & 0 \\ v_i & \delta_{ij} & 0 \\ \frac{1}{2}v^2 & v_j & 1 \end{bmatrix} = \begin{bmatrix} 1 & & & & \\ v_1 & 1 & & & \\ v_2 & 0 & 1 & & \\ v_3 & 0 & 0 & 1 & \\ \frac{1}{2}v^2 & v_1 & v_2 & v_3 & 1 \end{bmatrix}, \tag{8.33}$$

and from (8.31), we have $A_k$ for $k = 1, 2, 3$,

$$A_1 = \begin{bmatrix} 0 & & & & \\ 1 & 0 & & & \\ 0 & 0 & 0 & & \\ 0 & 0 & 0 & 0 & \\ 0 & 1 & 0 & 0 & 0 \end{bmatrix} \quad A_2 = \begin{bmatrix} 0 & & & & \\ 0 & 0 & & & \\ 1 & 0 & 0 & & \\ 0 & 0 & 0 & 0 & \\ 0 & 0 & 1 & 0 & 0 \end{bmatrix} \quad A_3 = \begin{bmatrix} 0 & & & & \\ 0 & 0 & & & \\ 0 & 0 & 0 & & \\ 1 & 0 & 0 & 0 & \\ 0 & 0 & 0 & 1 & 0 \end{bmatrix}.$$

Note that $X(\boldsymbol{v})$ and $A_k$ are all lower triangular matrices. □

**Exercise 8.2.3** From (8.26), show that

$$X(\boldsymbol{v})X(\boldsymbol{u}) = X(\boldsymbol{u})X(\boldsymbol{v}), \qquad X(\boldsymbol{v})^{-1} = X(-\boldsymbol{v}),$$

for any $\boldsymbol{v}, \boldsymbol{u} \in I\!R^3$.

## 8.3 System of Moment Equations

We have already mentioned some material models in mathematical physics that exhibit the elegant structure of symmetric hyperbolic systems (for some others, see [13]). Nevertheless, the most systematic models are the theory of moments akin to Grad's theory of moments in the kinetic theory of gases [24]. Our discussion of general balance equations will be centered on the structure of systems of fields consisting of moment densities of various orders.

We shall use the following notation: Round brackets indicate symmetrization of all indices within the brackets, i.e., terms are summed over all permutations of $n$ indices within the brackets and divided by $n!$, irrespective of whether the indices are distinct or not, e.g.,

$$v_{(i}A_{j)k} = \frac{1}{2}(v_i A_{jk} + v_j A_{ik}),$$
$$v_{(i}A_{ij)k} = \frac{1}{3}(v_i A_{ijk} + v_i A_{jik} + v_j A_{iik}).$$

## Definition of Moment Densities

In the kinetic theory of gases, physical quantities, such as density, stress, energy, and energy flux are associated with moments of a phase-density function $f(\boldsymbol{x}, \mathbf{c}, t)$. The phase density gives the number density of molecules with velocity $\mathbf{c}$ at place $\boldsymbol{x}$ and time $t$.

Macroscopic thermodynamic quantities can be defined, from the kinetic theory, as expectation values of a function $\psi(\boldsymbol{x}, \mathbf{c}, t)$,

$$\overline{\psi}(\boldsymbol{x}, t) = \int \psi(\boldsymbol{x}, \mathbf{c}, t) f(\boldsymbol{x}, \mathbf{c}, t)\, d\mathbf{c}. \tag{8.34}$$

In particular, the densities of mass, momentum and momentum flux are defined, respectively, with $\psi = m$, $\psi = mc_i$ and $\psi = mc_i c_j$,

$$F_0 = \overline{m}, \qquad F_i = \overline{mc_i}, \qquad F_{ij} = \overline{mc_i c_j},$$

where $m$ is the molecular mass. They are also called the zeroth-, the first- and the second-order moment densities. More generally, we define the moment density of order $n$ as

$$F_{i_1 \cdots i_n} = \overline{mc_{i_1} \cdots c_{i_n}}. \tag{8.35}$$

The mean velocity $\boldsymbol{v}$ of the gas can be defined as

$$v_i = \frac{F_i}{F_0}.$$

The velocity of a molecule relative to the mean velocity of the gas is called the peculiar velocity of the molecule,

$$C_i = c_i - v_i. \tag{8.36}$$

In the same manner, we may define moments with peculiar velocity,

$$\rho_{i_1 \cdots i_n} = \overline{mC_{i_1} \cdots C_{i_n}}, \tag{8.37}$$

called the internal moment densities of order $n$. We remark that the internal moment $\rho_{i_1 \cdots i_n}$ is an objective tensor quantity owing to the fact that the relative velocity is an objective vector (see Sect. 1.7.1 for verification).

With simple algebraic manipulations, it is easy to verify the following relations:

$$
\begin{aligned}
&F_0 = \rho, \\
&F_i = \rho_i + \rho v_i, \\
&F_{ij} = \rho_{ij} + 2\rho_{(i}v_{j)} + \rho v_i v_j, \\
&F_{ijk} = \rho_{ijk} + 3\rho_{(ij}v_{k)} + 3\rho_{(i}v_j v_{k)} + \rho v_i v_j v_k.
\end{aligned}
\tag{8.38}
$$

It is not a surprise to see that the right-hand sides look like a binomial expansion because of (8.36). Indeed, one can easily see that the following general relation holds:

$$
F_{i_1\cdots i_n} = \sum_{k=0}^{n} \binom{n}{k} v_{(i_1} \cdots v_{i_k} \rho_{i_{k+1}\cdots i_n)}, \tag{8.39}
$$

where

$$
\binom{n}{k} = \frac{n!}{k!(n-k)!},
$$

and we have adopted the following conventions:

$$
\begin{aligned}
&\text{for } k = n \qquad \rho_{i_{k+1}\cdots i_n} = \rho, \\
&\text{for } k = 0 \qquad F_{i_1\cdots i_k} = F_0, \quad v_{i_1}\cdots v_{i_k} = 1.
\end{aligned}
$$

### Moment Equations

We shall consider a system of balance equations (8.1) for the densities of moment $\boldsymbol{F} = (F_0, F_{i_1}, F_{i_1 i_2}, F_{i_1 i_2 i_3}, \cdots) \in I\!R^N$, $N \geq 5$,

$$
\frac{\partial \boldsymbol{F}}{\partial t} + \frac{\partial \boldsymbol{H}_k}{\partial x_k} = \boldsymbol{g}. \tag{8.40}
$$

We denote the non-convective part of the flux by $\boldsymbol{G}_k$,

$$
\boldsymbol{H}_k = \boldsymbol{F} v_k + \boldsymbol{G}_k. \tag{8.41}
$$

We shall denote the internal parts of moments, their fluxes, and production densities by

$$
\begin{aligned}
\tilde{\boldsymbol{F}} &= (\rho, \rho_{i_1}, \rho_{i_1 i_2}, \rho_{i_1 i_2 i_3}, \cdots), \\
\tilde{\boldsymbol{G}}_k &= (p_k, p_{i_1 k}, p_{i_1 i_2 k}, p_{i_1 i_2 i_3 k}, \cdots), \\
\tilde{\boldsymbol{g}} &= (\pi_0, \pi_{i_1}, \pi_{i_1 i_2}, \pi_{i_1 i_2 i_3}, \cdots).
\end{aligned}
\tag{8.42}
$$

Taking the relation (8.39) between $F_{i_1\cdots i_n}$ and $\rho_{i_1\cdots i_n}$ for granted, one can determine the $N \times N$ matrix $X(\boldsymbol{v})$ by comparison of (8.39) and $(8.25)_1$. The matrix, too cumbersome to write out here, is a lower triangular matrix with identity diagonal elements (a special case for $N = 5$ is given in (8.33)). For general discussions on the properties of these matrices for systems of moment equations see [57, 64].

Since the moments, the non-convective fluxes, and the production densities have the same structure given in (8.25), without explicitly using the matrix $X(\boldsymbol{v})$, we can easily obtain the following expressions:

$$\begin{aligned}
F_{i_1\cdots i_n} &= \sum_{k=0}^{n} \binom{n}{k} v_{(i_1} \cdots v_{i_k} \rho_{i_{k+1}\cdots i_n)}, \\
G_{i_1\cdots i_n j} &= \sum_{k=0}^{n} \binom{n}{k} v_{(i_1} \cdots v_{i_k} p_{i_{k+1}\cdots i_n)j}, \\
g_{i_1\cdots i_n} &= \sum_{k=0}^{n} \binom{n}{k} v_{(i_1} \cdots v_{i_k} \pi_{i_{k+1}\cdots i_n)}.
\end{aligned} \tag{8.43}$$

Note that the above structure implies that $F_{i_1\cdots i_n}$ and $g_{i_1\cdots i_n}$ are completely symmetric tensors, while $G_{i_1\cdots i_n j}$ is symmetric in the first $n$ indices only. Since $\rho_{i_1\cdots i_n}$ is completely symmetric by kinetic definition, we shall also assume that $\pi_{i_1\cdots i_n}$ is completely symmetric, and $p_{i_1\cdots i_n j}$ is symmetric in the first $n$ indices.

We can rewrite the balance equations (8.40) in terms of the internal parts of moments $\rho_{i_1\cdots t_n}$. It reads

$$\begin{aligned}
&\dot{\rho}_{i_1\cdots i_n} + \rho_{i_1\cdots i_n} \frac{\partial v_j}{\partial x_j} + \frac{\partial p_{i_1\cdots i_n j}}{\partial x_j} \\
&\quad + n \frac{\partial v_{(i_1}}{\partial x_j} p_{i_2\cdots i_n)j} + n \rho_{(i_1\cdots i_{n-1}} \dot{v}_{i_n)} = \pi_{i_1\cdots i_n},
\end{aligned} \tag{8.44}$$

for $n = 0, 1, 2, \cdots$, where the dot denotes the material time derivative. The proof of (8.44) using the relations (8.43) is straightforward algebraic calculations (see [39]) and will be left for the reader to verify.

### Hierarchy of Systems of Moments

The first five equations of the moment equations can be identified with conservation laws of mass, momentum, and energy, by introducing the following conventional quantities:

$$\begin{aligned}
&\rho && \text{mass density,} \\
&\varepsilon = \frac{1}{2\rho} \rho_{ii} && \text{specific internal energy,} \\
&T_{ij} = -p_{ij} && \text{stress tensor,} \\
&q_i = \frac{1}{2} p_{ijj} && \text{heat flux vector,}
\end{aligned} \tag{8.45}$$

and require

$$\rho_i = 0, \quad p_i = 0, \quad \pi_0 = 0, \quad \pi_i = 0, \quad \pi_{ii} = 0. \tag{8.46}$$

They read

$$\begin{aligned}
&\frac{\partial \rho}{\partial t} + \frac{\partial}{\partial x_j}(\rho v_j) = 0, \\
&\frac{\partial}{\partial t}(\rho v_i) + \frac{\partial}{\partial x_j}(\rho v_i v_j - T_{ij}) = 0, \\
&\frac{\partial}{\partial t}(\rho \varepsilon + \frac{1}{2}\rho v^2) + \frac{\partial}{\partial x_j}\Big((\rho \varepsilon + \frac{1}{2}\rho v^2)v_j - T_{ij}v_i + q_j\Big) = 0.
\end{aligned} \tag{8.47}$$

From (8.40) we have a system of equations for $N$ moments of increasing order. With the conditions (8.46), explicit expressions of (8.43) for moments up to the fourth-order are given in the following:

Moment densities:

$$\begin{aligned}
F_0 &= \rho, \\
F_i &= \rho v_i, \\
F_{ij} &= \rho_{ij} + \rho v_i v_j, \\
F_{ijk} &= \rho_{ijk} + 3\rho_{(ij}v_{k)} + \rho v_i v_j v_k.
\end{aligned}$$

Non-convective fluxes of moments:

$$\begin{aligned}
G_i &= 0, \\
G_{ij} &= p_{ij}, \\
G_{ijk} &= p_{ijk} + 2v_{(i}p_{j)k}, \\
G_{ijkl} &= p_{ijkl} + 3v_{(i}p_{jk)l} + 3v_{(i}v_j p_{k)l}.
\end{aligned}$$

Production densities of moments:

$$\begin{aligned}
g_0 &= 0, \\
g_i &= 0, \\
g_{ij} &= \pi_{\langle ij \rangle}, \\
g_{ijk} &= \pi_{ijk} + 3\pi_{(ij}v_{k)}.
\end{aligned}$$

We have introduced the notation,

$$A_{\langle ij \rangle} = A_{ij} - \frac{1}{3}A_{kk}\delta_{ij},$$

to represent the traceless part of a symmetric tensor $A_{ij}$.

We shall consider systems consist of $N$ moment equations with increasing tensorial order. The simplest one, $N = 5$, is a system consisting of conservation laws of mass, momentum, and energy given by (8.47).

In the framework of extended thermodynamics, a theory based on this system of five fields $(F, F_i, F_{ii})$, or $(\rho, v_i, \varepsilon)$, must be completed with constitutive equations:

$$T_{ij} = T_{ij}(\rho, v_i, \varepsilon),$$
$$q_j = q_j(\rho, v_i, \varepsilon).$$

Since the internal parts of moment and flux of moment are assumed to be objective, by the principle of material objectivity the above constitutive equations can not depend on $v_i$, and from the representation theorem (Sect. 4.3) they must reduce to

$$T_{ij} = -p(\rho, \varepsilon)\delta_{ij}, \qquad q_j = 0.$$

Therefore, the system (8.47) is identical to the Euler equations (8.13). In other words, an extended theory of five fields is necessarily a theory of an elastic fluid.

To account for a more accurate description of material behaviors, more basic field variables are needed and systems of balance equations involving higher moments will be required. Let

$$F_A = (F_0, F_i, F_{ij}, F_{ijk}, \cdots) \in I\!R^N.$$

The constitutive equations for the non-convective fluxes and the productions are functions of $F_A$,

$$\begin{aligned} G_{Ak} &= \tilde{G}_{Ak}(F_B), \\ g_A &= \tilde{g}_A(F_B), \end{aligned} \qquad A, B = 1, 2, \cdots, N. \tag{8.48}$$

Equivalently, in terms of internal parts of moments,

$$\rho_\alpha = (\rho, \rho_{ij}, \rho_{ijk}, \cdots) \in I\!R^{N-3},$$

we can replace (8.48) by the constitutive equations for internal parts of fluxes and productions,

$$\begin{aligned} p_{Ak} &= \tilde{p}_{Ak}(\rho_\alpha), \qquad & A &= 1, 2, \cdots, N, \\ \pi_A &= \tilde{\pi}_A(\rho_\alpha), \qquad & \alpha &= 1, 2, \cdots, N-3. \end{aligned} \tag{8.49}$$

The possible dependence on the velocity $v_i$ is not allowed by the principle of material objectivity. Moreover, since $\rho_\alpha$, $p_{Ak}$, and $\pi_A$ are all objective quantities, the constitutive functions $\tilde{p}_{Ak}$ and $\tilde{\pi}_A$ must be isotropic functions.

A severe restriction is placed on extended theories of moment equations by material symmetry. Recall that material symmetry is an invariance requirement of constitutive functions with respect to certain volume-preserving transformation of reference configurations (see Sect. 3.5). Since the quantities $\rho_\alpha$, $p_{Ak}$, and $\pi_A$ are invariant with respect to any such transformations, the constitutive functions $\tilde{p}_{Ak}$ and $\tilde{\pi}_A$ are invariant under any volume-preserving change of reference configurations. Therefore by the definition (3.46) of a fluid, we conclude that extended theories, based on the moment equations considered here, are applicable to *fluids* (or gases) only.[1]

We have seen that for $N = 5$ it is a theory of elastic fluids. In general, we have a hierarchy of fluid models depending on the number of fields $N$. For examples:

$$\begin{array}{ll}
N = 5 & F_A = (F_0, F_i, F_{ii}), \\
N = 10 & F_A = (F_0, F_i, F_{ij}), \\
N = 13 & F_A = (F_0, F_i, F_{ij}, F_{iij}), \\
N = 14 & F_A = (F_0, F_i, F_{ij}, F_{iij}, F_{iijj}), \\
N = \cdots & F_A = \cdots.
\end{array}$$

For a theory of fluids with viscosity and heat conduction, the minimum number of fields in the hierarchy is $N = 13$, which consists, in addition to the densities of mass, momentum, and energy, the fluxes of momentum and energy, so that the viscous stress and the heat flux are among the basic field variables.

More higher-order moments can be added to the basic fields, so as to obtain a better description of fluid behaviors. The higher-order moments do not usually have immediate physical interpretations, however, the presence of such fields may contribute some additional improvements to the fluid models.

**Exercise 8.3.1** From the relations $(8.25)_1$ and (8.39), determine the matrices $X(\boldsymbol{v})$ for the system of moments $F_A = (F_0, F_i, F_{ij}, F_{ijk})$; and by (8.31) show that $A_k$ is given by

$$A_k = \begin{bmatrix} 0 & & & \\ \delta_i^k & 0 & & \\ 0 & 2\delta_{(i}^k\delta_{j)}^p & 0 & \\ 0 & 0 & 3\delta_{(i}^k\delta_j^p\delta_{l)}^q & 0 \end{bmatrix}, \qquad k = 1, 2, 3. \tag{8.50}$$

## 8.4 Closure Problem

We consider a material state characterized by the fields $\boldsymbol{u} \in I\!R^N$ with their balance equations of the form (8.1),

$$\frac{\partial \boldsymbol{u}}{\partial t} + \operatorname{div} \boldsymbol{H}(\boldsymbol{u}) = \boldsymbol{g}(\boldsymbol{u}) \tag{8.51}$$

without external supplies. In order to be able to determine the state functions $\boldsymbol{u}$ for some initial boundary value problems, we need the constitutive equations for the flux $\boldsymbol{H}(\boldsymbol{u})$ and the production density $\boldsymbol{g}(\boldsymbol{u})$, so that (8.51) will

[1] Attempts have been made to generalize moment theories of extended thermodynamics to solids in [39, 40].

become a closed system of partial differential equations for the fields $\boldsymbol{u}(\boldsymbol{x}, t)$. Completion of balance equations with proper constitutive equations to yield a field theory is one of the main objectives of rational thermodynamics. This problem is also referred to as the "closure problem", a term frequently used in the kinetic theory of gases. The basic strategy of this problem has already been addressed earlier in Sect. 8.2. In the following sections, the closure procedure will be treated in more detail by the use of Lagrange multipliers.

### 8.4.1 Entropy Principle

Besides the Galilean invariance discussed in the preceding sections, the entropy principle is also imposed on the system of balance equations.

**Entropy principle.** *The system of balance equations* (8.51) *must be consistent with the entropy inequality,*

$$\frac{\partial h(\boldsymbol{u})}{\partial t} + \operatorname{div} \boldsymbol{\Pi}(\boldsymbol{u}) = \Sigma(\boldsymbol{u}) \geq 0, \tag{8.52}$$

*and the entropy density* $h(\boldsymbol{u})$ *is assumed to be a concave function.*

From the analysis in Sect. 8.2, the entropy principle implies the existence of the Lagrange multipliers $\boldsymbol{\Lambda}(\boldsymbol{u})$, which is invertible so that we can interchange $\boldsymbol{u}$ and $\boldsymbol{\Lambda}$ as variables. The existence of the Lagrange multipliers for the entropy–entropy flux integrability relations (8.4) imposes great restrictions on the constitutive functions. Moreover, in order to take into account the formal structure of the system and the principle of material objectivity, the entropy–entropy flux integrability relations (8.4) will be reformulated.

From the formal structure of balance equations, it is evident that the system depends on the velocity in a very explicit manner. The actual calculations in the closure procedure obviously can take this specific feature into account by separating velocity-dependent expressions in order to obtain the objective constitutive equations in terms of velocity-independent variables. This can be done by rewriting the entropy–entropy flux integrability relations (8.4) in terms of velocity-independent variables.

With the separation of the fields $\boldsymbol{u}$ into the velocity and the objective fields $\boldsymbol{w}$ in (8.21),

$$\boldsymbol{u} = \boldsymbol{F}(\boldsymbol{v}, \boldsymbol{w}) \in I\!R^N, \qquad \boldsymbol{w} \in I\!R^{N-3},$$

and the fluxes into convective and non-convective parts (8.22),

$$\boldsymbol{H}(\boldsymbol{u}) = \boldsymbol{F}(\boldsymbol{v}, \boldsymbol{w}) \otimes \boldsymbol{v} + \boldsymbol{G}(\boldsymbol{v}, \boldsymbol{w}), \tag{8.53}$$

we shall also express the entropy flux $\boldsymbol{\Pi}$ in a similar manner,

$$\boldsymbol{\Pi}(\boldsymbol{u}) = h(\boldsymbol{w})\boldsymbol{v} + \boldsymbol{\Phi}(\boldsymbol{w}). \tag{8.54}$$

In this expression, both the entropy density $h$ and the (internal) entropy flux $\boldsymbol{\Phi}$ are not allowed to depend on the velocity, as a consequence of the principle of material objectivity because they are regarded as objective quantities.

By the use of $(8.25)_1$, the relation $(8.4)_1$ gives

$$dh = \Lambda_A \, dF_A = \Lambda_A \, d(X_{AB}\tilde{F}_B) = \Lambda_A X_{AB} \, d\tilde{\boldsymbol{F}}_B + \Lambda_A \frac{\partial X_{AB}}{\partial v_k} \tilde{F}_B \, dv_k.$$

Since the entropy $h$ is independent of the velocity $v_i$, it follows that

$$dh = \tilde{\Lambda}_B d\tilde{F}_B, \qquad \Lambda_A \frac{\partial X_{AB}}{\partial v_k} \tilde{F}_B = 0, \tag{8.55}$$

where

$$\tilde{\boldsymbol{\Lambda}} = X(\boldsymbol{v})^T \boldsymbol{\Lambda} \tag{8.56}$$

is called the internal Lagrange multiplier. Since, from $(8.26)_2$, $X(0) = \mathit{1}$, it follows that

$$\tilde{\boldsymbol{\Lambda}}(\boldsymbol{w}) = \boldsymbol{\Lambda}(0, \boldsymbol{w}),$$

is the velocity-independent part of $\boldsymbol{\Lambda}$.

Similarly, the relation $(8.4)_2$ with (8.54) and (8.53) gives

$$\begin{aligned} d\Pi_i &= d(hv_i + \Phi_i) = v_i \, dh + h \, dv_i + d\Phi_i \\ &= \Lambda_A \, dH_{A_i} = \Lambda_A \, d(F_A v_i + G_{Ai}) \\ &= \Lambda_A X_{AB} \, d(\tilde{F}_B v_i + \tilde{G}_{Bi}) + \Lambda_A (\tilde{F}_B v_i + \tilde{G}_{Bi}) \, dX_{AB} \\ &= \tilde{\Lambda}_B \, d\tilde{G}_{Bi} + v_i \tilde{\Lambda}_B \, d\tilde{F}_B + \tilde{\Lambda}_B \tilde{F}_B \, dv_i + \Lambda_A (\tilde{F}_B v_i + \tilde{G}_{Bi}) \, dX_{AB}, \end{aligned}$$

and it follows that

$$d\Phi_i = \tilde{\Lambda}_B \, d\tilde{G}_{Bi}, \quad (\tilde{\Lambda}_B \tilde{F}_B - h)\delta_{ik} + \Lambda_A (\tilde{F}_B v_i + \tilde{G}_{Bi}) \frac{\partial X_{AB}}{\partial v_k} = 0. \tag{8.57}$$

Therefore, the above results can be stated in the following

**Proposition 8.4.1** *If the system of balance equations (8.51) is Galilean invariant, then the material objectivity and the entropy principle imply that there exists an internal Lagrange multiplier* $\tilde{\boldsymbol{\Lambda}}$ *satisfying*

$$\tilde{\boldsymbol{\Lambda}} \cdot A_k \tilde{\boldsymbol{F}} = 0, \qquad (\tilde{\boldsymbol{\Lambda}} \cdot \tilde{\boldsymbol{F}} - h)\delta_{ik} + \tilde{\boldsymbol{\Lambda}} \cdot A_k \tilde{\boldsymbol{G}}_i = 0, \tag{8.58}$$

*such that*

$$dh = \tilde{\boldsymbol{\Lambda}} \cdot d\tilde{\boldsymbol{F}}, \qquad d\boldsymbol{\Phi} = \tilde{\boldsymbol{\Lambda}} \cdot d\tilde{\boldsymbol{G}}, \tag{8.59}$$

*hold and the entropy production is given by*

$$\Sigma = \tilde{\boldsymbol{\Lambda}} \cdot \tilde{\boldsymbol{g}} \geq 0. \tag{8.60}$$

*Moreover,* $h(\tilde{\boldsymbol{F}})$, $\boldsymbol{\Phi}(\tilde{\boldsymbol{F}})$ *and* $\Sigma(\tilde{\boldsymbol{F}})$ *as well as* $\tilde{\boldsymbol{G}}(\tilde{\boldsymbol{F}})$ *and* $\tilde{\boldsymbol{g}}(\tilde{\boldsymbol{F}})$ *are isotropic functions.*

*Proof.* The relations (8.58) follow from $(8.55)_2$ and $(8.57)_2$ evaluated at $\boldsymbol{v} = 0$ and the definition of the matrix $A_k$ defined in (8.31). The entropy production (8.60) is a trivial consequence of $(8.3)_3$, $(8.25)_3$, and (8.56). □

We shall call the relations (8.58) the *Lagrange multiplier's identities* and (8.59) the *entropy–entropy flux integrability relations.* The integrability relation can be written in an equivalent form through a Legendre transformation similar to (8.6). We define $\widehat{h}$ and $\widehat{\boldsymbol{\Phi}}$ as

$$\widehat{h} = \tilde{\boldsymbol{\Lambda}} \cdot \tilde{\boldsymbol{F}} - h, \qquad \widehat{\boldsymbol{\Phi}} = \tilde{\boldsymbol{\Lambda}} \cdot \tilde{\boldsymbol{G}} - \boldsymbol{\Phi}, \tag{8.61}$$

then the integrability relations (8.59) become

$$d\widehat{h} = \tilde{\boldsymbol{F}} \cdot d\tilde{\boldsymbol{\Lambda}}, \qquad d\widehat{\boldsymbol{\Phi}} = \tilde{\boldsymbol{G}} \cdot d\tilde{\boldsymbol{\Lambda}}. \tag{8.62}$$

Note that both $\widehat{h}(\tilde{\boldsymbol{\Lambda}})$ and $\widehat{\boldsymbol{\Phi}}(\tilde{\boldsymbol{\Lambda}})$ are isotropic functions of $\tilde{\boldsymbol{\Lambda}}$.

### 8.4.2 Formal Procedures

The extended theory, in the framework of rational thermodynamics, was first formulated by Liu and Müller in [46], in which the closure procedure was based on the entropy–entropy flux integrability relations (8.59),

$$dh = \tilde{\boldsymbol{\Lambda}} \cdot d\tilde{\boldsymbol{F}}, \qquad d\boldsymbol{\Phi} = \tilde{\boldsymbol{\Lambda}} \cdot d\tilde{\boldsymbol{G}}, \tag{8.63}$$

by using the constitutive representations of $h$, $\boldsymbol{\Phi}$ and $\tilde{\boldsymbol{G}}$ in terms of $\tilde{\boldsymbol{F}}$ in a rather complicated calculation (also see [55]).

A different procedure, based on the relation (8.62)

$$d\widehat{h} = \tilde{\boldsymbol{F}} \cdot d\tilde{\boldsymbol{\Lambda}}, \qquad d\widehat{\boldsymbol{\Phi}} = \tilde{\boldsymbol{G}} \cdot d\tilde{\boldsymbol{\Lambda}}, \tag{8.64}$$

was proposed later (see [57]). It is clear that the fluxes $\tilde{\boldsymbol{G}}(\tilde{\boldsymbol{\Lambda}})$, as functions of the internal Lagrange multiplier $\tilde{\boldsymbol{\Lambda}}$, are completely determined by the knowledge of the vector function $\widehat{\boldsymbol{\Phi}}(\tilde{\boldsymbol{\Lambda}})$. On the other hand, the knowledge of the scalar function $\widehat{h}(\tilde{\boldsymbol{\Lambda}})$ gives the functions $\tilde{\boldsymbol{F}}(\tilde{\boldsymbol{\Lambda}})$, from which the Lagrange multiplier $\tilde{\boldsymbol{\Lambda}}$ in terms of $\tilde{\boldsymbol{F}}$ can be obtained by a functional inversion. Then, by a simple substitution, the constitutive equations of $\tilde{\boldsymbol{G}}(\tilde{\boldsymbol{F}}) = \tilde{\boldsymbol{G}}(\tilde{\boldsymbol{\Lambda}}(\tilde{\boldsymbol{F}}))$ can be obtained. This procedure based on the pair of functions $(\widehat{h}, \widehat{\boldsymbol{\Phi}})$, which are equivalent to the generator functions $(\widehat{h}, \widehat{\boldsymbol{\Pi}})$, is very elegant. The inversion is relatively easy for a linear theory, however, for higher-order constitutive equations, the inversion of a system of nonlinear equations may become more difficult if it can be done at all.

Here, we shall employ an alternative procedure for which there is no need for a functional inversion of the Lagrange multipliers but still maintains the elegant feature of the generator functions. It relies on the hybrid pair of generator functions $(h(\tilde{\boldsymbol{F}}), \widehat{\boldsymbol{\Phi}}(\tilde{\boldsymbol{\Lambda}}))$, with their integrability relations $(8.63)_1$ and $(8.64)_2$,

$$dh = \tilde{\boldsymbol{\Lambda}} \cdot d\tilde{\boldsymbol{F}}, \qquad d\widehat{\boldsymbol{\Phi}} = \tilde{\boldsymbol{G}} \cdot d\tilde{\boldsymbol{\Lambda}}. \tag{8.65}$$

From which we can obtain

$$\tilde{\boldsymbol{\Lambda}} = \tilde{\boldsymbol{\Lambda}}(\tilde{\boldsymbol{F}}) = \frac{\partial h}{\partial \tilde{\boldsymbol{F}}}, \qquad \tilde{\boldsymbol{G}} = \tilde{\boldsymbol{G}}(\tilde{\boldsymbol{\Lambda}}) = \frac{\partial \widehat{\boldsymbol{\Phi}}}{\partial \tilde{\boldsymbol{\Lambda}}},$$

and by a simple substitution of the first relation into the second, we immediately obtain the constitutive equations of the flux $\tilde{\boldsymbol{G}}$ in terms of the field variables $\tilde{\boldsymbol{F}}$,

$$\tilde{\boldsymbol{G}} = \tilde{\boldsymbol{G}}(\tilde{\boldsymbol{\Lambda}}(\tilde{\boldsymbol{F}})).$$

This procedure will be illustrated by the theory of 13 fields for viscous heat-conducting fluids in the following section. However, for simplicity, only linear constitutive equations will be derived. The procedure can be applied to obtain higher-order constitutive equations, for which calculations are generally cumbersome but mostly straightforward (see [44]).

## 8.5 Thirteen-Moment Theory of Viscous Heat-Conducting Fluid

We consider the state of a viscous heat-conducting fluid to be characterized by the fields of 13 moments, $(F_0, F_i, F_{ij}, F_{iij})$, and from (8.42), we have

$$\tilde{\boldsymbol{F}} = (\rho, 0, \rho_{ij}, \rho_{iij}),$$
$$\tilde{\boldsymbol{G}} = (0, p_{ik}, p_{ijk}, p_{iijk}),$$
$$\tilde{\boldsymbol{g}} = (0, 0, \pi_{\langle ij\rangle}, \pi_{iij}).$$

The entropy principle implies the existence of the internal Lagrange multiplier denoted as

$$\tilde{\boldsymbol{\Lambda}} = (\Lambda, \Lambda_i, \Lambda_{ij}, \lambda_j),$$

for the integrability relations (8.59), which now take the forms:

$$dh = \Lambda\, d\rho + \Lambda_{ij}\, d\rho_{ij} + \lambda_j\, d\rho_{iij}, \tag{8.66}$$
$$d\Phi_k = \Lambda_i\, dp_{ik} + \Lambda_{ij}\, dp_{ijk} + \lambda_j\, dp_{iijk}. \tag{8.67}$$

The Lagrange multipliers must also satisfy the identities (8.58), which by the use of (8.50) can be written as

$$\Lambda_j\, \rho + 3\lambda_{(i}\, \rho_{ij)} = 0, \tag{8.68}$$

$$(\Lambda\, \rho + \Lambda_{ij}\, \rho_{ij} + \lambda_j\, \rho_{iij} - h)\delta_{lk} + 2\Lambda_{lj}\, p_{jk} + 3\lambda_{(j}\, p_{jl)k} = 0. \tag{8.69}$$

From (8.60), we also have the inequality for the entropy production density,

$$\Sigma = \Lambda_{\langle ij\rangle}\pi_{\langle ij\rangle} + \lambda_j\, \pi_{iij} \geq 0. \tag{8.70}$$

**Exercise 8.5.1** Verify the identities (8.68) and (8.69) by direct calculations for the velocity independence of $h$ and $\Phi_i$ from (8.4) by the use of explicit expressions for $F_A$ and $G_{Ai}$.

### Absolute Temperature and Non-Equilibrium Variables

We define equilibrium as a process with no entropy production. From (8.70), the entropy production has a minimum in equilibrium for any process with no productions, i.e., $\Sigma = 0$ when $\pi_{\langle ij\rangle} = 0$ and $\pi_{iij} = 0$.

Since we can regard $\Sigma$, $\pi_{\langle ij\rangle}$ and $\pi_{iij}$ as functions of $\tilde{\Lambda}$, therefore, if we assume the invertibility so that we can replace $(\Lambda_{\langle ij\rangle}, \lambda_j)$ by $(\pi_{\langle ij\rangle}, \pi_{iij})$ as variables in $\Sigma$, then from (8.70) the necessary conditions for the minimum of the entropy production imply that

$$\left.\frac{\partial \Sigma}{\partial \pi_{ij}}\right|_E = \Lambda_{\langle ij\rangle}|_E = 0, \qquad \left.\frac{\partial \Sigma}{\partial \pi_{iij}}\right|_E = \lambda_j|_E = 0, \qquad \Lambda_j|_E = 0.$$

The last condition follows from (8.68). In other words, the Lagrange multipliers $\Lambda_{\langle ij\rangle}$, $\lambda_j$ and $\Lambda_j$ must also vanish in equilibrium.

With the conventional physical quantities (8.45), $\varrho_{ii} = 2\rho\varepsilon$, and the specific entropy density $\eta = h/\rho$, in equilibrium the relation (8.66) reduces to

$$dh|_E = d(\rho\eta|_E) = \Lambda|_E\, d\rho + \frac{2}{3}\Lambda_{ii}|_E\, d(\rho\varepsilon). \tag{8.71}$$

By reference to the Gibb's relation of thermostatics $(5.30)_2$, it follows that

$$\frac{1}{\theta} = \frac{2}{3}\Lambda_{ii}|_E, \qquad \Lambda|_E = -\frac{1}{\theta}g|_E, \tag{8.72}$$

where $\theta$ is the absolute temperature, $g$ is the free enthalpy function,

$$g = \varepsilon - \theta\eta + \frac{p}{\rho}, \tag{8.73}$$

and $p = p_{ii}/3$ is the mean pressure.

We shall be content with formulating the theory for thermodynamic processes "near" equilibrium. Mathematically, this means that non-equilibrium variables will be regarded as small quantities of the first-order and constitutive functions will be represented in terms of them up to certain order.

Since the Lagrange multipliers $\Lambda_{\langle ij\rangle}$, $\lambda_j$, and $\Lambda_j$ vanish in equilibrium, they are non-equilibrium quantities. We can also decompose the Lagrange multipliers $\Lambda_{ii}$ and $\Lambda$ into the equilibrium and non-equilibrium parts,

$$\Lambda_{ii} = \frac{3}{2\theta} + \lambda_\varepsilon, \qquad \Lambda = -\frac{1}{\theta}g|_E + \lambda_\rho. \tag{8.74}$$

Therefore, $(\lambda_\rho, \lambda_\varepsilon, \Lambda_i, \Lambda_{\langle ij\rangle}, \lambda_i)$ forms the set of non-equilibrium variables of the Lagrange multipliers $\tilde{\boldsymbol{\Lambda}}$.

On the other hand, what are the non-equilibrium variables of the moments $\tilde{\boldsymbol{F}}$? To answer this, we start by regarding $\tilde{\boldsymbol{F}}$ as functions of $\tilde{\boldsymbol{\Lambda}}$, and hence, from the decomposition (8.74), we can write

$$\rho_{ij} = \rho_{ij}(\rho, \theta, \lambda_\rho, \lambda_\varepsilon, \Lambda_m, \Lambda_{\langle mn\rangle}, \lambda_m),$$
$$\rho_{iij} = \rho_{iij}(\rho, \theta, \lambda_\rho, \lambda_\varepsilon, \Lambda_m, \Lambda_{\langle mn\rangle}, \lambda_n).$$

Since all quantities in these expressions are objective, the principle of material objectivity requires that for any orthogonal tensor $Q$,

$$\rho_{ij}(Q_{mp}\Lambda_p,\ Q_{mp}Q_{nq}\Lambda_{\langle pq\rangle},\ Q_{mp}\lambda_p) = Q_{ik}Q_{jl}\,\rho_{kl}(\Lambda_m,\ \Lambda_{\langle mn\rangle},\ \lambda_n),$$
$$\rho_{iij}(Q_{mp}\Lambda_p,\ Q_{mp}Q_{nq}\Lambda_{\langle pq\rangle},\ Q_{mp}\lambda_p) = Q_{jl}\,\rho_{iil}(\Lambda_m,\ \Lambda_{\langle mn\rangle},\ \lambda_n),$$

in which the irrelevant scalar variables have not been included. After evaluating the above relations in equilibrium, we obtain

$$\rho_{ij}(0,0,0) = Q_{ik}Q_{jl}\,\rho_{kl}(0,0,0),$$
$$\rho_{iij}(0,0,0) = Q_{jl}\,\rho_{iil}(0,0,0),$$

i.e., for any orthogonal tensor $Q$,

$$\rho_{ij}|_E = Q_{ik}Q_{jl}\,\rho_{kl}|_E, \qquad \rho_{iij}|_E = Q_{jl}\,\rho_{iil}|_E.$$

Therefore, it follows that

$$\rho_{\langle ij\rangle}|_E = 0, \qquad \rho_{iij}|_E = 0. \tag{8.75}$$

The second one follows from the fact that not every orthogonal tensor has the same eigenvalue, while the first one follows from the commutation theorem (see p. 258), which asserts that $\rho_{ij}|_E$ must be a multiple of the identity tensor.

Therefore, we have the following two sets of non-equilibrium variables for the moments and the Lagrange multipliers, respectively,

$$\{\rho_{\langle ij\rangle}, \rho_{iij}\} \in \tilde{\boldsymbol{F}}, \quad \text{and} \quad \{\lambda_\rho, \lambda_\varepsilon, \Lambda_i, \Lambda_{\langle ij\rangle}, \lambda_i\} \in \tilde{\boldsymbol{\Lambda}}. \tag{8.76}$$

### Representations of Hybrid Generator Functions

From (8.66) we have the integrability relation,

$$dh = \Lambda\, d\rho + \frac{2}{3}\Lambda_{ii}\, d(\rho\varepsilon) + \Lambda_{\langle ij\rangle}\, d\rho_{\langle ij\rangle} + \lambda_j\, d\rho_{iij},$$

which, by the use of (8.71), (8.72), and (8.74), can be written as

$$\begin{aligned} dh = dh|_E &+ \Big(\lambda_\rho + \frac{2}{3}\lambda_\varepsilon\big(\theta\frac{\partial h|_E}{\partial \rho} + g|_E\big)\Big)d\rho \\ &+ \frac{2}{3}\lambda_\varepsilon\theta\frac{\partial h|_E}{\partial \theta}\, d\theta + \Lambda_{\langle ij\rangle}\, d\rho_{\langle ij\rangle} + \lambda_j\, d\rho_{iij}. \end{aligned} \tag{8.77}$$

According to (8.61), we now define

$$\widehat{\Phi}_k = \Lambda_i\, p_{ik} + \Lambda_{ij}\, p_{ijk} + \lambda_j\, p_{iijk} - \Phi_k, \tag{8.78}$$

and from (8.67), we have the relations,

$$d\widehat{\Phi}_k = -\frac{1}{\theta^2} q_k\, d\theta + \frac{2}{3} q_k\, d\lambda_\varepsilon + p_{ik}\, d\Lambda_i + p_{\langle ij\rangle k}\, d\Lambda_{\langle ij\rangle} + p_{iijk}\, d\lambda_j. \tag{8.79}$$

The two relations (8.77) and (8.79) are in the form of the hybrid pair (8.65), from which we can regard $h$ and $\widehat{\Phi}_k$ as functions of

$$\begin{aligned} h &= h\,(\rho, \theta, \rho_{\langle ij\rangle}, \rho_{iij}), \\ \widehat{\Phi}_k &= \widehat{\Phi}_k(\rho, \theta, \lambda_\varepsilon, \Lambda_i, \Lambda_{\langle ij\rangle}, \lambda_j). \end{aligned} \tag{8.80}$$

Of course, from (8.79) we have

$$\frac{\partial \widehat{\Phi}_k}{\partial \rho} = 0. \tag{8.81}$$

We remark that although, from (8.68),

$$\Lambda_i = -\frac{10}{3}\varepsilon\, \lambda_i - \frac{2}{\rho}\rho_{\langle ij\rangle}\lambda_j, \tag{8.82}$$

and hence $\Lambda_i$ can be eliminated from the list of independent variables for the function $\widehat{\Phi}_k$, we have decided not to do so in order to maintain the simplicity of the integrability relation (8.79).

From the requirement of material objectivity, the constitutive functions for $h$ and $\widehat{\Phi}_k$ are isotropic functions. Since $(\rho_{\langle ij\rangle}, \rho_{iij})$ and $(\lambda_\varepsilon, \Lambda_i, \Lambda_{\langle ij\rangle}, \lambda_j)$ are small quantities of the first-order, we can easily write down their representations up to the second-order terms:

$$h \;= h|_E + h_1\, \rho_{iij}\rho_{nnj} + h_2\, \rho_{\langle ij\rangle}\rho_{\langle ij\rangle} + o(3), \tag{8.83}$$

$$\widehat{\Phi}_k = \alpha\, \lambda_k + \beta\, \Lambda_k + a_1\, \lambda_\varepsilon \lambda_k + b_1\, \lambda_\varepsilon \Lambda_k + a_2\, \Lambda_{\langle kj\rangle}\lambda_j + b_2\, \Lambda_{\langle kj\rangle}\Lambda_j + o(3), \tag{8.84}$$

where all the coefficients are functions of $(\rho, \theta)$.

Our objective is to determine the constitutive equations of the fluxes $(q_k, p_{ik}, p_{\langle ij\rangle k}, p_{iijk})$ as functions of the fields $(\rho, \theta, \rho_{\langle ij\rangle}, \rho_{iij})$. For the linear constitutive equations, we shall need only the representations (8.83) and (8.84) up to the second-order terms. The entire calculations are based on the four relations, namely, the entropy–entropy flux integrability relations (8.77), (8.79), and the Lagrange multiplier's identities (8.68) (or (8.82)) and (8.69).

We shall introduce the following abbreviations to be used later,

$$A_1 = a_1 - \frac{10}{3}\varepsilon b_1, \qquad A_2 = a_2 - \frac{10}{3}\varepsilon b_2. \tag{8.85}$$

**Entropy–Entropy Flux Integrability Relations**

First we obtain the Lagrange multipliers from the integrability relation (8.77) and the expression (8.83),

$$\begin{aligned} \lambda_\varepsilon &= o(2), \\ \lambda_\rho &= o(2), \\ \lambda_j &= 2h_1\, \rho_{iij} + o(2), \\ \Lambda_{\langle ij\rangle} &= 2h_2\, \rho_{\langle ij\rangle} + o(2). \end{aligned} \tag{8.86}$$

Similarly, from (8.79) and (8.84) we have the the following expressions for the fluxes,

$$\begin{aligned} \frac{2}{3}q_k &= a_1\lambda_k + b_1\Lambda_k + o(2), \\ p_{ik} &= \beta\,\delta_{ik} + b_2\Lambda_{\langle ik\rangle} + o(2), \\ p_{\langle ij\rangle k} &= a_2\,\delta_{k\langle i}\lambda_{j\rangle} + b_2\,\delta_{k\langle i}\Lambda_{j\rangle} + o(2), \\ p_{iijk} &= \alpha\,\delta_{jk} + a_2\Lambda_{\langle jk\rangle} + o(2), \end{aligned} \tag{8.87}$$

and

$$-\frac{q_k}{\theta^2} = \frac{\partial\alpha}{\partial\theta}\lambda_k + \frac{\partial\beta}{\partial\theta}\Lambda_k + o(2), \tag{8.88}$$

$$0 = \frac{\partial\alpha}{\partial\rho}\lambda_k + \frac{\partial\beta}{\partial\rho}\Lambda_k + o(2). \tag{8.89}$$

**Lagrange Multiplier's Identities**

From (8.82), (8.86) leads to

$$\Lambda_j = -\frac{20}{3}\varepsilon h_1\, \rho_{iij} + o(2). \tag{8.90}$$

The identity (8.69), after substitution of the Lagrange multipliers, gives an identity in the field variables. In particular, the first-order terms of the trace part are identically satisfied, while the traceless part leads to

$$\frac{1}{\theta}\,p_{\langle kl\rangle} + 2p\,\Lambda_{\langle kl\rangle} = 0,$$

from which the relation (8.87) implies that

$$\beta = p, \qquad b_2 = -2p\theta. \tag{8.91}$$

From the above results, we can obtain, from (8.88) and (8.89),

$$\frac{\partial\alpha}{\partial\theta} - \frac{10}{3}\varepsilon\frac{\partial p}{\partial\theta} = -\frac{3}{2\theta^2}A_1, \tag{8.92}$$

$$\frac{\partial\alpha}{\partial\rho} - \frac{10}{3}\varepsilon\frac{\partial p}{\partial\rho} = 0. \tag{8.93}$$

By the use of (8.82), (8.93) permits the relation (8.81) to be further evaluated up to the second-order terms. It reads

$$\frac{2}{\rho}\frac{\partial\beta}{\partial\rho}\rho_{\langle kj\rangle}\lambda_j = \frac{\partial A_2}{\partial\rho}\Lambda_{\langle kj\rangle}\lambda_j, \quad \text{or} \quad \frac{\partial A_2}{\partial\rho} = \frac{1}{\rho h_2}\frac{\partial p}{\partial\rho}. \tag{8.94}$$

### Summary of Results

Summarizing the above results, we can now obtain the linear constitutive equations for the fluxes:

$$\begin{aligned} p_{ik} &= p\,\delta_{ik} - 4p\theta h_2\,\rho_{\langle ik\rangle} + o(2),\\ q_k &= 3\,A_1 h_1\,\rho_{iik} + o(2),\\ p_{\langle ij\rangle k} &= 2\,A_2 h_1\,\delta_{k\langle i}\,\rho_{j\rangle nn} + o(2),\\ p_{iijk} &= \alpha\,\delta_{jk} + 2\,a_2 h_2\,\rho_{\langle jk\rangle} + o(2), \end{aligned} \tag{8.95}$$

where

$$\begin{aligned} A_1 &= -\frac{2}{3}\theta^2\Big(\frac{\partial\alpha}{\partial\theta} - \frac{10}{3}\,\varepsilon\frac{\partial p}{\partial\theta}\Big),\\ A_2 &= a_2 + \frac{20}{3}\,\theta p\varepsilon,\\ h_2 &= \frac{1}{\rho}\frac{\partial p}{\partial\rho}\Big(\frac{\partial A_2}{\partial\rho}\Big)^{-1}. \end{aligned} \tag{8.96}$$

On inspection of these constitutive equations, we conclude that besides the equilibrium state functions $\varepsilon(\rho,\theta)$ and $p(\rho,\theta)$, only three unknown material functions, $h_1(\rho,\theta)$, $a_2(\rho,\theta)$, and $\alpha(\rho,\theta)$, are needed. Note that the function $\alpha$ can be determined from the equilibrium state functions to within a function of $\theta$ by integrating (8.93).

### Concavity of Entropy Function

In the entropy principle, the entropy density is assumed to be a concave functions of $\boldsymbol{u}$. By evaluating at $\boldsymbol{v} = 0$ and $\delta\boldsymbol{v} = 0$, the condition (8.12) can now be expressed as

$$\delta\rho\,\delta\Lambda + \delta\rho_{ij}\,\delta\Lambda_{ij} + \delta\rho_{iij}\,\delta\lambda_{nnj} < 0, \tag{8.97}$$

for any variations of $(\rho, \theta, \rho_{\langle ij\rangle}, \rho_{iij})$.

In particular, If we consider a variation with $\delta\rho \neq 0$, $\delta\theta \neq 0$ and $\delta\rho_{\langle ij\rangle} = 0$, $\delta\rho_{iij} = 0$, then by the use of the relations (8.71) through (8.74), we obtain from (8.97)

$$\frac{1}{\rho\theta}\frac{\partial p}{\partial\rho}(\delta\rho)^2 + \frac{\varrho}{\theta^2}\frac{\partial\varepsilon}{\partial\theta}(\delta\theta)^2 > 0, \tag{8.98}$$

which implies that

$$\frac{\partial p}{\partial\rho} > 0, \qquad \frac{\partial\varepsilon}{\partial\theta} > 0. \tag{8.99}$$

The first inequality states that the isothermal compressibility is positive, while the second one ensures the positiveness of the specific heat at constant volume. These conditions are equivalent to the conditions of thermodynamic stability considered in Sect. 5.5.

Similarly, if we consider a variation with $\delta\rho = 0$, $\delta\theta = 0$ and $\delta\rho_{\langle ij\rangle} \neq 0$, $\delta\rho_{iij} \neq 0$, then from (8.86) it follows that

$$2h_2\delta\rho_{\langle ij\rangle}\delta\rho_{\langle ij\rangle} + 2h_1\delta\rho_{iij}\delta\rho_{nnj} < 0,$$

which leads to

$$h_1 < 0, \qquad h_2 < 0. \tag{8.100}$$

Of course, there are other restrictions imposed by the inequality (8.97), but they are of no immediate interest here.

**Exercise 8.5.2** Verify the inequality (8.98).

### Production Densities

The production densities can also be given by linear representations:

$$\begin{aligned} \pi_{ij} &= -\frac{1}{\tau_s}\rho_{\langle ij\rangle} + o(2), \\ \pi_{iij} &= -\frac{1}{\tau_q}\rho_{iij} + o(2), \end{aligned} \tag{8.101}$$

since they are isotropic functions and vanish in equilibrium. Both $\tau_s(\rho,\theta)$ and $\tau_q(\rho,\theta)$ are called relaxation times. Accordingly, from the relation (8.70) the entropy production density is given by

$$\Sigma = -2\,\frac{h_2}{\tau_s}\,\rho_{\langle ij\rangle}\rho_{\langle ij\rangle} - 2\,\frac{h_1}{\tau_q}\,\rho_{iij}\rho_{iij} + o(3), \tag{8.102}$$

and since it is non-negative, from (8.100) we have

$$\tau_s \geq 0, \qquad \tau_q \geq 0. \tag{8.103}$$

## 8.5.1 Field Equations

The system of balance equations for 13 moments can be written in terms of internal moments from (8.44),

$$\begin{aligned} &\dot{\rho} + \rho\frac{\partial v_i}{\partial x_i} = 0, \\ &\rho\dot{v}_i + \frac{\partial p_{ij}}{\partial x_j} = 0, \\ &\dot{\rho}_{ij} + \rho_{ij}\frac{\partial v_k}{\partial x_k} + \frac{\partial p_{ijk}}{\partial x_k} + 2\frac{\partial v_{(i}}{\partial x_k}p_{j)k} = \pi_{ij}, \\ &\dot{\rho}_{iij} + \rho_{iij}\frac{\partial v_k}{\partial x_k} + \frac{\partial p_{iijk}}{\partial x_k} + 3\frac{\partial v_{(i}}{\partial x_k}p_{ij)k} + 3\rho_{(ii}\dot{v}_{j)} = \pi_{iij}. \end{aligned} \tag{8.104}$$

With the constitutive equations, one can get a system of field equations for the 13 fields $(\rho, v_i, \rho_{ij}, \rho_{iij})$. However, it is more convenient to replace $(\rho_{ij}, \rho_{iij})$ by the conventional variables $(\theta, S_{ij}, q_j)$, namely, the temperature, the viscous stress tensor, and the heat flux, respectively. The viscous stress tensor is defined as the traceless part of the stress tensor,

$$T_{ij} = -p_{ij} = -p\,\delta_{ij} + S_{ij}, \qquad S_{\langle ij\rangle} = 0. \tag{8.105}$$

From (8.95) and (8.101) we obtain

$$\begin{aligned}
\rho_{ii} &= 2\,\rho\varepsilon, \\
\rho_{\langle ij\rangle} &= \frac{1}{4p\theta h_2} S_{ij} + o(2), \\
\rho_{iij} &= \frac{1}{3\,A_1 h_1} q_j + o(2), \\
p_{\langle ij\rangle k} &= \frac{2}{3}\frac{A_2}{A_1}\,\delta_{k\langle i}\, q_{j\rangle} + o(2), \\
p_{iijk} &= \alpha\,\delta_{jk} + \frac{a_2}{2p\theta}\,S_{jk} + o(2),
\end{aligned} \tag{8.106}$$

and

$$\begin{aligned}
\pi_{ij} &= -\frac{1}{4p\theta h_2 \tau_s} S_{ij} + o(2), \\
\pi_{iij} &= -\frac{1}{3\,A_1 h_1 \tau_q} q_j + o(2).
\end{aligned}$$

On substitution of the above constitutive equations into (8.104), we obtain a system of quasi-linear first-order partial differential equations for the 13 field variables $(\rho, \theta, v_i, S_{ij}, q_j)$. This is a hyperbolic system of field equations for viscous heat-conducting fluids.

### Linearized Field Equations

In order to associate the three unknown functions $h_1$, $a_2$, and $\alpha$ and two other coefficients, $\tau_s$ and $\tau_q$, with more suggestive physical parameters, we shall linearize the field equations by leaving out all nonlinear terms, such as the products of $S_{ij}$ and $q_j$ with the derivatives of $\rho$, $\theta$ and $v_i$. They read

$$\begin{aligned}
&\dot{\rho} + \rho\frac{\partial v_i}{\partial x_i} = 0, \\
&\rho\dot{v}_i + \frac{\partial p}{\partial x_i} - \frac{\partial S_{ij}}{\partial x_j} = 0, \\
&\rho\frac{\partial \varepsilon}{\partial \theta}\dot{\theta} + \theta\frac{\partial p}{\partial \theta}\frac{\partial v_i}{\partial x_i} + \frac{\partial q_i}{\partial x_i} = 0, \\
&\tau_s \dot{S}_{ij} - 2\mu\frac{\partial v_{\langle i}}{\partial x_{j\rangle}} - 2\mu K\frac{\partial q_{\langle i}}{\partial x_{j\rangle}} = -S_{ij}, \\
&\tau_q \dot{q}_i + \kappa\frac{\partial \theta}{\partial x_i} - \kappa K\theta\frac{\partial S_{ij}}{\partial x_j} = -q_i.
\end{aligned} \tag{8.107}$$

In these equations, we have identified five material parameters that are measurable, at least in principle, namely, $\tau_s$ and $\tau_q$ for the relaxation times of viscous stress and heat flux, and

$$\mu = -4\theta p^2 h_2 \tau_s, \qquad \kappa = -\frac{9}{2}\frac{A_1^2}{\theta^2} h_1 \tau_q, \qquad K = \frac{1}{3p}\frac{A_2}{A_1}, \tag{8.108}$$

for the shear viscosity, the thermal conductivity and the thermal-viscous coupling coefficient, respectively. From (8.100) and (8.103), we conclude that

$$\mu \geq 0, \qquad \kappa \geq 0. \tag{8.109}$$

Therefore, the shear viscosity $\mu$ and the thermal conductivity $\kappa$ as well as relaxation times $\tau_s$ and $\tau_q$ are all non-negative quantities.

Note that if the relaxation times and the thermo-viscous coupling are neglected, the equations $(8.107)_{4,5}$ reduce to the usual Navier–Stokes and Fourier laws with vanishing bulk viscosity (see Sect. 7.2.2).

### 8.5.2 Entropy and Entropy Flux

The constitutive equations for the entropy density $\eta$ and the entropy flux $\boldsymbol{\Phi}$ can be obtained up to the second-order terms directly from the representations (8.83), (8.84) and the definition (8.78). With the material parameters defined in the previous section, they can be written as

$$\begin{aligned} \eta &= \eta|_E - \frac{\tau_s}{4\rho\mu\theta} S_{\langle ij\rangle} S_{\langle ij\rangle} - \frac{\tau_q}{2\rho\kappa\theta^2} q_j q_j + o(3), \\ \Phi_j &= \frac{1}{\theta}(q_j + K S_{\langle jk\rangle} q_k) + o(3). \end{aligned} \tag{8.110}$$

The relation $(8.110)_2$ asserts that the difference between the entropy flux $\boldsymbol{\Phi}$ and $\boldsymbol{q}/\theta$ is of second-order, due to the thermo-viscous interaction. Therefore, the assumption $(5.9)_1$ that the heat flux and the entropy flux are proportional in the Clausius–Duhem inequality, is generally not valid.

The linearized equations (8.107) and the constitutive relations (8.110) were derived by Müller in 1966 [49] in the first phenomenological extended theory of moments within the framework of the then prevailing theory of thermodynamics of irreversible processes (see, for example, [14]). Unlike the rational thermodynamics we have considered in this book, thermodynamics of irreversible processes rests upon some heuristic assumptions, notably, a modified Gibbs relation to include non-equilibrium variables. Although this approach has been widely used in many branches of physics (known as extended irreversible thermodynamics, see [31]), it lacks the rational structure of modern thermodynamics and the results concern, at most, linear constitutive equations.

## 8.6 Monatomic Ideal Gases

In the kinetic theory of ideal gases, the equations of moments are obtained from Maxwell's equation of transfer for ideal gases (please refer to any standard textbook [24, 34, 55, 70] for derivation), which, in the absence of an external force, can be expressed as

$$\frac{\partial \overline{\psi}}{\partial t} + \frac{\partial \overline{\psi c_j}}{\partial x_j} = \mathcal{P}(\psi), \tag{8.111}$$

where the overhead bar denotes the expectation value defined by (8.34). The last term denotes the production density due to molecular collisions. The explicit form of $\mathcal{P}(\psi)$ is irrelevant here. It suffices to know that $\mathcal{P}(\psi)$ is linear in $\psi$ and

$$\mathcal{P}(m) = 0, \qquad \mathcal{P}(mc_i) = 0, \qquad \mathcal{P}(mc^2) = 0, \tag{8.112}$$

which follow from the conservation of mass, linear momentum and energy in the collision process.

With the moment $F_{i_1 \cdots i_n}$ defined by (8.35), taking $\psi = mc_{i_1} \cdots c_{i_n}$ in (8.111), we obtain

$$\frac{\partial F_{i_1 \cdots i_n}}{\partial t} + \frac{\partial F_{i_1 \cdots i_n j}}{\partial x_j} = g_{i_1 \cdots i_n}, \tag{8.113}$$

where $g_{i_1 \cdots i_n} = \mathcal{P}(mc_{i_1 \cdots i_n})$ denotes the production density of the moments.

It is interesting to note that in this equation, the flux of the $n$-th moment $F_{i_1 \cdots i_n}$ is the same as the $(n+1)$-th moment $F_{i_1 \cdots i_{n+1}}$. This forward-linking property between the moment and its flux is a characteristic feature of ideal gases.

Comparing this relation with (8.40), we conclude that ideal gases is a special case of the general structure considered in Sect 8.3 with

$$F_{i_1 \cdots i_n} = H_{i_1 \cdots i_n},$$

or, from (8.41) and (8.42),

$$\rho_{i_1 \cdots i_n} = p_{i_1 \cdots i_n}. \tag{8.114}$$

In particular, $\rho_{ii} = p_{ii} = 3p$, and from the definitions (8.45), it follows that

$$\varepsilon = \frac{3}{2} \frac{p}{\rho}. \tag{8.115}$$

It is known that this relation is valid for monatomic ideal gases only. In other words, the simple structure of the forward-linking property of moment equations (8.113) is only applicable to noble gases, such as helium and argon.

For classical ideal gases, the equation of state is given by the usual ideal gas law,

$$p = R\rho\theta, \tag{8.116}$$

where $R = k/m$ is the gas constant, $m$ is the atomic mass and $k$ is the Boltzmann constant.

### 8.6.1 Thirteen-Moment Theory

With the relations (8.114) through (8.116), we can further evaluate the constitutive relations in the previous section. Surprisingly, all the unknown functions can be determined explicitly.

First, we integrate (8.93) to yield

$$\alpha = 5\,R^2\rho\theta^2.$$

In the integration with respect to $\rho$, we have set the arbitrary function of $\theta$ to zero for simplicity and we shall do so hereafter without further remarks.[2]

From $(8.96)_1$, we obtain

$$A_1 = -\frac{10}{3}\,R^2\rho\theta^3,$$

and since $\rho_{\langle ik\rangle} = p_{\langle ik\rangle}$ and $\rho_{iik} = 2q_k$, from $(8.95)_{1,2}$ we have $-4p\theta h_2 = 1$ and $6A_1h_1 = 1$, which lead to

$$h_1 = -\frac{1}{20}\,\frac{1}{R^2\rho\theta^3}, \qquad h_2 = -\frac{1}{4}\,\frac{1}{R\rho\theta^2}.$$

We can now integrate the relation $(8.96)_3$ for $A_2$,

$$A_2 = -4\,R^2\rho\theta^3,$$

which, from $(8.96)_2$ gives

$$a_2 = -14\,R^2\rho\theta^3.$$

The coefficients $a_1$ and $b_1$ are not individually determined, instead, they are related through the value of $A_1$. Hence, only one of them, say $b_1$, is undetermined. Nevertheless, the explicit value of $b_1$ is not needed in the linear constitutive equations. Therefore, all material functions for the fluxes have been explicitly determined.

Moreover, from (8.86), the first-order non-equilibrium Lagrange multipliers are given by

$$\begin{aligned} \lambda_\varepsilon &= 0, & \Lambda_i &= \frac{1}{R\rho\theta^2}\,q_i\,, \\ \lambda_\rho &= 0, & & \\ \lambda_i &= -\frac{1}{5}\,\frac{1}{R^2\rho\theta^3}\,q_i, & \Lambda_{\langle ij\rangle} &= \frac{1}{2}\,\frac{1}{R\rho\theta^2}\,S_{ij}\,. \end{aligned}$$

[2] By doing so, the results are consistent with those obtained from the kinetic theory.

### 8.6.2 Constitutive Equations

From the above results, the first-order constitutive equation for the fluxes $\rho_{ijk}$ and $\rho_{iijk}$ can now be calculated,

$$\begin{aligned} \rho_{ij} &= R\rho\theta\,\delta_{ij} - S_{ij}, \\ \rho_{ijk} &= \frac{2}{5}\,(q_i\,\delta_{jk} + q_j\,\delta_{ki} + q_k\,\delta_{ij}) + o(2), \\ \rho_{iijk} &= 5\,R^2\rho\theta^2\,\delta_{jk} - 7\,R\theta\,S_{jk} + o(2). \end{aligned} \tag{8.117}$$

Moreover, we can also determine the constitutive equations of the entropy density $\eta$ and the entropy flux $\Phi_k$ from (8.78),

$$\Phi_k = \Lambda_i\rho_{ik} + \Lambda_{ij}\rho_{ijk} + \lambda_j\rho_{iijk} - \widehat{\Phi}_k.$$

They are given by

$$\begin{aligned} \eta &= \eta|_E - \frac{1}{4R\rho^2\theta^2}\,S_{ij}S_{ij} - \frac{1}{5R^2\rho^2\theta^3}\,q_iq_i + o(3), \\ \Phi_k &= \frac{1}{\theta}\,q_k + \frac{2}{5R\rho\theta^2}\,S_{kj}q_j + o(3). \end{aligned} \tag{8.118}$$

Note that in the linear theory, they can be determined up to the second-order terms.

We remark that all the coefficients of the representations in the above constitutive equations are explicitly determined. These amazing results of the rational extended thermodynamics were first derived by Liu and Müller [46]. They are in complete agreement with the results from the kinetic theory of gases [15, 34].

## 8.7 Stationary Heat Conduction in Ideal Gases

Extended thermodynamics provides a hierarchy of hyperbolic systems of first-order partial differential equations. Applications for ideal gases, such as heat wave and shear wave propagations, shock structure, and light scattering, are some of the phenomena successfully treated within extended thermodynamics. The extended theories have shown great improvement of results beyond Navier–Stokes–Fourier theory in ordinary thermodynamics. Systems of higher hierarchy are usually required to achieve better results in comparison with experiments. Discussion of various aspects of the theory and applications can be found in the monograph by Müller and Ruggeri [57].

As a simple application of extended thermodynamics, we shall consider boundary value problems of stationary heat conduction in ideal gases. In thirteen-moment theory, we shall see that Fourier's law of heat conduction is no longer valid, in general.

### 8.7.1 Fourier's Law and Heat Conduction

For stationary heat conduction in an ideal gas at rest, the equations of equilibrium and energy are given by

$$-g^{ij}p_{,j} + S^{ij}{}_{,j} = 0, \qquad q^j{}_{,j} = 0,$$

where $g^{ij}$ is the contravariant component of the metric tensor and a subscript comma (,) denotes the covariant derivative (see Sect. A.2.4). In ordinary theories, Fourier's law of heat conduction is given by

$$q^j = -\kappa\, g^{jk}\theta_{,k}, \tag{8.119}$$

where the conductivity $\kappa$ is a constant for ideal gases. Therefore, the governing equation for heat conduction becomes

$$g^{jk}\theta_{,jk} = 0.$$

In other words, the temperature field $\theta$ satisfies the Laplace equation. On the other hand, the pressure $p$ is a constant, since the viscous stress $S^{ij} = 0$ in a Navier–Stokes fluid when the gas is at rest. If, at the boundary, either the temperature or the heat flux is prescribed, it is known that the boundary value problem is well-posed. For purposes of comparison, the following simple example will be considered.

#### Heat Conduction Between Coaxial Cylinders

The inner and outer radii of two cylinders are denoted by $r_{\mathrm{i}}$ and $r_{\mathrm{o}}$, respectively, and we consider an axial symmetric solution $\theta(r)$ where $r$ is the radial coordinate. In this case, the Laplacian of $\theta(r)$ becomes

$$\frac{d^2\theta}{dr^2} + \frac{1}{r}\frac{d\theta}{dr} = 0,$$

for $r_{\mathrm{i}} < r < r_{\mathrm{o}}$. We also prescribed the following boundary conditions:

$$q(r_{\mathrm{i}}) = q_{\mathrm{i}}, \qquad \theta(r_{\mathrm{o}}) = \theta_{\mathrm{o}}, \tag{8.120}$$

where $q$ is the $r$-component of the heat flux. The solution is given by

$$\theta(r) = \theta_{\mathrm{o}} - \frac{q_{\mathrm{i}} r_{\mathrm{i}}}{\kappa} \log\frac{r}{r_{\mathrm{o}}}. \tag{8.121}$$

### 8.7.2 Heat Conduction in Thirteen-Moment Theory

Extended thermodynamics of the lowest hierarchy for the study of heat conduction problem is the theory of thirteen moments, $(\rho, v^i, p^{ij}, \rho_i{}^{ij})$. We consider a stationary process at rest, $v^i = 0$, for monatomic ideal gases. In this case, the equation of mass balance is identically satisfied.

In the cylindrical coordinate system $(x^1, x^2, x^3) = (r, \vartheta, z)$, for the coordinate, $\vartheta$ is used to avoid confusion with the temperature field $\theta$ – with

(8.117) we have the following system of moment equations:

$$
\begin{aligned}
& g^{jk}\frac{\partial p}{\partial x^k} - \frac{\partial S^{jk}}{\partial x^k} - \Gamma_{k\ i}^{\ j}S^{ik} - \Gamma_{k\ i}^{\ k}S^{ji} = 0, \\
& \frac{\partial q^k}{\partial x^k} + \Gamma_{k\ i}^{\ k}q^i = 0, \\
& \frac{2}{5}\left(g^{ik}\frac{\partial q^j}{\partial x^k} + g^{jk}\frac{\partial q^i}{\partial x^k} + g^{ik}\Gamma_{k\ l}^{\ j}q^l + g^{jk}\Gamma_{k\ l}^{\ i}q^l\right) = \frac{1}{\tau}S^{ij}, \\
& 5Rg^{jk}\frac{\partial(p\theta)}{\partial x^k} - 7R\frac{\partial(\theta S^{jk})}{\partial x^k} - 7R\theta(\Gamma_{k\ i}^{\ j}S^{ik} + \Gamma_{k\ i}^{\ k}S^{ji}) = -\frac{2}{\tau}q^j .
\end{aligned}
\tag{8.122}
$$

The metric tensor and the Christoffel symbols for cylindrical coordinate system can be found in Sect. A.2.7 (b).

In these equations, the production densities are given by (8.101) with a constant relaxation time $\tau = \tau_s = \tau_q$, corresponding to the BGK model for molecular interaction – the simplest linear model in the kinetic theory [6].

### Heat Conduction Between Coaxial Cylinders

We now consider again the boundary value problem of heat conduction between two coaxial cylinders. The solution is assumed to be axially symmetric and hence all fields are functions of $r$ only. The heat flux is also assumed to be in the radial direction, $q^r = q(r)$ and $q^\vartheta = q^z = 0$.

From $(8.122)_2$ it follows that

$$
\frac{dq}{dr} + \frac{1}{r}q = 0,
$$

and the solution is given by

$$
q(r) = \frac{c}{r}, \tag{8.123}
$$

where $c = q_\mathrm{i} r_\mathrm{i}$ from the boundary condition $(8.120)_1$.

Equation $(8.122)_3$ then implies that

$$
\begin{aligned}
S^{rr} &= \frac{4}{5}\tau\frac{dq}{dr} = -\frac{4}{5}\frac{\tau c}{r^2}, \\
S^{\vartheta\vartheta} &= \frac{4}{5}\tau\frac{1}{r^3}q = \frac{4}{5}\frac{\tau c}{r^4},
\end{aligned}
\tag{8.124}
$$

and the other components of the stress are zero. Moreover, by the use of (8.124), we obtain $S^{jk}{}_{,k} = 0$, i.e.,

$$
\frac{\partial S^{jk}}{\partial x^k} + \Gamma_{k\ i}^{\ j}S^{ik} + \Gamma_{k\ i}^{\ k}S^{ji} = 0.
$$

Therefore, $(8.122)_1$ implies that the pressure $p$ is constant and the last equation of (8.122) leads to the following relation between the heat flux and the temperature gradient,

$$q = -\kappa\Big(1 - \frac{7}{5}\frac{S^{rr}}{p}\Big)\frac{d\theta}{dr}, \qquad \kappa = \frac{5}{2}\,R\tau p, \tag{8.125}$$

where $\kappa$ is the thermal conductivity. Note that the presence of the normal stress makes Fourier's law of heat conduction no longer valid in the 13-moment theory. In the Navier Stokes–Fourier theory, of course, $S^{ij}$ vanishes identically in a gas at rest.

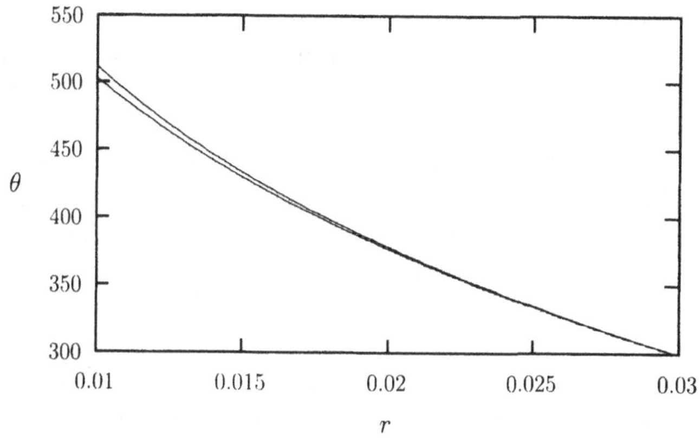


**Fig. 8.1.** Temperature $\theta(r)$ of 13-moment theory (*lower curve*) and Fourier theory (*upper curve*)

Finally, we can determine the temperature field. Substitution of (8.123) and (8.124) into (8.125) gives

$$\frac{d\theta}{dr} = -\frac{c}{\kappa}\,\frac{r}{r^2 + bc}, \qquad b = \frac{28}{25}\frac{\tau}{p}.$$

With the prescribed boundary conditions (8.120) of Example 8.7.1, the solution of 13-moment theory is

$$\theta(r) = \theta_o - \frac{q_i r_i}{2\kappa}\,\log\frac{r^2 + b\,q_i r_i}{r_o^2 + b\,q_i r_i}. \tag{8.126}$$

This result has been plotted together with the result (8.121) of Fourier theory for comparison in Fig. 8.1 for argon and the following data:

$r_i = 0.01\,\text{m}$, $r_o = 0.03\,\text{m}$, $q_i = 10^4\,\text{W/m}^2$, $\theta_o = 300\,\text{K}$,
$\tau = 10^{-5}\,\text{s}$, $p = 100\,\text{Pa}$, $\kappa = 0.5164\,\text{W/m/K}$.

Note that the temperature for the 13-moment theory is slightly lower than for the Fourier theory, and the largest difference is less than 1.80% at the inner surface.

### 8.7.3 Remark on Boundary Value Problems

Generally speaking, a system of field equations with more variables requires more boundary conditions in order to have unique solutions. However, in the heat conduction problem considered in this section, the usual boundary conditions (8.120) for the Fourier theory of heat conduction turn out to be sufficient for the determination of unique solutions within the 13-moment theory. This is generally not so in extended thermodynamics.

The presence of higher moments in extended thermodynamics presents an additional problem regarding the specification of their initial and boundary values, because usually higher moments do not have meaningful physical interpretations and can not be measured in conventional physical quantities, and hence we do not know how to prescribe their values. The problem with initial values seems to be of less concern, because in most of the applications, initial values of non-equilibrium higher moments can be set to zero if necessary. This has been done frequently in wave and shock problems [57]. However, with respect to boundary values of higher moments, the problem is more difficult. Mathematically they have to be assigned in order to have unique solutions. In other words, the dependence of such boundary values will result in the non-uniqueness of the solution. However, not only are they unconventional physical quantities, arbitrary assignment of additional boundary values is inconsistent with the nature of well-posedness of the corresponding physical problem in ordinary theories.

Therefore, it is necessary to establish some kind of criterion to determine the values of such quantities, so that the solution is unique and physically relevant in comparison with the solution of the corresponding problem in ordinary theories. In [67] Struchtrup and Weiss made the first attempt to consider such a criterion and they proposed to find the unique solution by requiring the entropy production to be as small as possible by the so-called Minimax Principle, similar to Prigogine's Principle of minimum entropy production [2, 60]. However, both principles do not always lead to reasonably consistent results. The boundary value problem in extended thermodynamics has come into discussion only recently and it is still largely an unexplored problem.

# A. Elementary Tensor Analysis

This appendix is intended to provide the mathematical preliminaries needed for a clear and rigorous presentation of the basic principles in continuum mechanics. The reader is expected to be familiar with some notions of vector spaces or matrix algebra. In the first part, we shall review some basic notions of vector spaces and linear transformations. At the same time, elementary properties of tensors as well as tensor notations will be introduced. All mathematical symbols are set in *italic*. Generally, scalars are represented by letters $(a, b, \alpha, \cdots)$ of normal typeface, vectors by lowercase boldface letters $(\boldsymbol{u}, \boldsymbol{v}, \cdots)$, and tensors by uppercase letters $(T, F, \cdots)$. The identity tensor is represented by *1* (italic "one").

## A.1 Linear Algebra

We shall consider finite-dimensional real vector spaces only. The field of real numbers is denoted by $\mathbb{R}$.

**Definition.** *A vector space $V$ is a set equipped with two operations:*

1) *$\boldsymbol{v} + \boldsymbol{u} \in V$, called addition of $\boldsymbol{v}$ and $\boldsymbol{u}$ in $V$,*
2) *$\alpha \boldsymbol{v} \in V$, called scalar multiplication of $\boldsymbol{v} \in V$ by $\alpha \in \mathbb{R}$,*

*which satisfy the following rules: for any $\boldsymbol{v}, \boldsymbol{u}, \boldsymbol{w} \in V$, and any $\alpha, \beta \in \mathbb{R}$,*

1) $\boldsymbol{v} + \boldsymbol{u} = \boldsymbol{u} + \boldsymbol{v}$.
2) $\boldsymbol{v} + (\boldsymbol{u} + \boldsymbol{w}) = (\boldsymbol{v} + \boldsymbol{u}) + \boldsymbol{w}$.
3) *There exists a null vector $0 \in V$, such that $\boldsymbol{v} + 0 = \boldsymbol{v}$.*
4) *For any $\boldsymbol{v} \in V$, there exist $-\boldsymbol{v} \in V$, such that $\boldsymbol{v} + (-\boldsymbol{v}) = 0$.*
5) $\alpha(\beta \boldsymbol{v}) = (\alpha\beta)\boldsymbol{v}$.
6) $(\alpha + \beta)\boldsymbol{v} = \alpha \boldsymbol{v} + \beta \boldsymbol{v}$.
7) $\alpha(\boldsymbol{v} + \boldsymbol{u}) = \alpha \boldsymbol{v} + \alpha \boldsymbol{u}$.
8) $1\boldsymbol{v} = \boldsymbol{v}$.

**Definition.** *A set of vectors $\{\boldsymbol{v}_1, \cdots, \boldsymbol{v}_n\}$ is said to be a basis of $V$, if*

1) *it is a linearly independent set, i.e., for any $a_1, \cdots, a_n \in \mathbb{R}$, if $a_1\boldsymbol{v}_1 + \cdots + a_n\boldsymbol{v}_n = 0$ then $a_1 = \cdots = a_n = 0$.*
2) *it spans the space $V$, i.e., for any $\boldsymbol{u} \in V$, the vector $\boldsymbol{u}$ can be expressed as a linear combination of $\{\boldsymbol{v}_1, \cdots, \boldsymbol{v}_n\}$.*

Let $\{\boldsymbol{e}_1, \cdots, \boldsymbol{e}_n\}$ be a basis of $V$, then any vector $\boldsymbol{u} \in V$ can be expressed as

$$\boldsymbol{u} = \sum_{i=1}^{n} u^i \boldsymbol{e}_i,$$

where $u^i$, called the components of $\boldsymbol{u}$, are uniquely determined relative to the basis $\{\boldsymbol{e}_i\}$.

A vector space can have many different bases, but all of them will have the same number of elements. The number of elements in a basis is called the *dimension* of the space, in this case, we have $\dim V = n$.

### A.1.1 Inner Product

We may think of a vector as a geometric object that has a length and points in a certain direction. To incorporate this notion we introduce an additional structure, inner product, into the vector space.

**Definition.** *An inner product is a map*

$$g : V \times V \to I\!R$$

*with the following properties: For any* $\boldsymbol{u}, \boldsymbol{v}, \boldsymbol{w} \in V$, *and* $\alpha \in I\!R$,

1) $g(\boldsymbol{u} + \alpha \boldsymbol{v}, \boldsymbol{w}) = g(\boldsymbol{u}, \boldsymbol{w}) + \alpha g(\boldsymbol{v}, \boldsymbol{w})$,
2) $g(\boldsymbol{u}, \boldsymbol{v}) = g(\boldsymbol{v}, \boldsymbol{u})$,
3) $g(\boldsymbol{u}, \boldsymbol{u}) > 0$, *if* $\boldsymbol{u} \neq 0$.

An inner product is a positive-definite symmetric bilinear function on $V$. We call $g(\boldsymbol{u}, \boldsymbol{v})$ the inner product of $\boldsymbol{u}$ and $\boldsymbol{v}$. The vector space equipped with an inner product is called an *inner product space.* Hereafter, all vector spaces considered are always inner product spaces.

**Notation.** $g(\boldsymbol{u}, \boldsymbol{v}) = \boldsymbol{u} \cdot \boldsymbol{v}$, if $g$ is given and fixed.

**Definition.** *The norm of a vector* $\boldsymbol{v} \in V$ *is defined as*

$$|\boldsymbol{v}| = \sqrt{(\boldsymbol{v} \cdot \boldsymbol{v})}.$$

*A vector space equipped with such a norm is called a Euclidean vector space.*

The notion of angle between two vectors can be defined based on the following Schwarz inequality:

$$|\boldsymbol{u} \cdot \boldsymbol{v}| \leq |\boldsymbol{u}|\, |\boldsymbol{v}|. \tag{A.1}$$

**Definition.** *For any non-zero* $\boldsymbol{u}, \boldsymbol{v} \in V$, *the angle between* $\boldsymbol{u}$ *and* $\boldsymbol{v}$, $\theta(\boldsymbol{u}, \boldsymbol{v}) \in [0, \pi]$, *is defined by*

$$\cos \theta(\boldsymbol{u}, \boldsymbol{v}) = \frac{\boldsymbol{u} \cdot \boldsymbol{v}}{|\boldsymbol{u}|\, |\boldsymbol{v}|}.$$

The vectors $\boldsymbol{u}$ and $\boldsymbol{v}$ are said to be *orthogonal* if $\theta(\boldsymbol{u},\boldsymbol{v}) = \pi/2$. Obviously, $\boldsymbol{u}$ and $\boldsymbol{v}$ are orthogonal if and only if $\boldsymbol{u}\cdot\boldsymbol{v} = 0$.

A vector $\boldsymbol{v}$ is called a *unit vector* if $|\boldsymbol{v}| = 1$. The *projection* of a vector $\boldsymbol{u}$ on the vector $\boldsymbol{v}$ can be defined as $|\boldsymbol{u}|\cos\theta(\boldsymbol{u},\boldsymbol{v})$, or as $(\boldsymbol{u}\cdot\boldsymbol{e})$, where $\boldsymbol{e} = \boldsymbol{v}/|\boldsymbol{v}|$ is the unit vector in the direction of $\boldsymbol{v}$. The vector $(\boldsymbol{u}\cdot\boldsymbol{e})\boldsymbol{e}$ is called the projection vector of $\boldsymbol{u}$ in the direction of $\boldsymbol{v}$.

Let $\{\boldsymbol{e}_i,\ i = 1,\cdots,n\}$ be a basis of $V$. Denote the inner product of $\boldsymbol{e}_i$ and $\boldsymbol{e}_j$ by $g_{ij}$,

$$g_{ij} = \boldsymbol{e}_i\cdot\boldsymbol{e}_j.$$

Clearly, $g_{ij}$ is symmetric, $g_{ij} = g_{ji}$. Let $\boldsymbol{u} = u^i\boldsymbol{e}_i$, $\boldsymbol{v} = v^j\boldsymbol{e}_j$ be arbitrary vectors in $V$ expressed in terms of the basis $\{\boldsymbol{e}_i\}$. Then

$$\begin{aligned}\boldsymbol{u}\cdot\boldsymbol{v} &= (u^i\boldsymbol{e}_i)\cdot(v^j\boldsymbol{e}_j)\\ &= u^iv^j(\boldsymbol{e}_i\cdot\boldsymbol{e}_j) = u^iv^jg_{ij},\end{aligned}$$

or

$$\boldsymbol{u}\cdot\boldsymbol{v} = g_{ij}u^iv^j. \tag{A.2}$$

Here we have used the following summation convention.

**Notation.** (Summation convention) *In the expression of a term, if an index is repeated once (and only once), a summation over the range of this index is assumed.*

For example,

$$u^i\boldsymbol{e}_i = \sum_{i=1}^{n} u^i\boldsymbol{e}_i,$$

$$g_{ij}u^iv^j = \sum_{i=1}^{n}\sum_{j=1}^{n} g_{ij}u^iv^j.$$

Note that in these expressions, we purposely write the indices in two different levels so that the repeated summation indices are always one superindex and one subindex. The reason for doing so will become clear in the next section.

### A.1.2 Dual Bases

Let $\{\boldsymbol{e}_1,\cdots,\boldsymbol{e}_n\}$ be a basis of $V$. There exists a non-zero vector orthogonal to the plane spanned by the $n-1$ vectors $\{\boldsymbol{e}_2,\cdots,\boldsymbol{e}_n\}$, and if, in addition, the projection of this vector on $\boldsymbol{e}_1$ is prescribed, then this vector is uniquely determined. In this manner, for any given basis $\{\boldsymbol{e}_1,\cdots,\boldsymbol{e}_n\}$, we can construct a set of vectors $\{\boldsymbol{e}^1,\cdots,\boldsymbol{e}^n\}$ such that

$$\boldsymbol{e}^i\cdot\boldsymbol{e}_j = \delta^i{}_j,$$

where ${\delta^i}_j$ is called the *Kronecker delta* defined by

$${\delta^i}_j = \begin{cases} 0, & \text{if} \quad i \neq j, \\ 1, & \text{if} \quad i = j. \end{cases}$$

From this construction, if $\boldsymbol{v} = v^i \boldsymbol{e}_i$ is a vector in $V$, then by taking the inner product with $\boldsymbol{e}^i$ we have

$$\boldsymbol{e}^i \cdot \boldsymbol{v} = \boldsymbol{e}^i \cdot (v^j \boldsymbol{e}_j) = v^j {\delta^i}_j = v^i.$$

Hence, the $i$-th component of $\boldsymbol{v}$ relative to the basis $\{\boldsymbol{e}_1, \cdots, \boldsymbol{e}_n\}$ is its inner product with the vector $\boldsymbol{e}^i$. Therefore, this set of vectors $\{\boldsymbol{e}^i\}$ associated with the basis $\{\boldsymbol{e}_i\}$ can be regarded as linear functions that map a vector to its components.[1]

We can easily show that this new set of vectors is a linearly independent set. Indeed, if for any linear combination $a_j \boldsymbol{e}^j = 0$, then it follows that $(a_j \boldsymbol{e}^j) \cdot \boldsymbol{e}_i = a_j {\delta^j}_i = a_i = 0$ for all $i$. Furthermore, it also spans the space $V$, for if $\boldsymbol{u} = u^i \boldsymbol{e}_i$ is a vector in $V$, then for any vector $\boldsymbol{v} = v^i \boldsymbol{e}_i$, from (A.2) and $v^j = \boldsymbol{e}^j \cdot \boldsymbol{v}$,

$$\boldsymbol{u} \cdot \boldsymbol{v} = g_{ij} u^i v^j = (g_{ij} u^i \boldsymbol{e}^j) \cdot \boldsymbol{v},$$

which implies that $\boldsymbol{u}$ can be expressed as $\boldsymbol{u} = u_i \boldsymbol{e}^i$ with

$$u_i = g_{ij} u^j.$$

Therefore we have proved that this set of vectors $\{\boldsymbol{e}^i\}$ is also a basis of $V$.

**Definition.** *Let $\beta = \{\boldsymbol{e}_i\}$ and $\beta^* = \{\boldsymbol{e}^i\}$ be two bases of $V$ related by the property*

$$\boldsymbol{e}^i \cdot \boldsymbol{e}_j = {\delta^i}_j.$$

*They are said to be a pair of dual bases for $V$, or $\beta^*$ is the dual basis of $\beta$.*

The dual bases are uniquely determined from each other. For this reason, we have used the same notation for their elements except the different level of indices to distinguish them. Clearly, if $\boldsymbol{u}$ is a vector in $V$, then we can express $\boldsymbol{u}$ in terms of components in two different ways relative to the dual bases,

$$\boldsymbol{u} = u^i \boldsymbol{e}_i = u_j \boldsymbol{e}^j,$$

where we have also employed different levels of component indices in order to be consistent with our summation convention, which sums over repeated

[1] In general, the space of all linear functions on $V$ is called the dual space of $V$ and is denoted by $V^*$. In this note, for simplicity, we shall not distinguish vectors in $V^*$ and $V$ when the space $V$ is equipped with an inner product.

indices in different levels. We call

$u^i$ the $i$-th *contravariant* component of $\boldsymbol{u}$,

$u_j$ the $j$-th *covariant* component of $\boldsymbol{u}$.

From the definition, it follows that

$$u^i = \boldsymbol{e}^i \cdot \boldsymbol{u}, \qquad u_j = \boldsymbol{e}_j \cdot \boldsymbol{u}, \tag{A.3}$$

and they are related by

$$u_i = g_{ij} u^j, \qquad u^i = g^{ij} u_j,$$

where we have denoted

$$g^{ij} = \boldsymbol{e}^i \cdot \boldsymbol{e}^j.$$

The two operations

$$g_{ij} : u^j \mapsto u_i, \qquad g^{ij} : u_j \mapsto u^i,$$

enable us to *lower* and *raise* the component index. One can also show that

$$\boldsymbol{e}_j = g_{ij} \boldsymbol{e}^i, \qquad \boldsymbol{e}^i = g^{ij} \boldsymbol{e}_j.$$

Therefore, lowering or raising the index for dual bases can be made in the same manner. It is easy to verify that

$$g^{ij} g_{jk} = \delta^i{}_k.$$

A basis $\{\boldsymbol{e}_i\}$ is called an *orthogonal basis* if all the elements of the basis are mutually orthogonal, i.e.,

$$\boldsymbol{e}_i \cdot \boldsymbol{e}_j = 0 \quad \text{if} \quad i \neq j.$$

If, in addition, $|\boldsymbol{e}_i| = 1$, for all $i$, it is called an *orthonormal basis.* Although in general, we carefully do our bookkeeping of super- and subindices, this becomes unnecessary if $\beta = \{\boldsymbol{e}_i\}$ is an orthonormal basis. Since then $g_{ij} = \delta_{ij}$, and

$$\boldsymbol{e}_i = g_{ij} \boldsymbol{e}^j = \delta_{ij} \boldsymbol{e}^j = \boldsymbol{e}^i.$$

That is, the basis $\beta$ is identical to its dual basis $\beta^*$. Hence, we do not have to distinguish contravariant and covariant components. In this case, we can write all the indices at the same level. for example,

$$\boldsymbol{v} = v_i \boldsymbol{e}_i.$$

Of course, according to our summation convention, we still sum over the repeated indices (now in the same level) in this situation.

**Exercise A.1.1** Let $\beta' = \{\mathbf{e}_1 = (1,0), \mathbf{e}_2 = (2,1)\}$ be a basis of $\mathbb{R}^2$, and $\boldsymbol{v} = (1,-1)$ be a vector in $\mathbb{R}^2$.

1) Find the dual basis $\{\mathbf{e}^1, \mathbf{e}^2\}$ of $\beta'$.
2) Determine the matrix representations $[g_{ij}]$ and $[g^{ij}]$ relative to $\beta'$.
3) Determine the contravariant and covariant components of $\boldsymbol{v}$ relative to the bases and make a graphic representation of the results.

### A.1.3 Tensor Product

The notion of matrix is related to linear functions on vector spaces. Let $U$ and $V$ be two vector spaces with inner product. A function $T : U \to V$, is called a *linear transformation* from $U$ to $V$, if for any $\boldsymbol{u}, \boldsymbol{v} \in U$ and $\alpha \in \mathbb{R}$,

$$T(\boldsymbol{u} + \alpha \boldsymbol{v}) = T(\boldsymbol{u}) + \alpha T(\boldsymbol{v}).$$

**Notation.** $\mathcal{L}(U, V) = \{T : U \to V \mid T \text{ is linear}\}$.

If $T$ and $S$ are two linear transformations in $\mathcal{L}(U, V)$, we can define the addition $T + S$ and the scalar multiplication $\alpha T$, as transformations in $\mathcal{L}(U, V)$, in the following manner, for all $\boldsymbol{v} \in U$,

$$(T + S)(\boldsymbol{v}) = T(\boldsymbol{v}) + S(\boldsymbol{v}),$$
$$(\alpha T)(\boldsymbol{v}) = \alpha T(\boldsymbol{v}).$$

With these operations the set $\mathcal{L}(U, V)$ becomes a vector space.

**Definition.** *For any vectors $\boldsymbol{v} \in V$ and $\boldsymbol{u} \in U$, the tensor product of $\boldsymbol{v}$ and $\boldsymbol{u}$, denoted by $\boldsymbol{v} \otimes \boldsymbol{u}$, is defined as a linear transformation from $U$ to $V$ such that*

$$(\boldsymbol{v} \otimes \boldsymbol{u})(\boldsymbol{w}) = (\boldsymbol{u} \cdot \boldsymbol{w})\boldsymbol{v}, \tag{A.4}$$

*for any $\boldsymbol{w} \in U$.*

The tensor product of two vectors is a linear transformation. We call such a linear transformation a *simple tensor.* Of course, not every linear transformation can be obtained as a tensor product of two vectors. However, we can show that, indeed, it can always be expressed as a linear combination of simple tensors.

**Proposition.** *Let $\{\mathbf{e}_i\}$, $i = 1, \cdots, n$ and $\{\mathbf{d}_\alpha\}$, $\alpha = 1, \cdots, m$ be bases of $V$ and $U$, respectively. Then the set $\{\mathbf{e}_i \otimes \mathbf{d}_\alpha\}$, $i = 1, \cdots, n$, $\alpha = 1, \cdots, m$, forms a basis of $\mathcal{L}(U, V)$.*

*Proof.* Let $\{\mathbf{e}^i\}$ be the dual basis of $\{\mathbf{e}_i\}$ and $\{\mathbf{d}^\alpha\}$ the dual basis of $\{\mathbf{d}_\alpha\}$. If $a^{i\alpha}\mathbf{e}_i \otimes \mathbf{d}_\alpha = 0$, then

$$a^{i\alpha}(\mathbf{e}_i \otimes \mathbf{d}_\alpha)(\mathbf{d}^\beta) = a^{i\alpha}(\mathbf{d}_\alpha \cdot \mathbf{d}^\beta)\mathbf{e}_i = a^{i\alpha}\delta^\beta_\alpha \mathbf{e}_i = a^{i\beta}\mathbf{e}_i = 0,$$

which implies that $a^{i\beta} = 0$, since $\{\boldsymbol{e}_i\}$ is a basis. Therefore, $\{\boldsymbol{e}_i \otimes \boldsymbol{d}_\alpha\}$ is a linearly independent set. Moreover, for any $T \in \mathcal{L}(U, V)$, let

$$\boldsymbol{e}^i \cdot T(\boldsymbol{d}^\alpha) = T^{i\alpha}.$$

Then, for any $\boldsymbol{v} \in V$ and any $\boldsymbol{u} \in U$,

$$\begin{aligned} \boldsymbol{v} \cdot T(\boldsymbol{u}) &= v_i \boldsymbol{e}^i \cdot T(u_\alpha \boldsymbol{d}^\alpha) \\ &= v_i u_\alpha \boldsymbol{e}^i \cdot T(\boldsymbol{d}^\alpha) = T^{i\alpha} v_i u_\alpha. \end{aligned}$$

On the other hand,

$$\begin{aligned} \boldsymbol{v} \cdot (\boldsymbol{e}_i \otimes \boldsymbol{d}_\alpha)(\boldsymbol{u}) &= v_j \boldsymbol{e}^j \cdot (\boldsymbol{e}_i \otimes \boldsymbol{d}_\alpha)(u_\beta \boldsymbol{d}^\beta) \\ &= v_j u_\beta (\boldsymbol{e}^j \cdot \boldsymbol{e}_i)(\boldsymbol{d}_\alpha \cdot \boldsymbol{d}^\beta) = v_i u_\alpha. \end{aligned}$$

Therefore, we have

$$\boldsymbol{v} \cdot T(\boldsymbol{u}) = T^{i\alpha} \boldsymbol{v} \cdot (\boldsymbol{e}_i \otimes \boldsymbol{d}_\alpha)(\boldsymbol{u}),$$

for any $\boldsymbol{v}$ and any $\boldsymbol{u}$, which leads to

$$T = T^{i\alpha} \boldsymbol{e}_i \otimes \boldsymbol{d}_\alpha.$$

That is, $\{\boldsymbol{e}_i \otimes \boldsymbol{d}_\alpha\}$ spans the space $\mathcal{L}(U, V)$. □

We may call $\mathcal{L}(U, V)$ the *tensor product space* of $V$ and $U$ and denote it by $V \otimes U$. Obviously, from this result, we have

$$\dim V \otimes U = (\dim V)(\dim U).$$

The basis $\{\boldsymbol{e}_i \otimes \boldsymbol{d}_\alpha\}$ is called a *product basis* of $V \otimes U$. Similarly, the sets $\{\boldsymbol{e}_i \otimes \boldsymbol{d}^\alpha\}$, $\{\boldsymbol{e}^i \otimes \boldsymbol{d}_\alpha\}$, and $\{\boldsymbol{e}^i \otimes \boldsymbol{d}^\alpha\}$ are also product bases of $V \otimes U$.

**Notation.** $V \otimes V = \mathcal{L}(V) = \mathcal{L}(V, V)$.

We shall call linear transformations in $\mathcal{L}(V)$ the *second-order tensors*. Let $\{\boldsymbol{e}_i\}$ and $\{\boldsymbol{e}^j\}$ be dual bases of $V$, a second-order tensor $T$ then has different component forms relative to the different product bases.

$$\begin{aligned} T &= T^{ij} \boldsymbol{e}_i \otimes \boldsymbol{e}_j = T^i{}_j \boldsymbol{e}_i \otimes \boldsymbol{e}^j \\ &= T_i{}^j \boldsymbol{e}^i \otimes \boldsymbol{e}_j = T_{ij} \boldsymbol{e}^i \otimes \boldsymbol{e}^j, \end{aligned}$$

where the various components are given by

$$\begin{aligned} T^{ij} &= \boldsymbol{e}^i \cdot T\boldsymbol{e}^j, \quad & T^i{}_j &= \boldsymbol{e}^i \cdot T\boldsymbol{e}_j, \\ T_{ij} &= \boldsymbol{e}_i \cdot T\boldsymbol{e}_j, \quad & T_i{}^j &= \boldsymbol{e}_i \cdot T\boldsymbol{e}^j. \end{aligned} \tag{A.5}$$

These components are called the *associated components* of the second-order tensor $T$. In classical tensor analysis, they are also called

| | |
|---|---|
| $T^{ij}$ | contravariant tensor of order 2, |
| $T_{ij}$ | covariant tensor of order 2, |
| $T^i{}_j,\ T_i{}^j$ | mixed tensor of order 2. |

They are related by

$$T^i{}_j = g_{kj}T^{ik} = g^{ik}T_{kj}, \quad \text{etc.,} \tag{A.6}$$

with the operations of raising or lowering the indices discussed in the previous section.

The matrices $[T^{ij}]$, $[T^i{}_j]$, $[T_i{}^j]$, $[T_{ij}]$ are called the *matrix representations* of $T$ relative to the corresponding product bases. Note that the first index refers to the row and the second index refers to the column of the matrix. It is important to distinguish the level as well as the position order of the component indices. In general, $T^i{}_j \neq T_j{}^i$, therefore it may cause some confusion to write $T^i_j$ with $i$ and $j$ at the same position one on top of the other. The relation (A.6) can be written in terms of matrix multiplication, in which the column of the first matrix is summed against the row of the second matrix,

$$\left[T^i{}_j\right] = \left[T^{ik}\right]\left[g_{kj}\right] = \left[g^{ik}\right]\left[T_{kj}\right].$$

Note that if $S, T \in \mathcal{L}(V)$, then the *composition* $S \circ T$, defined as $S \circ T(\boldsymbol{v}) = S(T(\boldsymbol{v}))$ for all $\boldsymbol{v} \in V$, is also in $\mathcal{L}(V)$. The composition $S \circ T$ will be more conveniently denoted by $ST$. In terms of components and matrix operation, we have

$$[(ST)^i{}_j] = [S^i{}_k T^k{}_j] = [S^i{}_k][T^k{}_j].$$

**Example A.1.1** The identity transformation, $\mathit{1}\boldsymbol{v} = \boldsymbol{v}$ for any $\boldsymbol{v}$ in $V$, has the components,

$$\mathit{1} = \delta^i{}_j \boldsymbol{e}_i \otimes \boldsymbol{e}^j = \delta_i{}^j \boldsymbol{e}^i \otimes \boldsymbol{e}_j = g^{ij} \boldsymbol{e}_i \otimes \boldsymbol{e}_j = g_{ij} \boldsymbol{e}^i \otimes \boldsymbol{e}^j, \tag{A.7}$$

since from (A.5), we have

$$\mathit{1}^i{}_j = \boldsymbol{e}^i \cdot \mathit{1}\boldsymbol{e}_j = \boldsymbol{e}^i \cdot \boldsymbol{e}_j = \delta^i{}_j,$$
$$\mathit{1}_{ij} = \boldsymbol{e}_i \cdot \mathit{1}\boldsymbol{e}_j = \boldsymbol{e}_i \cdot \boldsymbol{e}_j = g_{ij}.$$

Therefore, the Kronecker deltas are the mixed components of the identity tensor, while $g_{ij}$ and $g^{ij}$ are just its covariant and contravariant components.

Note that the identity transformation is denoted by $\mathit{1}$ (italic "one"). That is, $\mathit{1} \in \mathcal{L}(V)$, while the number, $1 \in I\!R$, is a scalar quantity. □

**Example A.1.2** For $\boldsymbol{v} = v^i \boldsymbol{e}_i$ and $\boldsymbol{u} = u^i \boldsymbol{e}_i$ in V, their tensor product has the component form:

$$\boldsymbol{v} \otimes \boldsymbol{u} = v^i u^j \boldsymbol{e}_i \otimes \boldsymbol{e}_j.$$

Let $\boldsymbol{v} = (v_1, v_2)$ and $\boldsymbol{u} = (u_1, u_2)$ be two vectors in $I\!R^2$, then, relative to the standard basis of $I\!R^2$, the matrix of $\boldsymbol{v} \otimes \boldsymbol{u}$ is given by

$$[(\boldsymbol{v} \otimes \boldsymbol{u})] = \begin{bmatrix} v_1 u_1 & v_1 u_2 \\ v_2 u_1 & v_2 u_2 \end{bmatrix}.$$

This product is sometimes referred to as the *dyadic product* of vectors $\boldsymbol{v}$ and $\boldsymbol{u}$. □

In general, the tensor products $\boldsymbol{v} \otimes \boldsymbol{u}$ and $\boldsymbol{u} \otimes \boldsymbol{v}$ belong to two different spaces, namely $V \otimes U$ and $U \otimes V$, respectively. Even in the case $V = U$, by definition, $\boldsymbol{v} \otimes \boldsymbol{u}$ and $\boldsymbol{u} \otimes \boldsymbol{v}$ are different, i.e., the tensor product is not symmetric.

**Definition.** *For $A \in V \otimes U$, the transpose of $A$, denoted by $A^T$, is defined as a tensor in $U \otimes V$ such that*

$$\boldsymbol{v} \cdot A\boldsymbol{u} = \boldsymbol{u} \cdot A^T \boldsymbol{v}, \tag{A.8}$$

*for any $\boldsymbol{v} \in V$ and any $\boldsymbol{u} \in U$.*

**Example A.1.3** For simple tensors, it follows that

$$(\boldsymbol{v} \otimes \boldsymbol{u})^T = \boldsymbol{u} \otimes \boldsymbol{v},$$

because for any $\boldsymbol{w}_1, \boldsymbol{w}_2 \in V$, we have

$$\begin{aligned} \boldsymbol{w}_1 \cdot (\boldsymbol{v} \otimes \boldsymbol{u})^T \boldsymbol{w}_2 &= \boldsymbol{w}_2 \cdot (\boldsymbol{v} \otimes \boldsymbol{u}) \boldsymbol{w}_1 \\ &= (\boldsymbol{w}_2 \cdot \boldsymbol{v})(\boldsymbol{u} \cdot \boldsymbol{w}_1) = \boldsymbol{w}_1 \cdot (\boldsymbol{u} \otimes \boldsymbol{v}) \boldsymbol{w}_2. \end{aligned}$$

□

**Example A.1.4** We have

$$A(\boldsymbol{u} \otimes \boldsymbol{v}) = A\boldsymbol{u} \otimes \boldsymbol{v}, \qquad (\boldsymbol{u} \otimes \boldsymbol{v})A = \boldsymbol{u} \otimes A^T \boldsymbol{v}.$$

Indeed, for any vector $\boldsymbol{w} \in V$, we obtain

$$A(\boldsymbol{u} \otimes \boldsymbol{v})\boldsymbol{w} = A\boldsymbol{u}(\boldsymbol{v} \cdot \boldsymbol{w}) = (A\boldsymbol{u} \otimes \boldsymbol{v})\boldsymbol{w},$$

and

$$(\boldsymbol{u} \otimes \boldsymbol{v})A\boldsymbol{w} = \boldsymbol{u}(\boldsymbol{v} \cdot A\boldsymbol{w}) = \boldsymbol{u}(A^T \boldsymbol{v} \cdot \boldsymbol{w}) = (\boldsymbol{u} \otimes A^T \boldsymbol{v})\boldsymbol{w}.$$

□

If $A$ is a second-order tensor in $\mathcal{L}(V)$, then the components of the transpose $A^T$ satisfy the following relations:

$$\begin{aligned} (A^T)^{ij} &= A^{ji}, \qquad & (A^T)_{ij} &= A_{ji}, \\ (A^T)^i{}_j &= A_j{}^i, \qquad & (A^T)_i{}^j &= A^j{}_i. \end{aligned} \tag{A.9}$$

We see from these relations that for contravariant or covariant tensors the matrix of $A^T$ is simply the transpose of the matrix of $A$. However, from the second group of the relations in (A.9) for mixed tensors, this is not valid in general, since the matrix transpose of $[A^i{}_j]$, by changing rows and columns, is $[A^j{}_i]$, instead of $[A_j{}^i]$.

A tensor $S \in \mathcal{L}(V)$ is called *symmetric* if $S^T = S$, and is called *skew-symmetric* if $S^T = -S$. In other words, $S$ is symmetric if $\boldsymbol{v} \cdot S\boldsymbol{u} = \boldsymbol{u} \cdot S\boldsymbol{v}$ and $S$ is skew-symmetric if $\boldsymbol{v} \cdot S\boldsymbol{u} = -\boldsymbol{u} \cdot S\boldsymbol{v}$, for all $\boldsymbol{u}, \boldsymbol{v} \in V$.

**Notation.** $\mathit{Sym}(V) = \{S \in \mathcal{L}(V) \mid S^T = S\}$ and
$\mathit{Skw}(V) = \{S \in \mathcal{L}(V) \mid S^T = -S\}$.

Note that both $\mathit{Sym}(V)$ and $\mathit{Skw}(V)$ are subspaces of $\mathcal{L}(V)$. If $S \in \mathit{Sym}(V)$, then its components satisfy

$$\begin{aligned} S^{ij} &= S^{ji}, \qquad S_{ij} = S_{ji}, \\ S^i{}_j &= S_j{}^i = g_{jk}\, g^{im} S^k{}_m. \end{aligned}$$

In terms of matrix representation we have

$$[S^{ij}] = [S^{ij}]^T, \qquad [S_{ij}] = [S_{ij}]^T.$$

Note that although $S$ is symmetric, the matrix $[S^i{}_j]$ is not symmetric, in general,

$$[S^i{}_j] \neq [S^i{}_j]^T.$$

A second-order tensor can also be regarded as a bilinear function in the following manner: For any $A \in \mathcal{L}(V)$, define the function on $V \times V$, also denoted by $A$,

$$A(\boldsymbol{u}, \boldsymbol{v}) = \boldsymbol{u} \cdot A\boldsymbol{v},$$

for any vectors $\boldsymbol{u}$ and $\boldsymbol{v}$ in $V$. In particular, for simple tensors, we have

$$(\boldsymbol{u} \otimes \boldsymbol{v})(\boldsymbol{u}', \boldsymbol{v}') = \boldsymbol{u}' \cdot (\boldsymbol{u} \otimes \boldsymbol{v})\boldsymbol{v}' = (\boldsymbol{u} \cdot \boldsymbol{u}')(\boldsymbol{v} \cdot \boldsymbol{v}').$$

By employing the notion of multilinear functions, we can generalize tensor products to higher orders. For example, we can define a tensor product of three vectors $\boldsymbol{u}$, $\boldsymbol{v}$, and $\boldsymbol{w}$ as a trilinear function on $V$ by

$$(\boldsymbol{u} \otimes \boldsymbol{v} \otimes \boldsymbol{w})(\boldsymbol{u}', \boldsymbol{v}', \boldsymbol{w}') = (\boldsymbol{u} \cdot \boldsymbol{u}')(\boldsymbol{v} \cdot \boldsymbol{v}')(\boldsymbol{w} \cdot \boldsymbol{w}')$$

for any vectors $\boldsymbol{u}'$, $\boldsymbol{v}'$ and $\boldsymbol{w}'$ in $V$. One can show as before, that if $\{\boldsymbol{e}_i\}$ is a basis of $V$, then $\{\boldsymbol{e}_i \otimes \boldsymbol{e}_j \otimes \boldsymbol{e}_k\}$ is a product basis for the space of all trilinear functions on V. We shall denote this space as $V \otimes V \otimes V$ and call it the space of third-order tensors. If $S$ is a third-order tensor, then

$$S = S^{ijk}\boldsymbol{e}_i \otimes \boldsymbol{e}_j \otimes \boldsymbol{e}_k = S^i{}_{jk}\boldsymbol{e}_i \otimes \boldsymbol{e}^j \otimes \boldsymbol{e}^k = \text{etc.}$$

There are several different component forms relative to the different product bases. In a similar manner, the tensor product of higher orders can be defined. We write

$$\overset{k}{\otimes} V = \overbrace{V \otimes \cdots \otimes V}^{k \text{ times}}$$

for tensors of order $k$. Clearly, $\dim \overset{k}{\otimes} V = (\dim V)^k$.

**Exercise A.1.2** Let $\beta' = \{\boldsymbol{e}_1 = (1,0), \boldsymbol{e}_2 = (2,1)\}$ be a basis of $\mathbb{R}^2$, and $T \in \mathcal{L}(\mathbb{R}^2)$ be defined by

$$T(x_1, x_2) = (3x_1 + x_2, x_1 + 2x_2), \qquad \forall\, (x_1, x_2) \in \mathbb{R}^2. \tag{A.10}$$

1) Show that $T$ is a symmetric transformation.
2) Determine the matrices of the associated components of $T$ relative to $\beta'$:

$$[T_{ij}],\ [T^{ij}],\ [T_i{}^j],\ [T^i{}_j].$$

Note that the last two matrices are not symmetric.

### A.1.4 Transformation Rules for Components

The components of a tensor relative to a basis are uniquely determined and their values depend on the basis. Therefore, if we make a change of basis, they must change accordingly. In this section, we shall establish the transformation rules for components of tensors under a change of basis.

Consider a change of basis from $\beta = \{\boldsymbol{e}_i\}$ to another basis $\bar{\beta} = \{\bar{\boldsymbol{e}}_i\}$ given by

$$\bar{\boldsymbol{e}}_k = M_k{}^j\, \boldsymbol{e}_j. \tag{A.11}$$

We call $M_k{}^j$ the transformation matrix for the change of basis from $\beta$ to $\bar{\beta}$. By the use of (A.3), we have

$$M_k{}^j = \bar{\boldsymbol{e}}_k \cdot \boldsymbol{e}^j,$$

from which we can also obtain the relation between the dual bases $\beta^* = \{\boldsymbol{e}^i\}$ and $\bar{\beta}^* = \{\bar{\boldsymbol{e}}^i\}$,

$$\boldsymbol{e}^j = M_k{}^j\, \bar{\boldsymbol{e}}^k. \tag{A.12}$$

The above two transformation relations (A.11) and (A.12) can be schematically represented by

$$\beta \xrightarrow{M} \bar{\beta}\,,$$

$$\beta^* \xleftarrow{M^T} \bar{\beta}^*.$$

In other words, if $M$ changes a basis $\beta$ to another basis $\bar{\beta}$, their corresponding dual bases $\beta^*$ and $\bar{\beta}^*$ are changed in the opposite direction through $M^T$.

The components of a vector transform in a similar manner. Indeed, let $\boldsymbol{v}$ be a vector in $V$, and

$$\boldsymbol{v} = v^i\,\boldsymbol{e}_i = \bar{v}^i\,\bar{\boldsymbol{e}}_i$$

$$= v_j\,\boldsymbol{e}^j = \bar{v}_j\,\bar{\boldsymbol{e}}^j.$$

One can easily verify that the transformation rules for the components are

$$\bar{v}_k = M_k{}^j\,v_j, \qquad v^j = M_k{}^j\,\bar{v}^k, \tag{A.13}$$

which look exactly like those for the change of basis (A.11) and (A.12). In matrix notations, we have

$$[\bar{v}_k] = [M_k{}^j]\,[v_j], \qquad [v^j] = [M_k{}^j]^T[\bar{v}^k]$$

or schematically

$$[v_i] \xrightarrow{M} [\bar{v}_i],$$

$$[v^i] \xleftarrow{M^T} [\bar{v}^i].$$

That is, the covariant components transform in the same direction as the change of basis by $M$, while the contravariant components transform in the opposite direction by $M^T$. This is why such components are called *covariant* and *contravariant* in classical tensor analysis, in which tensors are defined through their transformation properties.

For a second-order tensor $A$ in $\mathcal{L}(V)$,

$$A = A_{ij}\,\boldsymbol{e}^i \otimes \boldsymbol{e}^j = \bar{A}_{ij}\,\bar{\boldsymbol{e}}^i \otimes \bar{\boldsymbol{e}}^j$$
$$= A_i{}^j\,\boldsymbol{e}^i \otimes \boldsymbol{e}_j = \bar{A}_i{}^j\,\bar{\boldsymbol{e}}^i \otimes \bar{\boldsymbol{e}}_j.$$

We have the following transformation rules,

$$\begin{aligned} \bar{A}_{ij} &= A_{mn}\,M_i{}^m\,M_j{}^n, \\ \bar{A}_i{}^j &= A_m{}^n\,M_i{}^m\,\overset{-1}{M}{}_n{}^j, \end{aligned} \tag{A.14}$$

where the matrix $[\overset{-1}{M}{}_i{}^j]$ is the inverse matrix of $[M_i{}^j]$. In matrix notations, the transformation rules can be written as

$$[\bar{A}_{ij}] = [M_i{}^m]\,[A_{mn}]\,[M_j{}^n]^T,$$
$$[\bar{A}_i{}^j] = [M_i{}^m]\,[A_m{}^n]\,[M_n{}^j]^{-1}.$$

Transformation rules for other components and for tensors of higher orders are similar. The general rule can easily be obtained by composing the transformation rules for covariant and contravariant components, as shown in (A.13) or (A.14).

**Exercise A.1.3** Let $\beta = \{\boldsymbol{e}_1 = (1,0), \boldsymbol{e}_2 = (0,1)\}$ and $\bar{\beta} = \{\bar{\boldsymbol{e}}_1 = (1,0), \bar{\boldsymbol{e}}_2 = (2,1)\}$ be two bases of $I\!R^2$. Determine the transformation matrix of the change of basis from $\beta$ to $\bar{\beta}$ and also the transformation matrix from $\bar{\beta}^*$ to $\beta^*$. Let $T \in \mathcal{L}(\boldsymbol{R}^2)$ be defined in (A.10). Determine the various components of $T$ relative to the two different bases and verify the transformation rules.

**Exercise A.1.4** For any two bases $\beta = \{\boldsymbol{e}_i\}$ and $\bar{\beta} = \{\bar{\boldsymbol{e}}_i\}$ of $V$, there exists a linear transformation $A \in \mathcal{L}(V)$ such that $\bar{\boldsymbol{e}}_k = A\boldsymbol{e}_k$. Show that the transformation matrix $M$ for the change of basis from $\beta$ to $\bar{\beta}$ is given by $M_k{}^j = \boldsymbol{e}^j \cdot A\boldsymbol{e}_k$, that is, $[M_k{}^j] = [A^j{}_k]$.

### A.1.5 Determinant and Trace

In matrix algebra, the definition of the determinant of a square matrix is based on the notion of permutation. Let $(1, \cdots, n)$ be an ordered set of natural numbers. A reordering of the elements in $(1, \cdots, n)$ is called a permutation. More precisely, a permutation is a one-to-one mapping $\sigma : \{1, \cdots, n\} \to \{1, \cdots, n\}$ resulting in the ordered set $(\sigma(1), \cdots, \sigma(n))$. There are exactly $n!$ permutations of $(1, \cdots, n)$. A permutation by exchanging the order of two adjacent elements is called a transposition. It is known that any permutation can be obtained by merely subsequent transpositions, and although the number of such transpositions are not unique for a given permutation, the parity of this number is. Hence, a permutation is called even or odd according to the parity of the number of transpositions in order to restore the permutation back to the natural order and one can define the sign of a permutation, denoted $\operatorname{sign}\sigma$, as $+1$ if $\sigma$ is even and $-1$ if $\sigma$ is odd.

Let $[M_{ij}]$ be a square matrix. The first index denotes the row and the second the column (it does not matter whether they are superindices or subindices). The determinant of the matrix can be calculated by

$$\det [M_{ij}] = \sum_{\sigma} (\text{sign } \sigma)\, M_{\sigma(1)\,1} \cdots M_{\sigma(n)\,n}, \qquad (A.15)$$

where the summation is taken over all permutations of $(1, \cdots, n)$.

On the other hand, since the matrix representation of a linear transformation depends on the choice of basis, the question arises of whether it is meaningful to define the determinant of a linear transformation as the determinant of its matrix representation. In the following, we shall see that the notion of the determinant of a linear transformation can be defined in a natural way, independent of the choice of basis and we see how it is related to its matrix representations.

**Definition.** *Let $V$ be a vector space of dimension $n$. A function* $\omega : \overbrace{V \times \cdots \times V}^{n} \to \mathbb{R}$ *is said to be an alternating $n$-linear form if it is $n$-linear and for all $\boldsymbol{v}_1, \cdots, \boldsymbol{v}_n \in V$,*

$$\omega(\boldsymbol{v}_{\sigma(1)}, \cdots, \boldsymbol{v}_{\sigma(n)}) = (\text{sign } \sigma)\omega(\boldsymbol{v}_1, \cdots, \boldsymbol{v}_n). \qquad (A.16)$$

*$\omega$ is called non-trivial if there exist $\boldsymbol{u}_1, \cdots, \boldsymbol{u}_n \in V$, such that $\omega(\boldsymbol{u}_1, \cdots, \boldsymbol{u}_n) \neq 0$.*

It is obvious that if $\omega$ is alternating then

$$\omega(\cdots, \boldsymbol{u}, \cdots, \boldsymbol{v}, \cdots) = 0, \quad \text{if } \boldsymbol{u} = \boldsymbol{v}. \qquad (A.17)$$

More generally, if $\{\boldsymbol{v}_1, \cdots, \boldsymbol{v}_n\}$ is linearly dependent then $\omega(\boldsymbol{v}_1, \cdots, \boldsymbol{v}_n) = 0$. In other words, if $\omega(\boldsymbol{v}_1, \cdots, \boldsymbol{v}_n) \neq 0$ then $\{\boldsymbol{v}_1, \cdots, \boldsymbol{v}_n\}$ is a linearly independent set, and since the number of vectors in this set equals $\dim V$, $\{\boldsymbol{v}_1, \cdots, \boldsymbol{v}_n\}$ is also a basis of $V$.

**Theorem.** (uniqueness) *Let $\omega$ and $\omega'$ be two alternating $n$-linear forms and $\omega$ be non-trivial. Then there exists uniquely a $\lambda \in \mathbb{R}$, such that $\omega' = \lambda\omega$, i.e., $\forall\, \boldsymbol{v}_1, \cdots, \boldsymbol{v}_n \in V$,*

$$\omega'(\boldsymbol{v}_1, \cdots, \boldsymbol{v}_n) = \lambda\omega(\boldsymbol{v}_1, \cdots, \boldsymbol{v}_n).$$

*Proof.* Since $\omega$ is non-trivial, there exists a set of vectors, say $\{\boldsymbol{e}_1, \cdots, \boldsymbol{e}_n\}$, such that $\omega(\boldsymbol{e}_1, \cdots, \boldsymbol{e}_n) \neq 0$, and hence it is a basis of $V$. Let $\lambda$ be the number defined by

$$\lambda = \frac{\omega'(\boldsymbol{e}_1, \cdots, \boldsymbol{e}_n)}{\omega(\boldsymbol{e}_1, \cdots, \boldsymbol{e}_n)}.$$

Suppose that $\boldsymbol{v}_1, \cdots, \boldsymbol{v}_n \in V$, and

$$\boldsymbol{v}_a = v_a^i \boldsymbol{e}_i, \qquad a = 1, \cdots, n.$$

Then, using (A.16) and (A.17) one can easily obtain

$$\omega(\boldsymbol{v}_1, \cdots, \boldsymbol{v}_n) = \alpha\omega(\boldsymbol{e}_1, \cdots, \boldsymbol{e}_n),$$
$$\omega'(\boldsymbol{v}_1, \cdots, \boldsymbol{v}_n) = \alpha\omega'(\boldsymbol{e}_1, \cdots, \boldsymbol{e}_n),$$

where

$$\alpha = \sum_{\sigma} (\text{sign } \sigma) v_1^{\sigma(1)} \cdots v_n^{\sigma(n)}.$$

Theretore, we have

$$\omega'(\boldsymbol{v}_1, \cdots, \boldsymbol{v}_n) = \lambda\omega(\boldsymbol{v}_1, \cdots, \boldsymbol{v}_n).$$

Moreover, this relation also shows that $\lambda$ does not depend on the choice of basis. □

Let $T \in \mathcal{L}(V)$ be a linear transformation on $V$, and $\omega$ be a non-trivial alternating $n$-linear form on $V$. Define a map $T_\omega : V \times \cdots \times V \to I\!R$ by

$$T_\omega(\boldsymbol{v}_1, \cdots, \boldsymbol{v}_n) = \omega(T\boldsymbol{v}_1, \cdots, T\boldsymbol{v}_n). \tag{A.18}$$

Clearly it is alternating and $n$-linear, hence by the uniqueness theorem, there exists a unique $\lambda \in I\!R$, such that

$$T_\omega = \lambda\omega.$$

We can easily see that the scalar $\lambda$ so defined does not depend on the choice of $\omega$. For, if $\omega'$ is another non-trivial alternating $n$-linear form, then by the uniqueness theorem,

$$\omega' = \mu\omega, \qquad \mu \neq 0.$$

Therefore, we have

$$T_{\omega'} = \lambda'\omega' = \lambda'\mu\omega.$$

On the other hand, we have

$$\begin{aligned} T_{\omega'}(\boldsymbol{v}_1, \cdots, \boldsymbol{v}_n) &= \omega'(T\boldsymbol{v}_1, \cdots, T\boldsymbol{v}_n) = \mu\omega(T\boldsymbol{v}_1, \cdots, T\boldsymbol{v}_n) \\ &= \mu T_\omega(\boldsymbol{v}_1, \cdots, \boldsymbol{v}_n) = \mu\lambda\omega(\boldsymbol{v}_1, \cdots, \boldsymbol{v}_n), \end{aligned}$$

which implies that

$$T_{\omega'} = \mu\lambda\omega.$$

Consequently, $\lambda = \lambda'$. Therefore, $\lambda$ is uniquely determined by $T$ alone and we can lay down the following definition.

**Definition.** *$T \in \mathcal{L}(V)$, the determinant of $T$, $\det T \in I\!R$, is defined by the following relation,*

$$(\det T)\,\omega(\boldsymbol{v}_1, \cdots, \boldsymbol{v}_n) = \omega(T\boldsymbol{v}_1, \cdots, T\boldsymbol{v}_n), \tag{A.19}$$

*for any non-trivial alternating $n$-linear form $\omega$ and for any $\boldsymbol{v}_1, \cdots, \boldsymbol{v}_n \in V$.*

The function $\det : \mathcal{L}(V) \to I\!R$ has the following properties:

$$\begin{aligned} &1)\ \det \boldsymbol{u} \otimes \boldsymbol{v} = 0. \\ &2)\ \det(\alpha\mathit{1}) = \alpha^n. \\ &3)\ \det(ST) = (\det S)(\det T). \\ &4)\ \det S^T = \det S. \end{aligned} \tag{A.20}$$

The first two properties are almost trivial. Here, let us verify the property (3). By definition,

$$\begin{aligned} &\det(ST)\,\omega(\boldsymbol{v}_1, \cdots, \boldsymbol{v}_n) \\ &\quad = \omega(ST\boldsymbol{v}_1, \cdots, ST\boldsymbol{v}_n) = \omega(S(T\boldsymbol{v}_1), \cdots, S(T\boldsymbol{v}_n)) \\ &\quad = (\det S)\omega(T\boldsymbol{v}_1, \cdots, T\boldsymbol{v}_n) = (\det S)(\det T)\omega(\boldsymbol{v}_1, \cdots, \boldsymbol{v}_n). \end{aligned}$$

Since it holds for any $\omega(\boldsymbol{v}_1, \cdots, \boldsymbol{v}_n)$, relation (3) follows.

We can calculate the determinant of a linear transformation in term of its component matrix. Let $\{\boldsymbol{e}_i\}$ be a basis of $V$, and $T = T^i{}_j \boldsymbol{e}_i \otimes \boldsymbol{e}^j$. Then by definition,

$$
\begin{aligned}
&(\det T)\,\omega(\boldsymbol{e}_1, \cdots, \boldsymbol{e}_n) \\
&\quad = \omega(T\boldsymbol{e}_1, \cdots, T\boldsymbol{e}_n) = \omega(T^{i_1}{}_1 \boldsymbol{e}_{i_1}, \cdots, T^{i_n}{}_n \boldsymbol{e}_{i_n}) \\
&\quad = \sum_\sigma (\text{sign } \sigma) T^{\sigma(1)}{}_1 \cdots T^{\sigma(n)}{}_n \, \omega(\boldsymbol{e}_1, \cdots, \boldsymbol{e}_n).
\end{aligned}
$$

Hence, we obtain

$$
\det T = \sum_\sigma (\text{sign } \sigma) T^{\sigma(1)}{}_1 \cdots T^{\sigma(n)}{}_n,
$$

which assures that

$$
\det T = \det [T^i{}_j],
$$

i.e., $\det T$ is equal to the determinant of the component matrix $[T^i{}_j]$ according to the definition (A.15). Similarly, one can show that it is also equal to determinant of $[T_i{}^j]$. Therefore, we have

$$
\begin{aligned}
\det T &= \det[T^i{}_j] = \det[T_i{}^j] \\
&= \det[g^{ik} T_{kj}] = \det[g_{ik} T^{kj}].
\end{aligned}
$$

Note that $\det T$ is not equal to $\det[T_{ij}]$ nor to $\det[T^{ij}]$, unless $\det[g_{ij}] = 1$.

Similar to the determinant, another scalar can be associated with a linear transformation. Let $T \in \mathcal{L}(V)$, and $\omega$ be a non-trivial alternating $n$-linear form. Define a map $\widetilde{T}_\omega : V \times \cdots \times V \to I\!R$ by

$$
\widetilde{T}_\omega(\boldsymbol{v}_1, \cdots, \boldsymbol{v}_n) = \sum_{i=1}^n \omega(\boldsymbol{v}_1, \cdots, T\boldsymbol{v}_i, \cdots, \boldsymbol{v}_n).
$$

One can easily check that $\widetilde{T}_\omega$ is alternating and $n$-linear, hence by the uniqueness theorem, there exists a $\mu \in I\!R$, such that

$$
\widetilde{T}_\omega = \mu\omega.
$$

Moreover, $\mu$ does not depend on the choice of $\omega$. Therefore, we can make the following definition.

**Definition.** *$T \in \mathcal{L}(V)$, the trace of $T$, $\operatorname{tr} T \in I\!R$, is defined by the following relation*

$$
(\operatorname{tr} T)\,\omega(\boldsymbol{v}_1, \cdots, \boldsymbol{v}_n) = \sum_{i=1}^n \omega(\boldsymbol{v}_1, \cdots, T\boldsymbol{v}_i, \cdots, \boldsymbol{v}_n), \tag{A.21}
$$

*for any non-trivial alternating $n$-linear form $\omega$ and for any $\boldsymbol{v}_1, \cdots, \boldsymbol{v}_n \in V$.*

The function $\text{tr} : \mathcal{L}(V) \to \mathbb{R}$ has the following properties:

$$\begin{aligned}
&1)\ \text{tr}(\alpha S + T) = \alpha\ \text{tr}\, S + \text{tr}\, T. \\
&2)\ \text{tr}\, \mathit{1} = n. \\
&3)\ \text{tr}(\boldsymbol{v} \otimes \boldsymbol{u}) = \boldsymbol{v} \cdot \boldsymbol{u}. \\
&4)\ \text{tr}\, S^T = \text{tr}\, S. \\
&5)\ \text{tr}(ST) = \text{tr}(TS).
\end{aligned} \tag{A.22}$$

The property (1) states that trace is a linear function on $\mathcal{L}(V)$. Here, let us prove the property (3). Suppose that $\boldsymbol{v} = v^i \boldsymbol{e}_i$, then

$$\begin{aligned}
\text{tr}(\boldsymbol{v} \otimes \boldsymbol{u})\, \omega(\boldsymbol{e}_1, \cdots, \boldsymbol{e}_n) &= \sum_{i=1}^{n} \omega(\boldsymbol{e}_1, \cdots, (\boldsymbol{v} \otimes \boldsymbol{u})\boldsymbol{e}_i, \cdots, \boldsymbol{e}_n) \\
&= \sum_{i=1}^{n} (\boldsymbol{u} \cdot \boldsymbol{e}_i) \omega(\boldsymbol{e}_1, \cdots, \boldsymbol{v}, \cdots, \boldsymbol{e}_n) = \sum_{i=1}^{n} (\boldsymbol{u} \cdot \boldsymbol{e}_i) v^i \omega(\boldsymbol{e}_1, \cdots, \boldsymbol{e}_n),
\end{aligned}$$

which implies that

$$\text{tr}(\boldsymbol{v} \otimes \boldsymbol{u}) = \sum_{i=1}^{n} \boldsymbol{u} \cdot (v^i \boldsymbol{e}_i) = \boldsymbol{u} \cdot \boldsymbol{v}.$$

Hence (3) is proved.

In terms of components, let $T = T^i_{\ j} \boldsymbol{e}_i \otimes \boldsymbol{e}^j = T_i^{\ j} \boldsymbol{e}^i \otimes \boldsymbol{e}_j$, then

$$\text{tr}\, T = T^i_{\ j}\, \text{tr}(\boldsymbol{e}_i \otimes \boldsymbol{e}^j) = T^j_{\ j} = T_j^{\ j} = g_{ij} T^{ij} = g^{ij} T_{ij}.$$

That is, $\text{tr}\, T$ is equal to the sum of diagonal elements of the matrix $[T^i_{\ j}]$ or $[T_j^{\ i}]$, but, in general, is not equal to that of the matrix $[T_{ij}]$ or $[T^{ij}]$.

**Example A.1.5** Show that $\det(\mathit{1} + \boldsymbol{u} \otimes \boldsymbol{v}) = 1 + \boldsymbol{u} \cdot \boldsymbol{v}$.

By definition, we have

$$\begin{aligned}
&\det(\mathit{1} + \boldsymbol{u} \otimes \boldsymbol{v})\, \omega(\boldsymbol{e}_1, \cdots, \boldsymbol{e}_n) \\
&\quad = \omega((\mathit{1} + \boldsymbol{u} \otimes \boldsymbol{v})\boldsymbol{e}_1, \cdots, (\mathit{1} + \boldsymbol{u} \otimes \boldsymbol{v})\boldsymbol{e}_n) \\
&\quad = \omega(\boldsymbol{e}_1, \cdots, \boldsymbol{e}_n) + \sum_{i=1}^{n} \omega(\boldsymbol{e}_1, \cdots, (\boldsymbol{u} \otimes \boldsymbol{v})\boldsymbol{e}_i, \cdots, \boldsymbol{e}_n) + \cdots \\
&\quad = \omega(\boldsymbol{e}_1, \cdots, \boldsymbol{e}_n) + \text{tr}(\boldsymbol{u} \otimes \boldsymbol{v})\, \omega(\boldsymbol{e}_1, \cdots, \boldsymbol{e}_n),
\end{aligned}$$

where the dots represent terms involved with more than one factor of $(\boldsymbol{u} \otimes \boldsymbol{v})\boldsymbol{e}_i$ in $\omega$. Since $(\boldsymbol{u} \otimes \boldsymbol{v})\boldsymbol{e}_i = (\boldsymbol{v} \cdot \boldsymbol{e}_i)\boldsymbol{u}$, which is a vector in the direction of $\boldsymbol{u}$ for any index $i$, those terms must all equal zero because $\omega$ is an alternating form. □

Two non-trivial alternating $n$-linear forms $\omega_1$ and $\omega_2$ are said to be equivalent if $\omega_1 = \lambda\omega_2$ for some $\lambda > 0$. Clearly, this is an equivalence relation that decomposes the set of non-trivial alternating $n$-linear forms into two equivalent classes. Each of these classes is called an *orientation* of $V$. We call one of them, say $\Delta$, the *positive orientation.* A basis $\{\boldsymbol{e}_i\}$ of $V$ is called *positively oriented* if for any $\omega \in \Delta$,

$$\omega(\boldsymbol{e}_1, \cdots, \boldsymbol{e}_n) > 0,$$

and $A \in \mathcal{L}(V)$ is said to be *orientation preserving* if $A_\omega \in \Delta$, for any $\omega \in \Delta$. Here, $A_\omega$ is defined by (A.18). Since $A_\omega = (\det A)\omega$, $A$ preserves the orientation if and only if $\det A > 0$.

Let $\{\boldsymbol{e}_i\}$ and $\{\bar{\boldsymbol{e}}_i\}$ be two bases such that $A(\boldsymbol{e}_i) = \bar{\boldsymbol{e}}_i$. If $\det A > 0$ (or $< 0$), then $\{\boldsymbol{e}_i\}$ and $\{\bar{\boldsymbol{e}}_i\}$ are said to have the *same* (or the *opposite*) orientation.

Suppose that $V$ is a three-dimensional vector space and let $\{\boldsymbol{i}_1, \boldsymbol{i}_2, \boldsymbol{i}_3\}$ be a positively oriented orthonormal basis of $V$, then there exists a unique $e \in \Delta$, called the *volume element*, such that

$$e(\boldsymbol{i}_1, \boldsymbol{i}_2, \boldsymbol{i}_3) = 1.$$

Since $e \in \mathcal{L}(V \times V \times V, I\!R)$, it is a third-order tensor and can be represented as

$$e = \varepsilon_{ijk}\, \boldsymbol{i}_i \otimes \boldsymbol{i}_j \otimes \boldsymbol{i}_k,$$

where $\varepsilon_{ijk} = e(\boldsymbol{i}_i, \boldsymbol{i}_j, \boldsymbol{i}_k)$ are the components of $e$ relative to the basis $\{\boldsymbol{i}_k\}$. Obviously we have

$$\varepsilon_{ijk} = \begin{cases} 1 & \text{if } (i,j,k) \text{ is an even permutation of } (1,2,3), \\ -1 & \text{if } (i,j,k) \text{ is an odd permutation of } (1,2,3), \\ 0 & \text{otherwise,} \end{cases}$$

and we call it the *permutation symbol.* One can easily check the following identities:

$$\begin{aligned} &\varepsilon_{ijk}\varepsilon_{imn} = \delta_{jm}\delta_{kn} - \delta_{jn}\delta_{km}, \\ &\varepsilon_{ijk}\varepsilon_{ijn} = 2\,\delta_{kn}, \\ &\varepsilon_{ijk}\varepsilon_{ijk} = 6, \end{aligned} \tag{A.23}$$

where $\delta_{mn}$ is the Kronecker delta.

Let $\{\boldsymbol{e}_k\}$ be a basis and $A \in \mathcal{L}(V)$ be a change of basis from $\{\boldsymbol{i}_k\}$ to $\{\boldsymbol{e}_k\}$, i.e., $A\,\boldsymbol{i}_k = \boldsymbol{e}_k$. Then the covariant components of the volume element relative to $\{\boldsymbol{e}_k\}$ are

$$e_{ijk} = e(\boldsymbol{e}_i, \boldsymbol{e}_j, \boldsymbol{e}_k), \quad e = e_{ijk}\, \boldsymbol{e}^i \otimes \boldsymbol{e}^j \otimes \boldsymbol{e}^k.$$

From (A.19), it follows that

$$e_{ijk} = (\det A)\varepsilon_{ijk},$$

and also,

$$g_{ij} = \boldsymbol{e}_i \cdot \boldsymbol{e}_j = A\boldsymbol{i}_i \cdot A\boldsymbol{i}_j = (A^T A)_{ij},$$

which yields $g = (\det A)^2$, where $g = \det[g_{ij}]$. Therefore, we have

$$e_{ijk} = \sqrt{g}\,\varepsilon_{ijk}, \tag{A.24}$$

if $A$ preserves the orientation. Similarly, the contravariant components of the volume element are

$$e^{ijk} = e(\boldsymbol{e}^i, \boldsymbol{e}^j, \boldsymbol{e}^k), \quad e = e^{ijk}\boldsymbol{e}_i \otimes \boldsymbol{e}_j \otimes \boldsymbol{e}_k,$$

and

$$e^{ijk} = (\sqrt{g})^{-1}\varepsilon^{ijk},$$

where $\varepsilon^{ijk} = \varepsilon_{ijk}$. Moreover, the identities (A.23) can be written as

$$\begin{aligned} e^{ijk}e_{imn} &= \delta^j{}_m\delta^k{}_n - \delta^j{}_n\delta^k{}_m, \\ e^{ijk}e_{ijn} &= 2\,\delta^k{}_n, \\ e^{ijk}e_{ijk} &= 6. \end{aligned} \tag{A.25}$$

If $T \in \mathcal{L}(V)$ and $T = T^i{}_j\boldsymbol{e}_i \otimes \boldsymbol{e}^j$, then (A.19) leads to the following formula for the determinant of $T$,

$$e_{lmn}(\det T) = e_{ijk}T^i{}_lT^j{}_mT^k{}_n. \tag{A.26}$$

Multiplying by $e^{lmn}$ and using the last identity of (A.25), we obtain another formula for the determinant,

$$\det T = \frac{1}{6}\, e^{lmn}e_{ijk}T^i{}_l\,T^j{}_mT^k{}_n.$$

**Exercise A.1.5** Consider the tensor defined by (A.10) in the previous exercise. Calculate $\det T$ and $\operatorname{tr} T$ by means of definition and also by the use of component matrices relative to $\beta'$.

### A.1.6 Exterior Product and Vector Product

The usual vector product on a three-dimensional vector space can not be generalized directly to vector spaces in general. However, it can be associated with the skew-symmetric tensor product in a trivial manner.

**Definition.** *For any, $\boldsymbol{v}, \boldsymbol{u} \in V$, the exterior product of $\boldsymbol{v}$ and $\boldsymbol{u}$, denoted $\boldsymbol{v} \wedge \boldsymbol{u}$, is defined by*

$$\boldsymbol{v} \wedge \boldsymbol{u} = \boldsymbol{v} \otimes \boldsymbol{u} - \boldsymbol{u} \otimes \boldsymbol{v}.$$

It is obvious that the operation $\wedge : V \times V \longrightarrow V \otimes V$ is bilinear and skew-symmetric, i.e.,

$$\boldsymbol{v} \wedge \boldsymbol{u} = -\boldsymbol{u} \wedge \boldsymbol{v}.$$

The exterior product of two vectors $\boldsymbol{v} \wedge \boldsymbol{u}$ is a skew-symmetric tensor.

Suppose that $\{\boldsymbol{e}_i \otimes \boldsymbol{e}_j\}$, $i, j = 1, \cdots, n$ is a product basis of $V \otimes V$, then it is easy to verify that $\{\boldsymbol{e}_i \wedge \boldsymbol{e}_j\}$, $1 \leq i < j \leq n$ is a basis for $\mathit{Skw}(V)$. Therefore, we have the following proposition.

**Proposition.** *If* $\dim V = n$, *then* $\dim \mathit{Skw}(V) = n(n-1)/2$. *In particular, if* $n = 3$, *then* $\dim \mathit{Skw}(V) = 3$.

Now, suppose that $V$ is an oriented Euclidean three-dimensional vector space. Since the space of skew-symmetric tensors is also three-dimensional, we can define a linear map

$$\tau : \mathit{Skw}(V) \longrightarrow V$$

by the condition: for all $\boldsymbol{u}, \boldsymbol{v}, \boldsymbol{w} \in V$,

$$\tau(\boldsymbol{u} \wedge \boldsymbol{v}) \cdot \boldsymbol{w} = e(\boldsymbol{u}, \boldsymbol{v}, \boldsymbol{w}). \tag{A.27}$$

Here, $e$ is the volume element of $V$. This linear map, called the *duality map*, is one-to-one and onto and hence establishes a one-to-one correspondence between a skew-symmetric tensor and a vector. It is easy to verify that

$$\tau(\boldsymbol{e}_i \wedge \boldsymbol{e}_j) = e_{ijk} \boldsymbol{e}^k. \tag{A.28}$$

For an orthonormal basis $\{\boldsymbol{i}_k\}$ the duality map $\tau$ establishes the following correspondence,

$$\boldsymbol{i}_1 \wedge \boldsymbol{i}_2 \longmapsto \boldsymbol{i}_3,$$
$$\boldsymbol{i}_2 \wedge \boldsymbol{i}_3 \longmapsto \boldsymbol{i}_1,$$
$$\boldsymbol{i}_3 \wedge \boldsymbol{i}_1 \longmapsto \boldsymbol{i}_2.$$

For a skew-symmetric tensor $W$, let $\boldsymbol{w} = \tau(W)$ be the associated vector, which shall be denoted more conveniently by

$$\boldsymbol{w} = \langle W \rangle. \tag{A.29}$$

In component form, if

$$W = W^{ij} \boldsymbol{e}_i \otimes \boldsymbol{e}_j, \qquad W^{ij} = -W^{ji},$$

or

$$W = \frac{1}{2} W^{ij} \boldsymbol{e}_i \wedge \boldsymbol{e}_j.$$

Then, it follows from (A.28) that

$$\boldsymbol{w} = \frac{1}{2} e_{ijk} W^{ij} \boldsymbol{e}^k.$$

If the basis is orthonormal, it becomes

$$w_i = \frac{1}{2} \varepsilon_{ijk} W_{jk}, \qquad W_{ij} = \varepsilon_{ijk} w_k, \tag{A.30}$$

or, in matrix form,

$$[W_{ij}] = \begin{bmatrix} 0 & w_3 & -w_2 \\ -w_3 & 0 & w_1 \\ w_2 & -w_1 & 0 \end{bmatrix}.$$

**Remark.** It is worthwhile to point out that the vector associated with a skew-symmetric tensor behaves differently from usual vectors under linear transformations. To see this, let $\boldsymbol{u}, \boldsymbol{v} \in V$, then for any $\boldsymbol{w} \in V$ and $Q \in \mathcal{L}(V)$, it follows from the definition that

$$\begin{aligned} \langle Q\boldsymbol{u} \wedge Q\boldsymbol{v} \rangle \cdot Q\boldsymbol{w} &= e(Q\boldsymbol{u}, Q\boldsymbol{v}, Q\boldsymbol{w}) \\ &= (\det Q)\, e(\boldsymbol{u}, \boldsymbol{v}, \boldsymbol{w}) = (\det Q) \langle \boldsymbol{u} \wedge \boldsymbol{v} \rangle \cdot \boldsymbol{w}, \end{aligned}$$

which implies that

$$\langle Q\boldsymbol{u} \wedge Q\boldsymbol{v} \rangle = (\det Q) Q \langle \boldsymbol{u} \wedge \boldsymbol{v} \rangle.$$

In other words, as the vectors $\boldsymbol{u}$, $\boldsymbol{v}$ are transformed into $Q\boldsymbol{u}$, $Q\boldsymbol{v}$, respectively, the vector $\langle \boldsymbol{u} \wedge \boldsymbol{v} \rangle$ is transformed into $Q\langle \boldsymbol{u} \wedge \boldsymbol{v} \rangle$ only to within a scalar constant, or, into a vector that may point in one or the opposite sense of the same axial direction of $\langle Q\boldsymbol{u} \wedge Q\boldsymbol{v} \rangle$. For this reason, a vector associated with a skew-symmetric tensor is usually called an *axial vector*. □

The usual vector product, in the three-dimensional vector space, can now be defined from the exterior product in a similar manner.

**Definition.** *For any* $\boldsymbol{u}, \boldsymbol{v} \in V$, *the vector product of* $\boldsymbol{u}$ *and* $\boldsymbol{v}$, *denoted* $\boldsymbol{u} \times \boldsymbol{v}$, *is defined by*

$$\boldsymbol{u} \times \boldsymbol{v} = \langle \boldsymbol{u} \wedge \boldsymbol{v} \rangle. \tag{A.31}$$

Clearly the operation $\times : V \times V \longrightarrow V$ is bilinear and skew-symmetric. In components (A.31) gives

$$\boldsymbol{u} \times \boldsymbol{v} = e_{ijk} u^j v^k \boldsymbol{e}^i.$$

If the basis is orthonormal, say $\{\boldsymbol{i}_k\}$, then it becomes

$$\boldsymbol{u} \times \boldsymbol{v} = \varepsilon_{ijk} u_j v_k \boldsymbol{i}_i,$$

which is the usual definition of the vector product.

The relations (A.27) and (A.31) imply that

$$e(\boldsymbol{u}, \boldsymbol{v}, \boldsymbol{w}) = (\boldsymbol{u} \times \boldsymbol{v}) \cdot \boldsymbol{w}.$$

This is usually called the *triple product* of $\boldsymbol{u}$, $\boldsymbol{v}$, and $\boldsymbol{w}$. For convenience, we shall also use the notation,

$$[\boldsymbol{u}, \boldsymbol{v}, \boldsymbol{w}] = (\boldsymbol{u} \times \boldsymbol{v}) \cdot \boldsymbol{w}.$$

With this notation, we can rewrite the definitions (A.19) and (A.21) of the determinant and the trace in the following form

$$\begin{aligned} \det A &= \frac{[A\boldsymbol{e}_1, A\boldsymbol{e}_2, A\boldsymbol{e}_3]}{[\boldsymbol{e}_1, \boldsymbol{e}_2, \boldsymbol{e}_3]}, \\ \operatorname{tr} A &= \frac{[A\boldsymbol{e}_1, \boldsymbol{e}_2, \boldsymbol{e}_3] + [\boldsymbol{e}_1, A\boldsymbol{e}_2, \boldsymbol{e}_3] + [\boldsymbol{e}_1, \boldsymbol{e}_2, A\boldsymbol{e}_3]}{[\boldsymbol{e}_1, \boldsymbol{e}_2, \boldsymbol{e}_3]}. \end{aligned} \tag{A.32}$$

One may use the duality map to identify a skew-symmetric tensor with an axial vector, as well as the exterior product with the vector product. In other words, one may interpret the duality in either way, as far as the context requires.

**Exercise A.1.6** Verify the following relations, using index notations:
1) $W\boldsymbol{v} = -\boldsymbol{w} \times \boldsymbol{v}$.
2) $(\boldsymbol{u} \times \boldsymbol{v}) \times \boldsymbol{w} = (\boldsymbol{u} \cdot \boldsymbol{w})\boldsymbol{v} - (\boldsymbol{v} \cdot \boldsymbol{w})\boldsymbol{u}$.
3) $|\boldsymbol{u} \times \boldsymbol{v}|^2 = |\boldsymbol{u}|^2 |\boldsymbol{v}|^2 - |\boldsymbol{u} \cdot \boldsymbol{v}|^2$.
4) $|\boldsymbol{u} \times \boldsymbol{v}| = |\boldsymbol{u}|\,|\boldsymbol{v}| \sin\theta(\boldsymbol{u}, \boldsymbol{v})$.

### A.1.7 Second-Order Tensors

We shall review some of the important properties of linear transformations, i.e., the second-order tensors, mostly without proofs in this section. The proofs can be found in most standard books in linear algebra.

First, let us introduce an inner product of two second-order tensors. Let $A, B \in \mathcal{L}(V)$, we can define the *inner product* of $A$ and $B$ by

$$A \cdot B = \operatorname{tr}(AB^T),$$

which is obviously a bilinear, symmetric and positive-definite operation. We have

$$\mathit{1} \cdot A = \operatorname{tr} A,$$

where $\mathit{1}$ is the identity tensor, and for any $A, B, C \in \mathcal{L}(V)$,

$$AB \cdot C = B \cdot A^T C = A \cdot CB^T.$$

The *norm* of a tensor $A \in \mathcal{L}(V)$, can then be defined as

$$|A| = \sqrt{A \cdot A} = \sqrt{\operatorname{tr} AA^T}.$$

Note that if $A_{ij}$ are the components of $A$ relative to an orthonormal basis, then the norm of $A$ is simply

$$|A| = (A_{11}^2 + A_{12}^2 + \cdots + A_{nn}^2)^{1/2}.$$

Now, suppose that $A \in \mathcal{L}(V)$ is one-to-one (therefore, onto), then there is a unique $A^{-1} \in \mathcal{L}(V)$, called the *inverse* of $A$, such that

$$AA^{-1} = A^{-1}A = \mathit{1}.$$

If $A^{-1}$ exists, $A$ is said to be *invertible* or *non-singular*, otherwise, it is said to be *singular*. It can be proved that $A$ is invertible if and only if $\det A \neq 0$, and for any non-singular $A$ and $B$,

$$(AB)^{-1} = B^{-1}A^{-1},$$
$$(A^{-1})^T = (A^T)^{-1} = A^{-T}.$$

**Notation.** $Inv(V) = \{F \in \mathcal{L}(V) \mid F \text{ is invertible}\}$.

Recall that a set $G$ is called a *group* if it has the following properties:

1) If $A, B \in G$ then $AB \in G$.
2) If $A, B, C \in G$ then $A(BC) = (AB)C$.
3) There exists an identity element $\mathit{1} \in G$ such that $\mathit{1}A = A\mathit{1} = A$, for any $A \in G$.
4) For any $A \in G$, there exists $A^{-1} \in G$, such that $AA^{-1} = A^{-1}A = \mathit{1}$.

It is easy to verify that $Inv(V)$ forms a group under the operation of composition. It is usually known as the *general linear group* of $V$, denoted by $GL(V)$.

**Definition.** *$Q \in \mathcal{L}(V)$ is called an orthogonal transformation if it preserves the inner product of $V$. i.e., for all $\boldsymbol{u}, \boldsymbol{v} \in V$,*

$$Q\boldsymbol{u} \cdot Q\boldsymbol{v} = \boldsymbol{u} \cdot \boldsymbol{v}.$$

**Notation.** $\mathcal{O}(V) = \{Q \in \mathcal{L}(V) \mid Q \text{ is orthogonal}\}$.

The set $\mathcal{O}(V)$ forms a group and is called the *orthogonal group* of $V$. Orthogonal transformations have the following properties:

1) $Q^T = Q^{-1}$.
2) $|\det Q| = 1$.
3) $|Q\boldsymbol{v}| = |\boldsymbol{v}|$.
4) $\theta(Q\boldsymbol{v}, Q\boldsymbol{u}) = \theta(\boldsymbol{v}, \boldsymbol{u})$.

The last two relations assert that orthogonal transformations also preserve norms and angles. An orthogonal transformation $Q$ is said to be *proper* if $\det Q = 1$, and *improper* if $\det Q = -1$.

**Notation.** $\mathcal{O}^+(V) = \{Q \in \mathcal{O}(V) \mid \det Q = 1\}$.

The set $\mathcal{O}^+(V)$ also forms a group, called the *proper orthogonal group* of $V$. It is also called the *rotation group* since its elements are rotations. Note that the subset of $\mathcal{O}(V)$ with determinant equal to $-1$ does not form a group since it does not have an identity element.

**Notation.** $\mathcal{U}(V) = \{T \in \mathcal{L}(V) \mid |\det T| = 1\}$ and
$SL(V) = \{T \in \mathcal{L}(V) \mid \det T = 1\}$.

Elements of $\mathcal{U}(V)$ are called unimodular transformations and $\mathcal{U}(V)$ forms a group, called the *unimodular group* of $V$. $SL(V)$ also forms a group, called the *special linear group* of $V$. Clearly, we have the following relations:

$$\mathcal{O}^+(V) \subset \begin{matrix} SL(V) \\ \mathcal{O}(V) \end{matrix} \subset \mathcal{U}(V) \subset GL(V).$$

### A.1.8 Some Theorems of Linear Algebra

We shall mention some important theorems of linear algebra relevant to the study of mechanics. They are all related to the concept of eigenvalues and eigenvectors.

**Definition.** *Let $A \in \mathcal{L}(V)$. A scalar $\lambda \in I\!R$ is called an eigenvalue of $A$, if there exists a non-zero vector $\boldsymbol{v} \in V$, such that*

$$A\boldsymbol{v} = \lambda \boldsymbol{v}. \tag{A.33}$$

*$\boldsymbol{v}$ is called the eigenvector of $A$ associated with the eigenvalue $\lambda$.*

It follows from the definition that $\lambda$ is an eigenvalue if and only if

$$\det(A - \lambda \mathit{1}) = 0. \tag{A.34}$$

The left-hand side of (A.34) is a polynomial of degree $n$ in $\lambda$, where $n$ is the dimension of $V$. We may write it in the form

$$(-\lambda)^n + I_1(-\lambda)^{n-1} + \cdots + I_{n-1}(-\lambda) + I_n = 0.$$

It is called the *characteristic equation* of $A$. Its real roots are the eigenvalues of $A$. The coefficients $I_1, \cdots, I_n$ are scalar functions of $A$ and are called the *principal invariants* of $A$.

It can be shown that the characteristic equation is also satisfied by the tensor $A$ itself. We have the following

**Cayley–Hamilton Theorem.** *A second-order tensor $A \in \mathcal{L}(V)$ satisfies its own characteristic equation,*

$$(-A)^n + I_1(-A)^{n-1} + \cdots + I_{n-1}(-A) + I_n \mathit{1} = 0.$$

**Example A.1.6** For $\dim V = 3$ and $A \in \mathcal{V}$, we have

$$\det(A - \lambda \mathit{1}) = -\lambda^3 + I_A\lambda^2 - II_A\lambda + III_A. \tag{A.35}$$

The three principal invariants of $A$, more specifically denoted by $I_A$, $II_A$, and $III_A$ can be obtained from the following relations:

$$I_A = \operatorname{tr} A, \qquad II_A = \operatorname{tr} A^{-1} \det A, \qquad III_A = \det A. \tag{A.36}$$

Of course, the second relation is valid only when $A$ is non-singular.

*Proof.* From (A.32) we can write

$$\begin{aligned}
\det(A - \lambda \mathit{1})[\boldsymbol{e}_1, \boldsymbol{e}_2, \boldsymbol{e}_3] = {} & [(A - \lambda \mathit{1})\boldsymbol{e}_1, (A - \lambda \mathit{1})\boldsymbol{e}_2, (A - \lambda \mathit{1})\boldsymbol{e}_3] \\
= {} & -\lambda^3 [\boldsymbol{e}_1, \boldsymbol{e}_2, \boldsymbol{e}_3] \\
& + \lambda^2([A\boldsymbol{e}_1, \boldsymbol{e}_2, \boldsymbol{e}_3] + [\boldsymbol{e}_1, A\boldsymbol{e}_2, \boldsymbol{e}_3] + [\boldsymbol{e}_1, \boldsymbol{e}_2, A\boldsymbol{e}_3]) \\
& - \lambda ([\boldsymbol{e}_1, A\boldsymbol{e}_2, A\boldsymbol{e}_3] + [A\boldsymbol{e}_1, \boldsymbol{e}_2, A\boldsymbol{e}_3] + [A\boldsymbol{e}_1, A\boldsymbol{e}_2, \boldsymbol{e}_3]) \\
& + [A\boldsymbol{e}_1, A\boldsymbol{e}_2, A\boldsymbol{e}_3].
\end{aligned}$$

Comparing this with the right-hand side of (A.35), we obtain $(A.36)_{1,3}$ by the use of (A.32), as well as the following relation for the second invariant $II_A$,

$$II_A = \frac{[\boldsymbol{e}_1, A\boldsymbol{e}_2, A\boldsymbol{e}_3] + [A\boldsymbol{e}_1, \boldsymbol{e}_2, A\boldsymbol{e}_3] + [A\boldsymbol{e}_1, A\boldsymbol{e}_2, \boldsymbol{e}_3]}{[\boldsymbol{e}_1, \boldsymbol{e}_2, \boldsymbol{e}_3]}.$$

If $A \in Inv(V)$, then it implies the second relation of (A.36). In particular, if $\det A = 1$, we have $II_A = I_{A^{-1}}$. □

In general, the characteristic equation may not have real roots. However, it is known that if $A$ is symmetric all the roots are real and there exists a basis of $V$ consisting entirely of eigenvectors.

**Spectral Theorem.** *Let $S \in Sym(V)$, then there exists an orthonormal basis $\{\boldsymbol{e}_i\}$ of $V$, such that $S$ can be written in the form*

$$S = \sum_{i=1}^{n} s_i \boldsymbol{e}_i \otimes \boldsymbol{e}_i. \tag{A.37}$$

Such a basis is called a *principal basis* for $S$. Relative to this basis, the component matrix of $S$ is a diagonal matrix and the diagonal elements $s_i$ are the eigenvalues of $S$ associated with the eigenvectors $\boldsymbol{e}_i$, respectively. The eigenvalues $s_i$, $i = 1, \cdots, n$ may or may not be distinct.

**Definition.** *Let $\lambda$ be an eigenvalue of $S \in \mathcal{L}(V)$. We call $V_\lambda = \{\boldsymbol{v} \in V \mid S\boldsymbol{v} = \lambda\boldsymbol{v}\}$ the characteristic space of $S$ associated with $\lambda$.*

If $S$ is a symmetric tensor and suppose that $\boldsymbol{v} \in V_\lambda$, $\boldsymbol{u} \in V_\mu$, where $\lambda$ and $\mu$ are two distinct eigenvalues of $S$, then one can easily show that $\boldsymbol{v} \cdot \boldsymbol{u} = 0$, i.e., they are mutually orthogonal. Moreover, by the spectral theorem any vector $\boldsymbol{v}$ can be written in the form

$$\boldsymbol{v} = \sum_{\lambda} \boldsymbol{v}_\lambda, \qquad \boldsymbol{v}_\lambda \in V_\lambda, \tag{A.38}$$

where the summation is extended over all characteristic spaces of $S$.

**Commutation Theorem.** *Let $T \in \mathcal{L}(V)$ and $S \in Sym(V)$. Then*

$$ST = TS$$

*if and only if $T$ preserves all characteristic spaces of $S$, i.e., $T$ maps each characteristic space of $S$ into itself.*

*Proof.* Suppose that $S$ and $T$ commute, and $S\boldsymbol{v} = \lambda\boldsymbol{v}$. Then

$$S(T\boldsymbol{v}) = T(S\boldsymbol{v}) = \lambda(T\boldsymbol{v}),$$

so that both $\boldsymbol{v}$ and $T\boldsymbol{v}$ belong to the characteristic space $V_\lambda$.

To prove the converse, since $S$ is symmetric, for any $\boldsymbol{v} \in V$, let $\boldsymbol{v} = \sum_\lambda \boldsymbol{v}_\lambda$ be the decomposition relative to the characteristic spaces of $S$ as given in (A.38). If $T$ leaves each characteristic space $V_\lambda$ invariant, then $T\boldsymbol{v}_\lambda \in V_\lambda$ and

$$S(T\boldsymbol{v}_\lambda) = \lambda(T\boldsymbol{v}_\lambda) = T(\lambda\boldsymbol{v}_\lambda) = T(S\boldsymbol{v}_\lambda).$$

Therefore, from (A.38), we have

$$ST\boldsymbol{v} = \sum_\lambda ST\boldsymbol{v}_\lambda = \sum_\lambda TS\boldsymbol{v}_\lambda = TS\boldsymbol{v},$$

which shows that $ST = TS$. □

There is only one subspace of $V$ that is preserved by any rotation, namely $V$ itself. Therefore, we have the following

**Corollary.** *A symmetric $S \in \mathcal{L}(V)$ commutes with every orthogonal transformation if and only if $S = \lambda 1$, for some $\lambda \in \mathbb{R}$.*

**Definition.** *$S \in \mathcal{L}(V)$ is said to be positive definite (positive semi-definite) if for any $\boldsymbol{v} \in V$ and $\boldsymbol{v} \neq 0$, $\boldsymbol{v} \cdot S\boldsymbol{v} > 0$ $(\geq 0)$. Similarly, $S$ is said to be negative definite (negative semi-definite) if $\boldsymbol{v} \cdot S\boldsymbol{v} < 0$ $(\leq 0)$.*

One can easily see that if $S$ is symmetric, then it is positive definite if and only if all of its eigenvalues are positive. Consequently, for any symmetric positive definite transformation $S$, there is a *unique* symmetric positive definite transformation $T$ such that $T^2 = S$ and the eigenvalues of $T$ are the positive square roots of those of $S$ associated with the same eigenvectors. We denote $T = \sqrt{S}$ and call $T$ the *square root* of $S$. In other words, if $S$ is expressed by (A.37) in terms of the principal basis, then

$$T = \sqrt{S} = \sum_{i=1}^{n} \sqrt{s_i}\, \boldsymbol{e}_i \otimes \boldsymbol{e}_i.$$

**Example A.1.7** Let $S \in \mathcal{L}(\mathbb{R}^2)$ be given by $S(x, y) = (3x + \sqrt{2}y, \sqrt{2}x + 2y)$. Relative to the standard basis of $\mathbb{R}^2$, the matrix of $S$ is

$$[S_{ij}] = \begin{bmatrix} 3 & \sqrt{2} \\ \sqrt{2} & 2 \end{bmatrix},$$

which has the eigenvalues $s_1 = 4$ and $s_2 = 1$ and the corresponding principal basis $\boldsymbol{e}_1 = (\sqrt{2/3}, \sqrt{1/3})$ and $\boldsymbol{e}_2 = (-\sqrt{1/3}, \sqrt{2/3})$. Therefore, we have

$$T = \sqrt{S} = 2\boldsymbol{e}_1 \otimes \boldsymbol{e}_1 + \boldsymbol{e}_2 \otimes \boldsymbol{e}_2,$$

whose matrix, relative to the standard basis, becomes

$$[T_{ij}] = \frac{2}{3}\begin{bmatrix} 2 & \sqrt{2} \\ \sqrt{2} & 1 \end{bmatrix} + \frac{1}{3}\begin{bmatrix} 1 & -\sqrt{2} \\ -\sqrt{2} & 2 \end{bmatrix} = \frac{1}{3}\begin{bmatrix} 5 & \sqrt{2} \\ \sqrt{2} & 4 \end{bmatrix}.$$

One can easily verify that $[T_{ij}]^2 = [S_{ij}]$.

**Example A.1.8** Let $S$ be a positive definite symmetric tensor in a two-dimensional space, then

$$\sqrt{S} = \frac{1}{b}(S + a1),$$

where $a = \sqrt{\det S}$ and $b = \sqrt{2a + \operatorname{tr} S}$.

*Proof.* Let $A = \sqrt{S}$. By the Cayley–Hamilton theorem in the two-dimensional space, we have the identity

$$A^2 - (\operatorname{tr} A)A + (\det A)\mathit{1} = 0.$$

Since $A^2 = S$, if we let the eigenvalues of $A$ be $a_1$ and $a_2$, then $\det S = a_1^2 a_2^2$ and $\operatorname{tr} S = a_1^2 + a_2^2$. Therefore

$$a = \sqrt{a_1^2 a_2^2} = a_1 a_2 = \det A,$$

$$b = \sqrt{2a_1 a_2 + a_1^2 + a_2^2} = a_1 + a_2 = \operatorname{tr} A,$$

which, together with the above identity, prove the result. □

**Polar Decomposition Theorem.** *For any $F \in \operatorname{Inv}(V)$, there exist symmetric positive definite transformations $V$ and $U$ and an orthogonal transformation $R$ such that*

$$F = RU = VR.$$

*Moveover, the transformations $U$, $V$ and $R$ are uniquely determined in the above decompositions.*

*Proof.* We can easily verify that $FF^T$ and $F^TF$ are symmetric positive definite. Indeed, for any $\boldsymbol{v} \neq 0$, we have

$$(\boldsymbol{v} \cdot F^T F\boldsymbol{v}) = (F\boldsymbol{v} \cdot F\boldsymbol{v}) > 0,$$

since $F$ is non-singular.

To prove the theorem, let us define

$$U = \sqrt{F^TF}, \quad R = FU^{-1}, \quad V = RUR^T. \tag{A.39}$$

By definition, $U$ is symmetric positive definite and $R$ is orthogonal since

$$\begin{aligned} RR^T &= FU^{-1}(FU^{-1})^T = FU^{-1}U^{-T}F^T \\ &= FU^{-2}F^T = F(F^TF)^{-1}F^T = \mathit{1}. \end{aligned}$$

Moreover, from the definition (A.39) we also have

$$V^2 = RUR^T(RUR^T) = (RU)(RU)^T = FF^T.$$

Therefore, $V$ is the square root of $FF^T$ and hence is itself a symmetric positive definite transformation. Furthermore, the uniqueness follows from the definition of a square root. □

The polar decomposition theorem, which decomposes a non-singular transformation into a rotation and a positive definite tensor, is crucial in the development of continuum mechanics. The following decomposition of a tensor into its symmetric and skew-symmetric parts is also important in mechanics.

For any $T \in \mathcal{L}(V)$, let

$$A = \frac{1}{2}(T + T^T), \qquad B = \frac{1}{2}(T - T^T),$$

then

$$T = A + B, \qquad A \in Sym(V), \quad B \in Skw(V).$$

This is sometimes called the *Cartesian decomposition* of a tensor. Such a decomposition is also unique.

**Exercise A.1.7** Let $A \in \mathcal{L}(V)$ be such that $(\mathit{1} + A)$ is non-singular. Verify that

1) $(\mathit{1} + A)^{-1} = \mathit{1} - A(\mathit{1} + A)^{-1}$.
2) $(\mathit{1}+A)^{-1} = \mathit{1} - A + A^2 - \cdots + (-1)^n A^n + o(A^n)$ if $\lim_{|A| \to 0} \dfrac{o(A^n)}{|A|^n} = 0$.

**Exercise A.1.8** Let $\boldsymbol{u}, \boldsymbol{v} \in V$. Show that if $1 + \boldsymbol{u} \cdot \boldsymbol{v} \neq 0$ then

$$(\mathit{1} + \boldsymbol{u} \otimes \boldsymbol{v})^{-1} = \mathit{1} - \frac{\boldsymbol{u} \otimes \boldsymbol{v}}{1 + \boldsymbol{u} \cdot \boldsymbol{v}}.$$

**Exercise A.1.9** For $\dim V = 3$, let $A \in \mathcal{L}(V)$ and $B = \mathit{1} + A$. Show that

$$\begin{aligned} I_B &= 3 + I_A, \\ II_B &= 3 + 2I_A + II_A, \\ III_B &= 1 + I_A + II_A + III_A, \end{aligned}$$

and if $a = \det B \neq 0$, verify that

$$(\mathit{1} + A)^{-1} = \frac{1}{a}\Big((1 + I_A + II_A)\mathit{1} - (1 + I_A)A + A^2\Big).$$

**Exercise A.1.10** Prove the Cayley–Hamilton theorem for the special case that $A \in \mathcal{L}(V)$ is symmetric, by employing the spectral theorem.

**Exercise A.1.11** Let $\beta = \{(1,0,0), (0,1,0), (0,0,1)\}$ be the standard basis of $I\!R^3$ and the matrix representation of $F \in \mathcal{L}(I\!R^3)$ relative to $\beta$ be given by

$$F = \begin{bmatrix} \sqrt{3} & 1 & 0 \\ 0 & 2 & 0 \\ 0 & 0 & 1 \end{bmatrix}.$$

Suppose that $F = RU = VR$ is the polar decomposition of $F$. Find the matrix representation of $U$, $V$, and $R$ relative to the standard basis $\beta$.

## A.2 Tensor Calculus

In the second part of this appendix, we shall discuss some basic notions of calculus on Euclidean spaces: gradients and other differential operators of tensor functions.

### A.2.1 Euclidean Point Space

Let $\mathcal{E}$ be a set of points and $V$ be a Euclidean vector space of dimension $n$.

**Definition.** *$\mathcal{E}$ is called a Euclidean point space of dimension $n$, and $V$ is called the translation space of $\mathcal{E}$ if, for any pair of points $x, y \in \mathcal{E}$, there is a vector $\boldsymbol{v} \in V$, called the difference vector of $x$ and $y$, written as*

$$\boldsymbol{v} = y - x, \tag{A.40}$$

*with the following properties:*
1) $\forall\, x \in \mathcal{E}, \quad x - x = 0 \in V.$
2) $\forall\, x \in \mathcal{E}, \forall\, \boldsymbol{v} \in V$, *there exists a unique point $y \in \mathcal{E}$, such that* (A.40) *is satisfied. We write $y = x + \boldsymbol{v}$.*
3) $\forall\, x, y, z \in \mathcal{E}, \quad (x - y) + (y - z) = (x - z).$

Obviously, with (A.40) we can define the *distance* between $x$ and $y$ in $\mathcal{E}$, denoted $d(x, y)$, by

$$d(x, y) = |\boldsymbol{v}|,$$

or equivalently

$$d(x, y) = \sqrt{(x - y) \cdot (x - y)},$$

where the dot denotes the inner product on $V$.

**Notation.** $\mathcal{E}_x = \{\boldsymbol{v}_x = (x, \boldsymbol{v}) \mid \boldsymbol{v} = y - x, \ \forall\, y \in \mathcal{E}\}.$

$\mathcal{E}_x$ denotes the set of all difference vectors at $x$. It can be made into a Euclidean vector space in an obvious way, with the addition and scalar multiplication defined as

$$\boldsymbol{v}_x + \boldsymbol{u}_x = (\boldsymbol{v} + \boldsymbol{u})_x,$$
$$\alpha\, \boldsymbol{v}_x = (\alpha \boldsymbol{v})_x.$$

We call $\mathcal{E}_x$ the *tangent space* of $\mathcal{E}$ at $x$.

Clearly $\mathcal{E}_x$ is a copy of $V$, i.e., it is isomorphic to $V$. In other words, for any $x \in \mathcal{E}$, the map $i_x : V \to \mathcal{E}_x$, called the *Euclidean parallelism*, taking $\boldsymbol{v}$ to $\boldsymbol{v}_x$ trivially establishes a one-to-one correspondence between $\mathcal{E}_x$ and $V$. The composite map

$$\tau_{xy} = i_y \circ i_x^{-1} : \mathcal{E}_x \longrightarrow \mathcal{E}_y$$

taking

$$\boldsymbol{v}_x = (x, \boldsymbol{v}) \longmapsto \boldsymbol{v}_y = (y, \boldsymbol{v})$$

defines the *parallel translation* of vectors at $x$ to vectors at $y$ (Fig. A.1).

Therefore, although $\mathcal{E}_x$ and $\mathcal{E}_y$ for $x \neq y$, are two different tangent spaces, they can be identified through $V$ in an obvious manner,

$$\mathcal{E}_x \cong \mathcal{E}_y \cong V, \qquad \forall\, x, y \in \mathcal{E}.$$

In other words, $\boldsymbol{v}_x = (x, \boldsymbol{v}) \in \mathcal{E}_x$ and $\boldsymbol{u}_y = (y, \boldsymbol{u}) \in \mathcal{E}_y$ are regarded as the same vector if and only if $\boldsymbol{v} = \boldsymbol{u}$. In this manner, vectors at different tangent spaces can be added or subtracted as if they were in the same vector space.

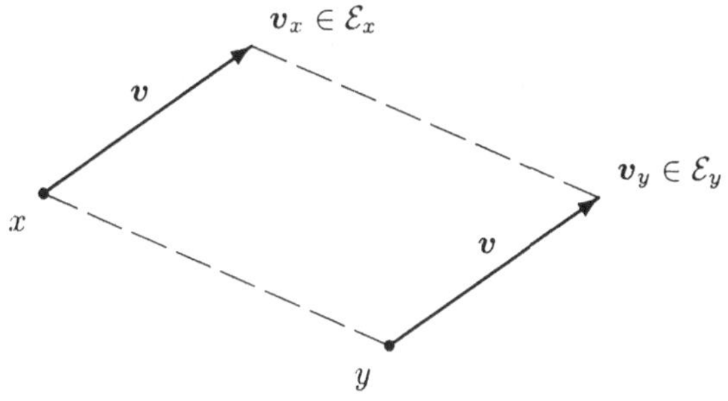


**Fig. A.1.** Parallel translation

### A.2.2 Differentiation

Before we define the derivative of tensor functions on Euclidean space in general, let us recall the definition of derivative of a real-valued function of a real variable. Let $f : (a, b) \to \mathbb{R}$ be a function on the interval $(a, b) \subset \mathbb{R}$. The derivative of $f$ at $t \in (a, b)$ is defined as

$$\frac{df(t)}{dt} = \lim_{h \to 0} \frac{1}{h} \Big( f(t+h) - f(t) \Big),$$

if the limit exists.

This definition can easily be extended to tensor-valued functions of a real variable. Let $W$ be a space equipped with a norm or a distance function. As examples, we have

$$\begin{array}{lll} I\!R & : & d(x,y) = |x-y|\,, \\ \mathcal{E} & : & d(x,y) = \sqrt{(x-y)\cdot(x-y)}, \\ V & : & |\boldsymbol{u}| = \sqrt{\boldsymbol{u}\cdot\boldsymbol{u}}, \\ \mathcal{L}(V), Sym(V), Skw(V) & : & |A| = \sqrt{\operatorname{tr} AA^T}. \end{array} \tag{A.41}$$

It makes sense to talk about limit and convergence in the space $W$ when a norm or a distance function is defined.

Let $\boldsymbol{f} : (a,b) \to W$ be a function defined on an interval $(a,b) \subset I\!R$. The derivative of $\boldsymbol{f}$ at $t \in (a,b)$ is defined as

$$\frac{d\boldsymbol{f}(t)}{dt} = \lim_{h\to 0} \frac{1}{h}\Big(\boldsymbol{f}(t+h) - \boldsymbol{f}(t)\Big). \tag{A.42}$$

The derivative of $\boldsymbol{f}$ at $t$ will also be denoted by $\dot{\boldsymbol{f}}(t)$. Obviously for any $t \in (a,b)$ we have $\dot{\boldsymbol{f}}(t) \in W$.

Note that if $\boldsymbol{f}$ is defined on a more general space, the expression on the right-hand side of the definition (A.42) may not make sense at all. However, we can rewrite the relation (A.42) in a different form.

For fixed $t$, let $D\boldsymbol{f}(t) : I\!R \to W$ be the linear transformation defined by

$$D\boldsymbol{f}(t)[h] = \dot{\boldsymbol{f}}(t)\,h.$$

Then (A.42) is equivalent to

$$\lim_{h\to 0} \frac{1}{|h|} |\boldsymbol{f}(t+h) - \boldsymbol{f}(t) - D\boldsymbol{f}(t)[h]| = 0.$$

In this form, the definition of the derivative can easily be generalized to other functions.

### Tensor Fields

Now we shall consider functions on a Euclidean point space $\mathcal{E}$. Let $\mathcal{D}$ be an open set in $\mathcal{E}$, and $\boldsymbol{f}$ be a tensor-valued function, $\boldsymbol{f} : \mathcal{D} \to W$. Such functions are usually called tensor fields, more specifically,

1) $W = I\!R$, $\boldsymbol{f}$ is called a scalar field on $\mathcal{D}$,

$$\boldsymbol{f} : x \in \mathcal{D} \longmapsto \boldsymbol{f}(x) \in I\!R.$$

2) $W = V$, $\boldsymbol{f}$ is called a vector field on $\mathcal{D}$,

$$\boldsymbol{f} : x \in \mathcal{D} \longmapsto \boldsymbol{f}(x) \in \mathcal{E}_x \cong V.$$

3) $W = \mathcal{L}(V)$, $\boldsymbol{f}$ is called a second-order tensor field on $\mathcal{D}$,

$$\boldsymbol{f} : x \in \mathcal{D} \longmapsto \boldsymbol{f}(x) \in \mathcal{E}_x \otimes \mathcal{E}_x \cong \mathcal{L}(V).$$

4) $W = \mathcal{E}$, $\boldsymbol{f}$ is called a point field on $\mathcal{D}$ or a deformation of $\mathcal{D}$,

$$\boldsymbol{f} : x \in \mathcal{D} \longmapsto \boldsymbol{f}(x) \in \mathcal{E}.$$

**Definition.** *A function $\boldsymbol{f} : \mathcal{D} \to W$ is said to be differentiable at $x \in \mathcal{D} \subset \mathcal{E}$, if there exists a linear transformation $D\boldsymbol{f}(x) \in \mathcal{L}(V, W)$ at $x$, such that for any $\boldsymbol{v} \in V$,*

$$\lim_{|\boldsymbol{v}| \to 0} \frac{1}{|\boldsymbol{v}|} \left| \boldsymbol{f}(x + \boldsymbol{v}) - \boldsymbol{f}(x) - D\boldsymbol{f}(x)[\boldsymbol{v}] \right| = 0. \tag{A.43}$$

The linear transformation $D\boldsymbol{f}(x)$ is uniquely determined by the above relation, and it is called the *gradient* (or *derivative*) of $\boldsymbol{f}$ at $x$, denoted by $\operatorname{grad} \boldsymbol{f}$, or $\nabla_x \boldsymbol{f}$, or simply $\nabla \boldsymbol{f}$. By definition, $\nabla \boldsymbol{f}(x)$ is a tensor in $W \otimes V$, or is a vector in $V$ if $W = I\!R$.

The condition (A.43) is equivalent to

$$\boldsymbol{f}(x + \boldsymbol{v}) - \boldsymbol{f}(x) = \nabla \boldsymbol{f}(x)[\boldsymbol{v}] + o(\boldsymbol{v}),$$

where $o(\boldsymbol{v})$ is a quantity containing terms such that

$$\lim_{|\boldsymbol{v}| \to 0} \frac{o(\boldsymbol{v})}{|\boldsymbol{v}|} = 0.$$

Moreover, if we substitute $t\boldsymbol{v}$ for $\boldsymbol{v}$ for some fixed $\boldsymbol{v}$ in $V$, (A.43) is also equivalent to

$$\begin{aligned} \nabla \boldsymbol{f}(x)[\boldsymbol{v}] &= \lim_{t \to 0} \frac{1}{t} \Big( \boldsymbol{f}(x + t\boldsymbol{v}) - \boldsymbol{f}(x) \Big) \\ &= \frac{d}{dt} \boldsymbol{f}(x + t\boldsymbol{v}) \Big|_{t=0}. \end{aligned} \tag{A.44}$$

The right-hand side of the above relation is usually known as the *directional derivative* of $\boldsymbol{f}$ relative to the vector $\boldsymbol{v}$. Note that for fixed $x$ and $\boldsymbol{v}$, $\boldsymbol{f}(x + t\boldsymbol{v})$ is a tensor-valued function of a real variable and its derivative can easily be determined from (A.42).

## Functions on Tensor Spaces

Let $W_1$ and $W_2$ be two spaces on which a norm or a distance function is defined, such as the spaces mentioned in (A.41), and let $\mathcal{D} \subset W_1$ be an open subset. The gradient of tensor functions on $\mathcal{D}$ can be defined in a similar manner.

**Definition.** *A function $\boldsymbol{F} : \mathcal{D} \to W_2$ is said to be differentiable at $X \in \mathcal{D} \subset W_1$, if there exists a linear transformation $D\boldsymbol{F}(X) \in \mathcal{L}(W_1, W_2)$ at $X$, such that $\forall\, Y \in \mathcal{D}$,*

$$\lim_{|Y| \to 0} \frac{1}{|Y|} |\boldsymbol{F}(X+Y) - \boldsymbol{F}(X) - D\boldsymbol{F}(X)[Y]| = 0.$$

The linear transformation $D\boldsymbol{F}(X)$ is uniquely determined by the above relation, and it is called the *gradient* of $\boldsymbol{F}$ with respect to $X$, denoted by $\partial_X \boldsymbol{F}$. We have $\partial_X \boldsymbol{F} \in W_2 \otimes W_1$. The definition is equivalent to the condition: for any $Y \in \mathcal{D}$, we have

$$\boldsymbol{F}(X+Y) - \boldsymbol{F}(X) = \partial_X \boldsymbol{F}(X)[Y] + o(Y), \tag{A.45}$$

or

$$\partial_X \boldsymbol{F}(X)[Y] = \frac{d}{dt} \boldsymbol{F}(X + tY)\Big|_{t=0}. \tag{A.46}$$

For $\phi \in W_2 \otimes W_1$, and $Y \in W_1$, the notation $\phi[Y]$ used in the above relations is self-evident: for $\phi = K \otimes X$,

$$(K \otimes X)[Y] = (X \cdot Y)\, K, \qquad \forall\, K \in W_2,\ X, Y \in W_1,$$

and for all $\boldsymbol{v}, \boldsymbol{u} \in V$ and $A, S \in \mathcal{L}(V)$, we have

$$\begin{aligned} \boldsymbol{v}[\boldsymbol{u}] &= \boldsymbol{v} \cdot \boldsymbol{u}, \\ A[\boldsymbol{u}] &= A\boldsymbol{u}, \\ A[S] &= A \cdot S = \operatorname{tr} AS^T, \\ (\boldsymbol{v} \otimes \boldsymbol{u})[S] &= \boldsymbol{v} \cdot S\boldsymbol{u}. \end{aligned}$$

Gradients can easily be computed directly from the definition (A.45) or (A.46). We demonstrate this procedure with some examples.

**Example A.2.1** Let $\phi : \mathcal{L}(V) \times V \to I\!R$ be defined by

$$\phi(A, \boldsymbol{v}) = \boldsymbol{v} \cdot A\boldsymbol{v}.$$

Then, from (A.45),

$$\begin{aligned} \phi(A, \boldsymbol{v} + \boldsymbol{u}) &= (\boldsymbol{v} + \boldsymbol{u}) \cdot A(\boldsymbol{v} + \boldsymbol{u}) \\ &= \boldsymbol{v} \cdot A\boldsymbol{v} + \boldsymbol{v} \cdot A\boldsymbol{u} + \boldsymbol{u} \cdot A\boldsymbol{v} + \boldsymbol{u} \cdot A\boldsymbol{u} \\ &= \phi(A, \boldsymbol{v}) + \partial_{\boldsymbol{v}}\phi[\boldsymbol{u}] + o(\boldsymbol{u}), \end{aligned}$$

so that

$$\begin{aligned} \partial_{\boldsymbol{v}}\phi[\boldsymbol{u}] &= \boldsymbol{v} \cdot A\boldsymbol{u} + \boldsymbol{u} \cdot A\boldsymbol{v} \\ &= A^T\boldsymbol{v} \cdot \boldsymbol{u} + A\boldsymbol{v} \cdot \boldsymbol{u} \\ &= (A^T + A)\boldsymbol{v}[\boldsymbol{u}]. \end{aligned}$$

Therefore, we obtain

$$\partial_{\boldsymbol{v}}\phi = (A + A^T)\boldsymbol{v}.$$

Moreover, we have

$$\phi(A + S, \boldsymbol{v}) = \boldsymbol{v} \cdot (A + S)\boldsymbol{v} = \boldsymbol{v} \cdot A\boldsymbol{v} + \boldsymbol{v} \cdot S\boldsymbol{v},$$

which implies

$$\partial_A\phi[S] = \boldsymbol{v} \cdot S\boldsymbol{v} = (\boldsymbol{v} \otimes \boldsymbol{v})[S],$$

so that

$$\partial_A\phi = \boldsymbol{v} \otimes \boldsymbol{v}.$$

□

**Example A.2.2** Let $\boldsymbol{u}, \boldsymbol{v} \in V$ be constant vectors, and let $\phi : \mathcal{L}(V) \to I\!R$ be defined by

$$\phi(A) = \boldsymbol{u} \cdot A\boldsymbol{v}.$$

From (A.46) we have

$$\partial_A\phi[S] = \frac{d}{dt}\Big(\boldsymbol{u} \cdot (A + tS)\boldsymbol{v}\Big)\Big|_{t=0} = \boldsymbol{u} \cdot S\boldsymbol{v} = (\boldsymbol{u} \otimes \boldsymbol{v})[S],$$

for all $S \in \mathcal{L}(V)$, and we obtain

$$\partial_A\phi = \boldsymbol{u} \otimes \boldsymbol{v}.$$

Now, suppose that $A$ is a symmetric tensor, hence the function $\phi$ is defined on the subspace $Sym(V)$ only,

$$\phi : Sym(V) \to I\!R,$$

and by definition $\partial_A\phi \in Sym(V)$ also. In this case, we have the same relation,

$$\partial_A\phi[S] = (\boldsymbol{u} \otimes \boldsymbol{v})[S],$$

but it holds only for all $S \in Sym(V)$. Therefore, we conclude that

$$\partial_A\phi = \frac{1}{2}(\boldsymbol{u} \otimes \boldsymbol{v} + \boldsymbol{v} \otimes \boldsymbol{u}),$$

after symmetrization.

Similarly, if $A$ is a skew-symmetric tensor, then $\partial_A\phi \in Skw(V)$ and the result must be skew-symmetrized,

$$\partial_A\phi = \frac{1}{2}(\boldsymbol{u} \otimes \boldsymbol{v} - \boldsymbol{v} \otimes \boldsymbol{u}).$$

□

**Example A.2.3** We consider trace and determinant functions. Since

$$\operatorname{tr}(A+S) = \operatorname{tr} A + \operatorname{tr} S = \operatorname{tr} A + \mathit{1} \cdot S,$$

so that, trivially, the gradient of the trace is the identity transformation,

$$\partial_A(\operatorname{tr} A) = \mathit{1}. \tag{A.47}$$

For the gradient of the determinant, we have

$$(\partial_A \det A)[S] = \det(A+S) - \det(A) + o(S).$$

Let $\omega$ be a non-trivial alternating $n$-linear form, then

$$\begin{aligned}&\omega(\boldsymbol{v}_1, \cdots, \boldsymbol{v}_n)(\partial_A \det A)[S] \\ &\qquad = \omega((A+S)\boldsymbol{v}_1, \cdots, (A+S)\boldsymbol{v}_n) - \omega(A\boldsymbol{v}_1, \cdots, A\boldsymbol{v}_n) + o(S).\end{aligned}$$

By the linearity of $\omega$, after removing all the higher-order terms into $o(S)$, the right-hand side becomes

$$\begin{aligned}&= \sum_{i=1}^{n} \omega(A\boldsymbol{v}_1, \cdots, S\boldsymbol{v}_i, \cdots, A\boldsymbol{v}_n) + o(S) \\ &= \sum_{i=1}^{n} \omega(A\boldsymbol{v}_1, \cdots, AA^{-1}S\boldsymbol{v}_i, \cdots, A\boldsymbol{v}_n) + o(S) \\ &= (\det A) \sum_{i=1}^{n} \omega(\boldsymbol{v}_1, \cdots, A^{-1}S\boldsymbol{v}_i, \cdots, \boldsymbol{v}_n) + o(S) \\ &= (\det A)(\operatorname{tr} A^{-1}S)\omega(\boldsymbol{v}_1, \cdots, \boldsymbol{v}_n) + o(S),\end{aligned}$$

from (A.21). Therefore, we have

$$(\partial_A \det A)[S] = (\det A)(\operatorname{tr} SA^{-1}) = (\det A)A^{-T}[S],$$

which implies the following formula,

$$\partial_A \det A = (\det A)A^{-T}. \tag{A.48}$$

□

In differential calculus, we frequently differentiate a composite function by the chain rule. This rule can be stated for composite tensor functions in general. Let $W_1$, $W_2$, $W_3$ be normed spaces of the type (A.41) and $\mathcal{D}_1 \subset W_1$, $\mathcal{D}_2 \subset W_2$ be open subsets, and let

$$\phi : \mathcal{D}_1 \to W_2, \qquad \psi : \mathcal{D}_2 \to W_3,$$

with $\phi(\mathcal{D}_1) \subset \mathcal{D}_2$. Then we have the following

**Chain Rule.** *Let $\phi$ be differentiable at $X \in \mathcal{D}_1$, and $\psi$ be differentiable at $Y = \phi(X) \in \mathcal{D}_2$. Then the composition $\boldsymbol{f} = \psi \circ \phi$ is differentiable at $X$ and*

$$D\boldsymbol{f}(X)[Z] = D\psi(\phi(X))[D\phi(X)[Z]], \tag{A.49}$$

*for any $Z \in W_1$ or simply*

$$D\boldsymbol{f}(X) = D\psi(Y) \circ D\phi(X).$$

**Example A.2.4** If $\phi$ is a scalar-valued function of a vector variable, $\boldsymbol{g}(x)$ is a vector field on $\mathcal{E}$, and $\boldsymbol{h}(\boldsymbol{v})$ is a vector-valued function of a vector variable, then

$$\nabla \boldsymbol{h}(\boldsymbol{g}(x)) = \partial_{\boldsymbol{v}} \boldsymbol{h}\Big|_{\boldsymbol{v}=\boldsymbol{g}(x)} (\nabla \boldsymbol{g}(x)),$$

$$\nabla \phi(\boldsymbol{g}(x)) = \left(\nabla \boldsymbol{g}(x)\right)^T \partial_{\boldsymbol{v}} \phi\Big|_{\boldsymbol{v}=\boldsymbol{g}(x)}.$$

Let us verify the last one in the above formulae. For any $\boldsymbol{u} \in V$, from (A.49),

$$\begin{aligned}\nabla \phi(\boldsymbol{g}(x))[\boldsymbol{u}] &= \partial_{\boldsymbol{v}} \phi\Big|_{\boldsymbol{v}=\boldsymbol{g}(x)} [\nabla \boldsymbol{g}(x)[\boldsymbol{u}]] = \partial_{\boldsymbol{v}} \phi\Big|_{\boldsymbol{v}=\boldsymbol{g}(x)} \cdot (\nabla \boldsymbol{g}(x))\boldsymbol{u} \\ &= \left(\nabla \boldsymbol{g}(x)\right)^T \partial_{\boldsymbol{v}} \phi\Big|_{\boldsymbol{v}=\boldsymbol{g}(x)} \cdot \boldsymbol{u} = \left(\nabla \boldsymbol{g}(x)\right)^T \partial_{\boldsymbol{v}} \phi\Big|_{\boldsymbol{v}=\boldsymbol{g}(x)} [\boldsymbol{u}],\end{aligned}$$

where in the third step we have used the definition of transpose (A.8). Note that $\nabla \boldsymbol{h}$, $\nabla \boldsymbol{g}$, and $\partial_{\boldsymbol{v}} \boldsymbol{h}$ are all second-order tensors, while $\partial_{\boldsymbol{v}} \phi$ is a vector quantity. □

Another important result in differentiation is the product rule. For tensor functions, in general, there are many different products available, for example, the product of a scalar and a vector, the inner product, the tensor product, the action of a tensor on a vector. These products have one property in common, namely, bilinearity. Therefore, in order to establish a product rule valid for all cases of interest, we consider the bilinear operation

$$\pi : W_1 \times W_2 \longrightarrow W_3,$$

which assigns to each $\phi \in W_1$, $\psi \in W_2$, the product $\pi(\phi, \psi) \in W_3$. If $\phi$, $\psi$ are two functions,

$$\phi : \mathcal{D} \to W_1, \qquad \psi : \mathcal{D} \to W_2,$$

where $\mathcal{D}$ is an open subset of some normed space $W$, then the product $\boldsymbol{f} = \pi(\phi, \psi)$ is the function defined by

$$\begin{aligned}&\boldsymbol{f} : \mathcal{D} \longrightarrow W_3 \\ &\boldsymbol{f}(X) = \pi(\phi(X), \psi(X)), \qquad \forall X \in \mathcal{D}.\end{aligned}$$

We then have the following

**Product Rule.** *Suppose that $\phi$ and $\psi$ are differentiable at $X \in \mathcal{D} \subset W$, then their product $\boldsymbol{f} = \pi(\phi, \psi)$ is differentiable at $X$ and*

$$D\boldsymbol{f}(X)[V] = \pi(D\phi(X)[V], \psi(X)) + \pi(\phi(X), D\psi(X)[V]), \tag{A.50}$$

*for all $V \in W$.*

In other words, the derivative of the product $\pi(\phi, \psi)$ is the derivative of $\pi$ holding $\psi$ fixed plus the derivative of $\pi$ holding $\phi$ fixed.

**Example A.2.5** Let $f$ be a scalar-valued, and $\boldsymbol{h}$, $\boldsymbol{q}$ be vector-valued functions on $\mathcal{D} \subset W$. For $W = I\!R$, we have

$$\begin{aligned}(f\boldsymbol{h})^{\cdot} &= \dot{f}\boldsymbol{h} + f\dot{\boldsymbol{h}},\\ (\boldsymbol{q}\cdot\boldsymbol{h})^{\cdot} &= \dot{\boldsymbol{q}}\cdot\boldsymbol{h} + \boldsymbol{q}\cdot\dot{\boldsymbol{h}}.\end{aligned} \tag{A.51}$$

For $W = \mathcal{E}$, we have

$$\begin{aligned}\nabla(f\boldsymbol{h}) &= \boldsymbol{h}\otimes\nabla f + f\nabla\boldsymbol{h},\\ \nabla(\boldsymbol{q}\cdot\boldsymbol{h}) &= (\nabla\boldsymbol{q})^T\boldsymbol{h} + (\nabla\boldsymbol{h})^T\boldsymbol{q}.\end{aligned} \tag{A.52}$$

For $W = V$, we have

$$\begin{aligned}\partial_{\boldsymbol{v}}(f\boldsymbol{h}) &= \boldsymbol{h}\otimes\partial_{\boldsymbol{v}}f + f\,\partial_{\boldsymbol{v}}\boldsymbol{h},\\ \partial_{\boldsymbol{v}}(\boldsymbol{q}\cdot\boldsymbol{h}) &= (\partial_{\boldsymbol{v}}\boldsymbol{q})^T\boldsymbol{h} + (\partial_{\boldsymbol{v}}\boldsymbol{h})^T\boldsymbol{q}.\end{aligned} \tag{A.53}$$

Unlike the simple formulae in (A.51), the relations in (A.52) and (A.53) do not look like the familiar product rules, because they have to be consistent with our notation conventions.

Let us demonstrate the first relation of (A.52). By the product rule (A.50), for any $\boldsymbol{w} \in V$, we have

$$\begin{aligned}\nabla(f\boldsymbol{h})[\boldsymbol{w}] &= (\nabla f[\boldsymbol{w}])\boldsymbol{h} + f(\nabla\boldsymbol{h}[\boldsymbol{w}]) = (\nabla f\cdot\boldsymbol{w})\boldsymbol{h} + f(\nabla\boldsymbol{h})\boldsymbol{w}\\ &= (\boldsymbol{h}\otimes\nabla f)\boldsymbol{w} + f(\nabla\boldsymbol{h})\boldsymbol{w} = \big(\boldsymbol{h}\otimes\nabla f + f(\nabla\boldsymbol{h})\big)[\boldsymbol{w}],\end{aligned}$$

where, in the third step, we have used the definition (A.4). □

If $\boldsymbol{f} : \mathcal{D} \subset U \to W$ is differentiable and its derivative $D\boldsymbol{f}$ is continuous in $\mathcal{D}$, we say that $\boldsymbol{f}$ is of class $C^1$. The derivative is again a function, $D\boldsymbol{f} : \mathcal{D} \to W \otimes U$, for which we can talk about the differentiability and continuity. We say that $\boldsymbol{f}$ is of class $C^2$, if $D\boldsymbol{f}$ is of class $C^1$, and so forth. Frequently, we say a function is *smooth* to mean that it is of class $C^k$ for some $k \geq 1$. We mention the following

**Inverse Function Theorem.** *Let $\mathcal{D} \subset W$ be an open set and $\boldsymbol{f} : \mathcal{D} \to W$ be a one-to-one function of class $C^k (k \geq 1)$. Assume that the linear transformation $D\boldsymbol{f}(X) : W \to W$ is invertible at each $X \in \mathcal{D}$, then $\boldsymbol{f}^{-1}$ exists and is of class $C^k$.*

**Example A.2.6** Let $\mathcal{D} \subset \mathcal{E}$ and $\phi : \mathcal{D} \to I\!R$ be of class $C^2$. Then the second gradient of $\phi$ is a symmetric tensor, that is, $\nabla(\nabla\phi) \in Sym(V)$.

Indeed, from the definition, we have

$$\nabla\phi(x+\boldsymbol{u}) - \nabla\phi(x) = \nabla(\nabla\phi)[\boldsymbol{u}] + o(\boldsymbol{u}).$$

Taking the inner product with $\boldsymbol{v}$, we obtain

$$\nabla\phi(x+\boldsymbol{u})[\boldsymbol{v}] - \nabla\phi(x)[\boldsymbol{v}] = \boldsymbol{v}\cdot\nabla(\nabla\phi)\boldsymbol{u} + o(\boldsymbol{u}),$$

which implies that

$$\begin{aligned}\boldsymbol{v}\cdot\nabla(\nabla\phi)\boldsymbol{u} &= \Big(\phi(x+\boldsymbol{u}+\boldsymbol{v}) - \phi(x+\boldsymbol{u})\Big) \\ &\quad - \Big(\phi(x+\boldsymbol{v}) - \phi(x)\Big) + o(\boldsymbol{u}) + o(\boldsymbol{v}).\end{aligned}$$

Since the right-hand side of the last relation is symmetric in $\boldsymbol{u}$ and $\boldsymbol{v}$, it follows that

$$\boldsymbol{v}\cdot\nabla(\nabla\phi)\boldsymbol{u} = \boldsymbol{u}\cdot\nabla(\nabla\phi)\boldsymbol{v},$$

which proves that the second gradient of $\phi$ is symmetric. □

**Exercise A.2.1** Show that if $Q : I\!R \to \mathcal{O}(V)$ is differentiable, then $\dot{Q}Q^T$ is skew-symmetric.

**Exercise A.2.2** Let $\boldsymbol{h}(\boldsymbol{v}, A) = (\boldsymbol{v}\cdot A\boldsymbol{v})A^2\boldsymbol{v}$ be a vector function of a vector $\boldsymbol{v}$ and a second-order tensor $A$. Compute $\partial_{\boldsymbol{v}}\boldsymbol{h}$ and $(\partial_A\boldsymbol{h})[S]$ for any $S \in \mathcal{L}(V)$.

**Exercise A.2.3** If $A \in \mathcal{L}(V)$ is invertible, show that
1) $(\partial_A A^{-1})[S] = -A^{-1}SA^{-1}$, for any $S \in \mathcal{L}(V)$,
2) $\partial_A \operatorname{tr}(A^{-1}) = -(A^{-2})^T$.

**Exercise A.2.4** Let $A$ be a second-order tensor. Show that
1) For any positive integer $k$,

$$\partial_A \operatorname{tr} A^k = k(A^{k-1})^T.$$

2) For principal invariants $I_A, II_A, III_A$,

$$\begin{aligned}\partial_A I_A &= 1, \\ \partial_A II_A &= (I_A 1 - A)^T, \\ \partial_A III_A &= (II_A 1 - I_A A + A^2)^T.\end{aligned} \tag{A.54}$$

Hint: Calculate $\partial_A \det(A + \lambda 1) = \partial_A(\lambda^3 + I_A\lambda^2 + II_A\lambda + III_A)$.

### A.2.3 Coordinate System

Tensor functions can be expressed in terms of components relative to smooth fields of bases in the Euclidean point space $\mathcal{E}$ associated with a coordinate system.

**Definition.** *Let $\mathcal{D} \subset \mathcal{E}$ be an open set. A coordinate system on $\mathcal{D}$ is a smooth one-to-one mapping*

$$\psi : \mathcal{D} \longrightarrow U,$$

*where $U$ is an open set in $I\!R^n$, such that $\psi^{-1}$ is also smooth.*

Let $x \in \mathcal{D}$,

$$\psi : x \longmapsto (x^1, \cdots, x^n) = \psi(x).$$

$(x^1, \cdots, x^n)$ is called the (*curvilinear*) *coordinate* of $x$, and the functions

$$\begin{aligned} &\chi^i : \mathcal{D} \longrightarrow I\!R \\ &\chi^i(x) = x^i, \qquad i = 1, \cdots, n, \end{aligned} \tag{A.55}$$

are called the $i$-th *coordinate function* of $\psi$. For convenience, we call $(x^i)$ a coordinate system on $\mathcal{D}$.

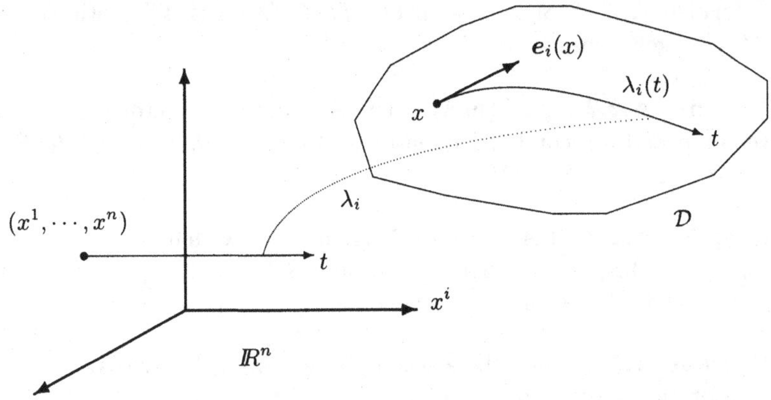


**Fig. A.2.** Coordinate curve

Let $\chi = \psi^{-1}$, then

$$x = \chi(x^1, \cdots, x^n). \tag{A.56}$$

For $x^1, \cdots, x^n$ fixed, the mapping (Fig. A.2)

$$\begin{aligned} &\lambda_i : I\!R \longrightarrow \mathcal{D} \\ &\lambda_i(t) = \chi(x^1, \cdots, x^i + t, \cdots, x^n), \end{aligned} \tag{A.57}$$

is a curve in $\mathcal{D}$ passing through $x$ at $t = 0$, called the $i$-th *coordinate curve* at $x$. We denote the tangent of this curve at $x$ by $\mathbf{e}_i(x)$.

$$\mathbf{e}_i(x) = \dot{\lambda}_i(t)\Big|_{t=0} = \frac{\partial \chi}{\partial x^i}\Big|_{(x^1,\cdots,x^n)}. \tag{A.58}$$

**Proposition.** *The set* $\{\mathbf{e}_i(x), \quad i = 1, \cdots, n\}$ *forms a basis for the tangent space* $\mathcal{E}_x$.

*Proof.* For any vector $\boldsymbol{v} \in \mathcal{E}_x$, we can define a curve through $x$ by

$$\lambda(t) = x + t\boldsymbol{v}.$$

Let

$$\lambda(t) = \chi(\lambda^1(t), \cdots, \lambda^n(t)),$$

where $\lambda^i(t)$ are the coordinates of $\lambda(t)$ given by

$$\lambda^i(t) = \chi^i(x + t\boldsymbol{v}). \tag{A.59}$$

Then, from (A.58), the tangent vector becomes

$$\boldsymbol{v} = \dot{\lambda}(t)\Big|_{t=0} = \frac{\partial \chi}{\partial x^i}\Big|_x \frac{d\lambda^i}{dt}\Big|_{t=0} = \frac{d\lambda^i}{dt}\Big|_{t=0} \mathbf{e}_i(x),$$

In other words, $\{\mathbf{e}_i(x)\}$ spans the space $\mathcal{E}_x$. □

The set $\{\mathbf{e}_i(x)\}$ is a basis of $\mathcal{E}_x$ for each $x$. This field of bases is called the *natural basis* of the coordinate system $(x^i)$ for $V$, the translation space of $\mathcal{E}$. The corresponding dual basis of this natural basis is denoted by $\{\mathbf{e}^i(x)\}$.

Combining (A.55) and (A.56), we have

$$x^i = \chi^i(\chi(x^1, \cdots, x^n)),$$

which implies

$$\frac{\partial x^i}{\partial x^j} = \delta^i_j = (\nabla \chi^i) \cdot \frac{\partial \chi}{\partial x^j} = (\nabla \chi^i) \cdot \mathbf{e}_j(x),$$

by the use of (A.58). Therefore, the two natural bases of the coordinate system $(x^i)$ are given by the following relations:

$$\mathbf{e}_i(x) = \frac{\partial \chi}{\partial x^i}\Big|_x, \qquad \mathbf{e}^i(x) = \nabla \chi^i(x). \tag{A.60}$$

The inner products,

$$g_{ij}(x) = \mathbf{e}_i(x) \cdot \mathbf{e}_j(x), \qquad g^{ij}(x) = \mathbf{e}^i(x) \cdot \mathbf{e}^j(x),$$

are called the *metric tensors* of the coordinate system.

Now let us consider a change of coordinate systems. Let $(x^i)$ and $(\bar{x}^i)$ be two coordinate systems on $\mathcal{D}$, and $\{\mathbf{e}_i(x)\}$, $\{\bar{\mathbf{e}}_i(x)\}$ be the corresponding natural bases. Suppose that the coordinate transformations are given by

$$x^i = x^i(\bar{x}^1, \cdots, \bar{x}^n),$$
$$\bar{x}^k = \bar{x}^k(x^1, \cdots, x^n).$$

Then, by taking the gradients, one immediately obtains the change of the corresponding natural bases given by

$$\mathbf{e}^i(x) = \frac{\partial x^i}{\partial \bar{x}^k}\bar{\mathbf{e}}^k(x), \qquad \mathbf{e}_i(x) = \frac{\partial \bar{x}^k}{\partial x^i}\bar{\mathbf{e}}_k(x). \tag{A.61}$$

Comparing the change of bases considered in Sect. A.1.4, $[\partial x^i/\partial \bar{x}^k]$ plays the role of the transformation matrix $[M_k{}^i]$ in (A.12), and hence, the transformation rules (A.14) for the components of an arbitrary tensor in the change of coordinate system becomes

$$\bar{A}^i{}_j = A^k{}_l \frac{\partial \bar{x}^i}{\partial x^k}\frac{\partial x^l}{\partial \bar{x}^j}. \tag{A.62}$$

For other components of tensors, in general, the transformation rules are similar.

**Example A.2.7** Let us consider a deformation $\kappa : \mathcal{D} \to \mathcal{E}$,

$$\kappa(x) = \tilde{x}.$$

Let $(x^i)$ be a coordinate system on $\mathcal{D}$, and $(\tilde{x}^\alpha)$ be a coordinate system on $\kappa(\mathcal{D})$,

$$x = \chi(x^1, \cdots, x^n), \qquad \tilde{x} = \tilde{\chi}(\tilde{x}^1, \cdots, \tilde{x}^n).$$

The deformation $\kappa$ is usually expressed explicitly in the form,

$$\tilde{x}^\alpha = \kappa^\alpha(x^1, \cdots, x^n), \qquad \alpha = 1, \cdots, n. \tag{A.63}$$

Using the chain rule, we obtain, with $x^i = \chi^i(x)$,

$$\nabla\kappa(x) = \frac{\partial \tilde{\chi}}{\partial \tilde{x}^\alpha}\bigg|_{\tilde{x}} \frac{\partial \kappa^\alpha}{\partial x^i}\bigg|_x \nabla\chi^i(x),$$

which, from (A.60) becomes

$$\nabla\kappa(x) = \frac{\partial \kappa^\alpha}{\partial x^i}\bigg|_x \tilde{\mathbf{e}}_\alpha(\kappa(x)) \otimes \mathbf{e}^i(x).$$

This is the component form of the deformation gradient $\nabla\kappa(x)$ in terms of two different coordinate systems $(x^i)$ and $(\tilde{x}^\alpha)$. With respect to these two natural bases at two different points, namely, $x$ and $\kappa(x)$, the components of the deformation gradient are just the partial derivatives of the deformation function (A.63), which can most easily be calculated. Other component forms of $\nabla\kappa$ can be obtained through the metric tensors and by the change of bases relative to the coordinate systems. □

### A.2.4 Covariant Derivatives

We shall now consider the component form of the gradient of a tensor field, in general, relative to the natural basis of a coordinate system. Let $(x^i)$ be a coordinate system on $\mathcal{D} \subset \mathcal{E}$, and $\{\mathbf{e}_i(x)\}$, $\{\mathbf{e}^i(x)\}$ be its natural bases.

To begin with, let us consider a scalar field, $f : \mathcal{D} \to I\!R$, the gradient of $f$ is then a vector field. From (A.44), (A.57), and (A.58) we have

$$\begin{aligned}(\nabla f(x)) \cdot \mathbf{e}_i(x) &= \lim_{t\to 0} \frac{1}{t}\left(f(x + t\mathbf{e}_i) - f(x)\right) \\ &= \lim_{t\to 0} \frac{1}{t}\left(f(\chi(x^1, \cdots, x^i + t, \cdots, x^n)) - f(\chi(x^1, \cdots, x^n))\right) \\ &= \left.\frac{\partial(f \circ \chi)}{\partial x^i}\right|_{(x^1,\cdots,x^n)},\end{aligned}$$

which are the covariant components of $\nabla f$.

Usually, we shall write $f(\chi(x^1, \cdots, x^n))$ as $f(x^1, \cdots, x^n)$ for simplicity. Therefore, the component form of the gradient of $f(x)$ becomes

$$\nabla f(x) = \left.\frac{\partial f}{\partial x^i}\right|_x \mathbf{e}^i(x). \tag{A.64}$$

In other words, for the gradient of a scalar field $f$, its covariant component relative to the natural basis, $(\nabla f)_i$, is just the partial derivative relative to the coordinate $x^i$.

Now, let us consider the gradients of natural bases themselves. For each $i$ fixed, $\{\mathbf{e}_i\}$ and $\{\mathbf{e}^i\}$ can be regarded as vector fields on $\mathcal{D}$,

$$\mathbf{e}_i : x \in \mathcal{D} \longmapsto \mathbf{e}_i(x) \in \mathcal{E}_x.$$

Let us denote the gradients of natural bases by

$$\begin{aligned}\Gamma_i(x) &= \nabla \mathbf{e}_i(x) \in \mathcal{E}_x \otimes \mathcal{E}_x, \\ \Gamma^i(x) &= \nabla \mathbf{e}^i(x) \in \mathcal{E}_x \otimes \mathcal{E}_x.\end{aligned} \tag{A.65}$$

We write

$$\Gamma_i = \Gamma_{i\ k}^{\ j} \mathbf{e}_j \otimes \mathbf{e}^k, \qquad \Gamma^i = \Gamma^i_{\ jk} \mathbf{e}^j \otimes \mathbf{e}^k. \tag{A.66}$$

The components $\Gamma_{i\ k}^{\ j}$ and $\Gamma^i_{\ jk}$ are called the *Christoffel symbols.* Note that $\Gamma_{i\ k}^{\ j}$ and $\Gamma^i_{\ jk}$ are not the associated components of a third-order tensor.

By taking the gradient of $(\mathbf{e}^i(x) \cdot \mathbf{e}_j(x))$, one can obtain the relation,

$$\Gamma_{j\ k}^{\ i} = -\Gamma^i_{\ jk}. \tag{A.67}$$

Moreover, since $\Gamma^i = \nabla(\nabla\chi^i(x))$ by $(\mathrm{A}.60)_1$ and the second gradient is a symmetric tensor, we have the following symmetry conditions,

$$\Gamma^i_{\ jk} = \Gamma^i_{\ kj}, \qquad \Gamma_{j\ k}^{\ i} = \Gamma_{k\ j}^{\ i}. \tag{A.68}$$

Since both Christoffel symbols are related in such a simple manner, usually only one is in use, namely, $\Gamma_{j\ k}^{\ i}$, and it is called the Christoffel symbol of the second kind in classical tensor analysis.

Now let us calculate the gradient of a vector field in terms of the coordinate system. Suppose that $\boldsymbol{v}(x)$ is a vector field and

$$\boldsymbol{v}(x) = v^i(x)\boldsymbol{e}_i(x) = v_i(x)\boldsymbol{e}^i(x).$$

Then from $(A.52)_1$, (A.64), (A.65), and (A.66), we have

$$\begin{aligned}
\nabla\boldsymbol{v} &= \nabla(v^i\boldsymbol{e}_i) \\
&= \boldsymbol{e}_i \otimes \nabla v^i + v^i \nabla \boldsymbol{e}_i \\
&= \boldsymbol{e}_i \otimes \frac{\partial v^i}{\partial x^k}\boldsymbol{e}^k + v^i \Gamma_{i\ k}^{\ j}\boldsymbol{e}_j \otimes \boldsymbol{e}^k \\
&= \left(\frac{\partial v^j}{\partial x^k} + v^i \Gamma_{i\ k}^{\ j}\right)\boldsymbol{e}_j \otimes \boldsymbol{e}^k.
\end{aligned}$$

Hence, the gradient of $\boldsymbol{v}(x)$ has the component form,

$$\nabla\boldsymbol{v} = v^j_{\ ,k}\boldsymbol{e}_j \otimes \boldsymbol{e}^k,$$

where

$$v^j_{\ ,k} = \frac{\partial v^j}{\partial x^k} + v^i \Gamma_{i\ k}^{\ j}. \tag{A.69}$$

Similarly, we also have

$$\nabla\boldsymbol{v} = v_{j,k}\boldsymbol{e}^j \otimes \boldsymbol{e}^k,$$

where

$$v_{j,k} = \frac{\partial v_j}{\partial x^k} - v_i \Gamma_{j\ k}^{\ i}. \tag{A.70}$$

Here, the relation (A.67) has been used.

$v^j_{\ ,k}$ and $v_{j,k}$ are the mixed and the covariant components of $\nabla\boldsymbol{v}$. The comma stands for the operation called the *covariant derivative*, since it increases the order of the covariant components by one.

More generally, suppose that $A$ is a second-order tensor field, then $\nabla A$ is a third-order tensor field that has the following component form,

$$\nabla A = A^i_{\ j,k}\, \boldsymbol{e}_i \otimes \boldsymbol{e}^j \otimes \boldsymbol{e}^k,$$

where

$$A^i{}_{j,k} = \frac{\partial A^i{}_j}{\partial x^k} + A^l{}_j \Gamma_l{}^i{}_k - A^i{}_l \Gamma_j{}^l{}_k. \qquad (A.71)$$

Covariant derivatives of other components can easily be written using the same recipes for covariant and contravariant components, respectively.

We have seen in (A.7) that the components of the metric tensor, $g_{ij}(x)$ and $g^{ij}(x)$, are also the components of the identity tensor, therefore their covariant derivatives must vanish,

$$g_{ij,k} = 0, \qquad g^{ij}{}_{,k} = 0. \qquad (A.72)$$

Consequently, from (A.24), the covariant derivatives of the volume tensor also vanish,

$$e_{ijk,l} = 0, \qquad e^{ijk}{}_{,l} = 0.$$

In other words, the components of the metric tensor and the volume tensor behave like constant tensors in covariant derivation, although they are, in general, functions of $x$.

From $(A.72)_1$, we can derive a formula for the determination of the Christoffel symbols in terms of the metric tensor. From (A.71) we have

$$\frac{\partial g_{ij}}{\partial x^k} = g_{lj}\Gamma_i{}^l{}_k + g_{il}\Gamma_j{}^l{}_k.$$

Rotating the indices $(i, j, k)$ of this relation, then adding two of the three resulting equations and subtracting the remaining one, we get

$$2\, g_{lj}\Gamma_i{}^l{}_k = \left(\frac{\partial g_{jk}}{\partial x^i} + \frac{\partial g_{ij}}{\partial x^k} - \frac{\partial g_{ik}}{\partial x^j}\right).$$

Hence, we have the following formula:

$$\Gamma_i{}^j{}_k = \frac{1}{2} g^{jl}\left(\frac{\partial g_{li}}{\partial x^k} + \frac{\partial g_{lk}}{\partial x^i} - \frac{\partial g_{ik}}{\partial x^l}\right). \qquad (A.73)$$

The Christoffel symbols are not components of a third-order tensor. For two coordinate systems $(x^i)$ and $(\bar{x}^i)$, they have the following transformation rules:

$$\bar{\Gamma}_i{}^j{}_k = \Gamma_r{}^s{}_t \frac{\partial x^r}{\partial \bar{x}^i}\frac{\partial \bar{x}^j}{\partial x^s}\frac{\partial x^t}{\partial \bar{x}^k} + \frac{\partial^2 x^r}{\partial \bar{x}^i \partial \bar{x}^k}\frac{\partial \bar{x}^j}{\partial x^r}.$$

### A.2.5 Other Differential Operators

Divergence and curl of a vector field can be defined in the usual way and their definitions can be adopted also for tensor fields.

**Definition.** *The divergence of a vector field $\boldsymbol{u}$ is a scalar field defined by*

$$\operatorname{div}\boldsymbol{u} = \operatorname{tr}(\nabla \boldsymbol{u}). \qquad (A.74)$$

In component form,

$$\operatorname{div}\boldsymbol{u} = u^i{}_{,i}.$$

**Definition.** *The curl (or rotation) of* $\boldsymbol{u}$ *is a vector field defined by*

$$\operatorname{curl} \boldsymbol{u} = \langle \nabla \boldsymbol{u}^T - \nabla \boldsymbol{u} \rangle.$$

In component form,

$$\operatorname{curl} \boldsymbol{u} = e^{ijk} u_{k,j} \boldsymbol{e}_i.$$

Here, the duality map defined in (A.29) is employed and, according to (A.30), $\operatorname{curl} \boldsymbol{u}$ is the axial vector of the skew-symmetric part of the gradient of $(-2\boldsymbol{u})$. One can easily verify the following condition:

$$\boldsymbol{v} \cdot \operatorname{curl} \boldsymbol{u} = \operatorname{div}(\boldsymbol{u} \times \boldsymbol{v}),$$

for any constant vector field $\boldsymbol{v}$. This condition can be used as the definition for the curl operator. In a similar manner, we can define the divergence of a second-order tensor in terms of the divergence of a vector.

**Definition.** *The divergence of a second-order tensor field* $S$ *is a vector field defined by the condition: for any constant vector field* $\boldsymbol{v}$,

$$\boldsymbol{v} \cdot \operatorname{div} S = \operatorname{div}(S^T \boldsymbol{v}). \tag{A.75}$$

In component form, we have

$$\operatorname{div} S = S^{ij}{}_{,j} \boldsymbol{e}_i.$$

**Definition.** *The Laplacian of a scalar (or vector) field* $\phi$, *denoted by* $\nabla^2 \phi$, *is a scalar (or vector) field defined by*

$$\nabla^2 \phi = \operatorname{div}(\nabla \phi).$$

In component form, if $\phi$ is a scalar field,

$$\nabla^2 \phi = g^{jk} (\phi_{,j})_{,k} = g^{jk} \phi_{,jk}.$$

If $\phi = \boldsymbol{h}$ is a vector field,

$$\nabla^2 \boldsymbol{h} = g^{jk} h^i{}_{,jk} \boldsymbol{e}_i.$$

In the above expressions, the comma denotes the covariant derivative.

**Example A.2.8** Let $f$, $S$, and $\boldsymbol{u}$, $\boldsymbol{v}$ be scalar, tensor, and vector fields, respectively. Then we can show the following relations:

$$\begin{aligned}
&\operatorname{div}(f\boldsymbol{u}) = \boldsymbol{u} \cdot \nabla f + f \operatorname{div} \boldsymbol{u}, \\
&\operatorname{div}(S\boldsymbol{u}) = \boldsymbol{u} \cdot \operatorname{div} S^T + \operatorname{tr}(S \nabla \boldsymbol{u}), \\
&\operatorname{div}(\boldsymbol{u} \times \boldsymbol{v}) = \boldsymbol{v} \cdot \operatorname{curl} \boldsymbol{u} - \boldsymbol{u} \cdot \operatorname{curl} \boldsymbol{v}, \\
&\nabla^2(\boldsymbol{u} \cdot \boldsymbol{v}) = \nabla^2 \boldsymbol{u} \cdot \boldsymbol{v} + 2 \nabla \boldsymbol{u} \cdot \nabla \boldsymbol{v} + \boldsymbol{u} \cdot \nabla^2 \boldsymbol{v}.
\end{aligned} \tag{A.76}$$

Let us verify the first relation.

$$\begin{aligned}\operatorname{div}(f\boldsymbol{u}) &= \operatorname{tr}(\nabla(f\boldsymbol{u}))\\ &= \operatorname{tr}\Big(\boldsymbol{u}\otimes\nabla f + f(\nabla\boldsymbol{u})\Big)\\ &= \operatorname{tr}(\boldsymbol{u}\otimes\nabla f) + f\ \operatorname{tr}(\nabla\boldsymbol{u}),\end{aligned}$$

which gives $(A.76)_1$. In this calculation, we have used the definition (A.74), the relation $(A.52)_1$, and the linearity of the trace operator.

Verification of the other relations in (A.76) may not be so straightforward in *direct notation*. And more annoyingly, these relations, as well as the relations (A.52) and (A.53), are not easy to memorize. Nevertheless, if we express all of these relations in *index notation*, they all become trivially simple. Indeed, (A.76) may be written out directly as:

$$\begin{aligned}&(fu^i)_{,i} = f_{,i}u^i + fu^i{}_{,i},\\ &(S^{ij}u_j)_{,i} = S^{ij}{}_{,i}u_j + S^{ij}u_{j,i},\\ &(g^{il}e_{ljk}u^jv^k)_{,i} = g^{il}e_{ljk}u^j{}_{,i}v^k + g^{il}e_{ljk}u^jv^k{}_{,i},\\ &g^{jk}(u^iv_i)_{,jk} = g^{jk}u^i{}_{,jk}v_i + 2\,g^{jk}u^i{}_{,k}v_{i,j} + g^{jk}u^iv_{i,jk},\end{aligned}$$

which are merely the usual product rules of differentiating scalar functions and the symmetry of second gradient. The only difference here is that the comma denotes the covariant derivative instead of the usual partial derivative. □

**Remark**. From the observation made in the above example, the use of index notation is often encouraged, especially when complicated calculations are involved. In arbitrary curvilinear coordinate systems, contravariant and covariant indices must be carefully distinguished and the pair of repeated indices, for which the summation convention is applied, must always appear in different levels. An index can be raised or lowered to its proper level with the metric tensor $g_{ij}$ or $g^{ij}$. Moreover, since the gradients of the metric tensor and the volume tensor vanish, in covariant differentiation, the metric tensor $g_{ij}$ as well as the components of the volume element $e_{ijk}$ can be treated as constants. Furthermore, if the Cartesian coordinate system is used, there is no difference between contravariant and covariant components and hence all the indices can be written at the same level, and more conveniently, the covariant derivative becomes the partial derivative, and $g_{ij} = \delta_{ij}$, $e_{ijk} = \varepsilon_{ijk}$ are constants.

It is important to note that given an expression in index notation, one can always turn it into an expression in direct notation or *vice versa*. Therefore, in handling calculations, the choice of using direct notation or index notation, or even using Cartesian index notation is totally up to one's taste and convenience.

We shall also mention some important theorems of integral calculus often used in mechanics.

**Divergence Theorem.** *Let $\mathcal{R}$ be a bounded regular region*[2] *in $\mathcal{E}$, and let $\phi : \mathcal{R} \to I\!R$, $\boldsymbol{h} : \mathcal{R} \to V$, $S : \mathcal{R} \to \mathcal{L}(V)$ be smooth fields. Then*

$$\int_{\partial\mathcal{R}} \phi \boldsymbol{n}\, da = \int_{\mathcal{R}} \nabla\phi\, dv,$$

$$\int_{\partial\mathcal{R}} \boldsymbol{v} \cdot \boldsymbol{n}\, da = \int_{\mathcal{R}} \operatorname{div} \boldsymbol{v}\, dv, \tag{A.77}$$

$$\int_{\partial\mathcal{R}} S\boldsymbol{n}\, da = \int_{\mathcal{R}} \operatorname{div} S\, dv,$$

*where $\boldsymbol{n}$ is the outward unit normal field on $\partial\mathcal{R}$.*

*Proof.* The relations $(A.77)_{1,2}$ are well-known classical results. To show $(A.77)_3$, let $\boldsymbol{v}$ be an arbitrary constant vector. Then

$$\begin{aligned}\boldsymbol{v} \cdot \int_{\partial\mathcal{R}} S\boldsymbol{n}\, da &= \int_{\partial\mathcal{R}} \boldsymbol{v} \cdot S\boldsymbol{n}\, da = \int_{\partial\mathcal{R}} S^T\boldsymbol{v} \cdot \boldsymbol{n}\, da \\ &= \int_{\mathcal{R}} \operatorname{div}(S^T\boldsymbol{v})\, dv = \int_{\mathcal{R}} \boldsymbol{v} \cdot \operatorname{div} S\, dv \\ &= \boldsymbol{v} \cdot \int_{\mathcal{R}} \operatorname{div} S\, dv,\end{aligned}$$

where we have used $(A.77)_2$ and the definition (A.75). □

**Proposition.** *Let $\phi : \mathcal{D} \to W$ be a continuous function on an open set $\mathcal{D}$ in $\mathcal{E}$. If*

$$\int_{\mathcal{N}} \phi\, dv = 0,$$

*for any $\mathcal{N} \subset \mathcal{D}$, then $\phi$ is identically zero in $\mathcal{D}$,* i.e.,

$$\phi(x) = 0, \qquad \forall x \in \mathcal{D}.$$

*Proof.* Suppose that $\phi(x_\circ) \neq 0$ for some $x_\circ \in \mathcal{D}$, then since $\phi$ is continuous, there exists a small neighborhood $\mathcal{N} \subset \mathcal{D}$ containing $x_\circ$, such that $\phi(x) \neq 0$, $\forall x \in \mathcal{N}$. Therefore, by the mean value theorem of integral calculus,

$$\int_{\mathcal{N}} \phi\, dv = K\phi(\bar{x}) \neq 0,$$

[2] A regular region, roughly speaking, is a closed region with a piecewise smooth boundary.

for some $\bar{x} \in \mathcal{N}$, where $K$ denotes the volume of $\mathcal{N}$. This contradicts the hypothesis. □

This proposition and the divergence theorem enable us to deduce local field equations from the integral balance laws.

**Exercise A.2.5** Let $f$, $\boldsymbol{u}$, $\boldsymbol{v}$, and $S$ be smooth scalar, vector, and second-order tensor fields. Verify the following identities:

1) $\operatorname{div}(S\boldsymbol{u}) = \boldsymbol{u} \cdot \operatorname{div} S^T + \operatorname{tr}(S\nabla\boldsymbol{u})$,
2) $\operatorname{div}(fS) = S\,\nabla f + f \operatorname{div} S$,
3) $\operatorname{div}(\boldsymbol{u} \otimes \boldsymbol{v}) = (\nabla\boldsymbol{u})\boldsymbol{v} + \boldsymbol{u} \operatorname{div} \boldsymbol{v}$,
4) $\operatorname{div}(\nabla\boldsymbol{u})^T = \nabla(\operatorname{div} \boldsymbol{u})$.

**Exercise A.2.6** Let $f$ and $\boldsymbol{v}$ be smooth scalar and vector fields, respectively. Show that

1) $\operatorname{curl} \nabla f = 0$,
2) $\operatorname{div} \operatorname{curl} \boldsymbol{v} = 0$,
3) If $\operatorname{div} \boldsymbol{v} = 0$ and $\operatorname{curl} \boldsymbol{v} = 0$, then $\nabla^2 \boldsymbol{v} = 0$.

**Exercise A.2.7** Let $\boldsymbol{v}$ and $S$ be smooth vector and tensor field, on a bound regular region $\mathcal{R}$, respectively. Show that

1) $\displaystyle\int_{\partial\mathcal{R}} \boldsymbol{v} \otimes \boldsymbol{n}\, da = \int_{\mathcal{R}} \nabla\boldsymbol{v}\, dv$,

2) $\displaystyle\int_{\partial\mathcal{R}} \boldsymbol{v} \otimes S\boldsymbol{n}\, da = \int_{\mathcal{R}} \left(\boldsymbol{v} \otimes \operatorname{div} S + (\nabla\boldsymbol{v})S^T\right) dv$.

### A.2.6 Physical Components

Let $(x^i)$ be a coordinate system on $\mathcal{E}$ and $\{\boldsymbol{e}_i(x)\}$ and $\{\boldsymbol{e}^i(x)\}$ be its natural bases. The system $(x^i)$ is called an *orthogonal coordinate system* if the metric tensor

$$g_{ij}(x) = 0, \quad \text{for } i \neq j, \ \forall x \in \mathcal{E}.$$

For an orthogonal coordinate system, we can define a field of orthonormal basis, denoted by $\{\boldsymbol{e}_{\langle i \rangle}(x)\}$, by normalizing the natural basis,

$$\boldsymbol{e}_{\langle i \rangle} = \frac{\boldsymbol{e}_i}{|\boldsymbol{e}_i|} \qquad \text{(no sum)}.$$

In this expression, the summation notation is not invoked as indicated explicitly. Since

$$|\boldsymbol{e}_i| = \sqrt{\boldsymbol{e}_i \cdot \boldsymbol{e}_i} = \sqrt{g_{ii}} \qquad \text{(no sum)},$$

therefore,

$$\begin{aligned} \boldsymbol{e}_{\langle i\rangle} &= \frac{\boldsymbol{e}_i}{\sqrt{g_{ii}}} = \frac{\boldsymbol{e}^i}{\sqrt{g^{ii}}} \\ &= \sqrt{g^{ii}}\, \boldsymbol{e}_i = \sqrt{g_{ii}}\, \boldsymbol{e}^i \qquad \text{(no sum).} \end{aligned}$$

Here, we have noted that normalization of the two dual natural bases of an orthogonal coordinate system gives rise to the same orthonormal basis.

The components of a tensor field relative to the orthonormal basis $\{\boldsymbol{e}_{\langle i\rangle}(x)\}$ are called the *physical components* in the coordinate system $(x^i)$. For a vector field $\boldsymbol{v}$,

$$\boldsymbol{v} = v^i \boldsymbol{e}_i = v_i \boldsymbol{e}^i = v_{\langle i\rangle} \boldsymbol{e}_{\langle i\rangle}.$$

The physical components $v_{\langle i\rangle}$ are given by

$$v_{\langle i\rangle} = \sqrt{g_{ii}}\, v^i = \frac{v_i}{\sqrt{g_{ii}}}. \qquad \text{(no sum)} \tag{A.78}$$

For a second-order tensor field $T$,

$$\begin{aligned} T &= T^{ij} \boldsymbol{e}_i \otimes \boldsymbol{e}_j = T_{ij} \boldsymbol{e}^i \otimes \boldsymbol{e}^j = T^i{}_j \boldsymbol{e}_i \otimes \boldsymbol{e}^j \\ &= T_{\langle ij\rangle} \boldsymbol{e}_{\langle i\rangle} \otimes \boldsymbol{e}_{\langle j\rangle}. \end{aligned}$$

The physical components $T_{\langle ij\rangle}$ are given by

$$T_{\langle ij\rangle} = \sqrt{g_{ii}}\sqrt{g_{jj}}\, T^{ij} = \frac{T_{ij}}{\sqrt{g_{ii}}\sqrt{g_{jj}}} = \frac{\sqrt{g_{ii}}}{\sqrt{g_{jj}}} T^i{}_j \qquad \text{(no sum).} \tag{A.79}$$

In particular, we have $g_{\langle ij\rangle} = \delta_{ij}$.

The advantage of using physical components is obvious in practical applications. Since the norms of the basis vectors of the natural basis, in general, vary from point to point in $\mathcal{E}$, hence it is inconvenient for the measurement of physical quantities relative to this basis.

### A.2.7 Orthogonal Coordinate Systems

We now consider three orthogonal coordinate systems most commonly used: the Cartesian, the cylindrical, and the spherical coordinate systems and derive their basic characteristics.

#### a) Cartesian Coordinate System

Fix a point $o$ in $\mathcal{E}$. Let $\{\boldsymbol{i}_1, \boldsymbol{i}_2, \boldsymbol{i}_3\}$ be an orthonormal basis of $V$. For any $x \in \mathcal{E}$, then $x - o \in V$. We write

$$x - o = x_i\, \boldsymbol{i}_i.$$

Clearly, this defines a coordinate system

$$x \longmapsto (x_1, x_2, x_3),$$

with $\{\boldsymbol{i}_1, \boldsymbol{i}_2, \boldsymbol{i}_3\}$ as its natural basis, which is, of course, independent of $x \in \mathcal{E}$. We call such a system a *Cartesian coordinate system.*

For a Cartesian coordinate system, we have

$$g_{ij}(x) = \delta_{ij}, \qquad \forall\, x \in \mathcal{E},$$

and hence, from (A.73)

$$\Gamma_j{}^i{}_k(x) = 0.$$

It is also a custom to write the basis $\{\boldsymbol{i}_1, \boldsymbol{i}_2, \boldsymbol{i}_3\}$ as $\{\boldsymbol{e}_x, \boldsymbol{e}_y, \boldsymbol{e}_z\}$ and the coordinate $(x_1, x_2, x_3)$ as $(x, y, z)$ for a Cartesian coordinate system.

## b) Cylindrical Coordinate System

The cylindrical coordinate system $(r, \theta, z)$ is defined as

$$x = \chi(r, \theta, z),$$

by the following coordinate transformation (see Fig. A.3 (a)),

$$\begin{aligned} x_1 &= r\,\cos\theta, & r &> 0 \\ x_2 &= r\,\sin\theta, & 0 &< \theta < 2\pi \\ x_3 &= z, \end{aligned} \tag{A.80}$$

where $x = (x_1, x_2, x_3)$ is the Cartesian coordinate system.

The natural bases are denoted by $\{\boldsymbol{e}_r, \boldsymbol{e}_\theta, \boldsymbol{e}_z\}$ and $\{\boldsymbol{e}^r, \boldsymbol{e}^\theta, \boldsymbol{e}^z\}$. From (A.80) and $(A.60)_2$, we can determine the basis in terms of the Cartesian components.

$$\begin{aligned} \boldsymbol{e}_r &= \frac{\partial\chi}{\partial r} = \cos\theta\,\boldsymbol{i}_1 + \sin\theta\,\boldsymbol{i}_2, \\ \boldsymbol{e}_\theta &= \frac{\partial\chi}{\partial\theta} = -r\,\sin\theta\,\boldsymbol{i}_1 + r\,\cos\theta\,\boldsymbol{i}_2, \\ \boldsymbol{e}_z &= \frac{\partial\chi}{\partial z} = \boldsymbol{i}_3. \end{aligned}$$

Therefore, we obtain the matrix representations of the metric tensor in the cylindrical coordinate system,

$$[g_{ij}] = \begin{bmatrix} 1 & & \\ & r^2 & \\ & & 1 \end{bmatrix}, \qquad [g^{ij}] = \begin{bmatrix} 1 & & \\ & r^{-2} & \\ & & 1 \end{bmatrix},$$

and the Christoffel symbols from (A.73),

$$\Gamma_{r}{}^{\theta}{}_{\theta} = \Gamma_{\theta}{}^{\theta}{}_{r} = \frac{1}{r},$$

$$\Gamma_{\theta}{}^{r}{}_{\theta} = -r,$$

$$\text{others} = 0.$$

Moreover, we have

$$\boldsymbol{e}_r = \boldsymbol{e}^r, \quad \boldsymbol{e}_\theta = r^2 \boldsymbol{e}^\theta, \quad \boldsymbol{e}_z = \boldsymbol{e}^z,$$

and

$$\boldsymbol{e}_{\langle r\rangle} = \cos\theta\, \boldsymbol{i}_1 + \sin\theta\, \boldsymbol{i}_2,$$

$$\boldsymbol{e}_{\langle \theta\rangle} = -\sin\theta\, \boldsymbol{i}_1 + \cos\theta\, \boldsymbol{i}_2,$$

$$\boldsymbol{e}_{\langle z\rangle} = \boldsymbol{i}_3.$$

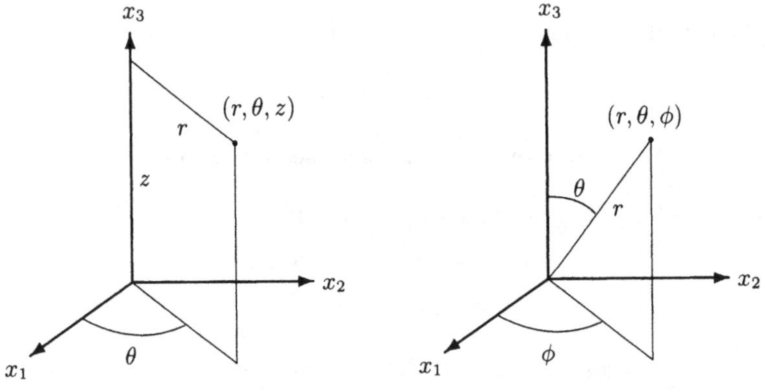


(a) Cylindrical (b) Spherical

**Fig. A.3.** Coordinate systems

### c) Spherical Coordinate System

The spherical coordinate system $(r, \theta, \phi)$ is defined as

$$x = \chi(r, \theta, \phi),$$

by the following coordinate transformation (see Fig. A.3 (b)),

$$x_1 = r \sin\theta \cos\phi, \qquad r > 0$$

$$x_2 = r \sin\theta \sin\phi, \qquad 0 < \theta < \pi$$

$$x_3 = r \cos\theta, \qquad 0 < \phi < 2\pi$$

where $x = (x_1, x_2, x_3)$ is the Cartesian coordinate system.

The natural bases are denoted by $\{\boldsymbol{e}_r, \boldsymbol{e}_\theta, \boldsymbol{e}_\phi\}$ and $\{\boldsymbol{e}^r, \boldsymbol{e}^\theta, \boldsymbol{e}^\phi\}$. We have

$$\begin{aligned}
\boldsymbol{e}_r &= \sin\theta\,\cos\phi\,\boldsymbol{i}_1 + \sin\theta\,\sin\phi\,\boldsymbol{i}_2 + \cos\theta\,\boldsymbol{i}_3,\\
\boldsymbol{e}_\theta &= r\,\cos\theta\,\cos\phi\,\boldsymbol{i}_1 + r\,\cos\theta\,\sin\phi\,\boldsymbol{i}_2 - r\,\sin\theta\,\boldsymbol{i}_3,\\
\boldsymbol{e}_\phi &= -r\,\sin\theta\,\sin\phi\,\boldsymbol{i}_1 + r\,\sin\theta\,\cos\phi\,\boldsymbol{i}_2,
\end{aligned}$$

and

$$\boldsymbol{e}_r = \boldsymbol{e}^r, \quad \boldsymbol{e}_\theta = r^2\,\boldsymbol{e}^\theta, \quad \boldsymbol{e}_\phi = r^2\sin^2\theta\,\boldsymbol{e}^\phi.$$

The matrix representations of the metric tensor have the forms

$$[g_{ij}] = \begin{bmatrix} 1 & & \\ & r^2 & \\ & & r^2\sin^2\theta \end{bmatrix}, \qquad [g^{ij}] = \begin{bmatrix} 1 & & \\ & r^{-2} & \\ & & (r\,\sin\theta)^{-2} \end{bmatrix},$$

and the Christoffel symbols are

$$\begin{aligned}
&\Gamma_{r\ \theta}^{\ \theta} = \Gamma_{\theta\ r}^{\ \theta} = \Gamma_{r\ \phi}^{\ \phi} = \Gamma_{\phi\ r}^{\ \phi} = \frac{1}{r},\\
&\Gamma_{\theta\ \theta}^{\ r} = -r,\\
&\Gamma_{\phi\ \phi}^{\ r} = -r\,\sin^2\theta,\\
&\Gamma_{\theta\ \phi}^{\ \phi} = \Gamma_{\phi\ \theta}^{\ \phi} = \cot\theta,\\
&\Gamma_{\phi\ \phi}^{\ \theta} = -\sin\theta\,\cos\theta,\\
&\text{others} = 0.
\end{aligned}$$

Moreover, the orthonormal basis for the physical components are

$$\begin{aligned}
\boldsymbol{e}_{\langle r\rangle} &= \sin\theta\,\cos\phi\,\boldsymbol{i}_1 + \sin\theta\,\sin\phi\,\boldsymbol{i}_2 + \cos\theta\,\boldsymbol{i}_3,\\
\boldsymbol{e}_{\langle\theta\rangle} &= \cos\theta\,\cos\phi\,\boldsymbol{i}_1 + \cos\theta\,\sin\phi\,\boldsymbol{i}_2 - \sin\theta\,\boldsymbol{i}_3,\\
\boldsymbol{e}_{\langle\phi\rangle} &= -\sin\phi\,\boldsymbol{i}_1 + \cos\phi\,\boldsymbol{i}_2.
\end{aligned}$$

**Remark**. More frequently, we would like to express quantities in these coordinate systems in terms of their physical components. A simple way to do this is to derive the expressions first in terms of contravariant or covariant components and then convert them into physical components using relations like (A.78) and (A.79).

**Example A.2.9** Let us calculate the Laplacian of a scalar field $\varPhi$ in the spherical coordinate system. We have

$$\begin{aligned}
\varPhi_{,j} &= \frac{\partial\varPhi}{\partial x^j},\\
\varPhi_{,jk} &= \frac{\partial^2\varPhi}{\partial x^j\partial x^k} - \frac{\partial\varPhi}{\partial x^i}\Gamma_{j\ k}^{\ i},
\end{aligned}$$

from which we obtain the following covariant components:

$$\Phi_{,rr} = \frac{\partial^2 \Phi}{\partial r^2},$$
$$\Phi_{,\theta\theta} = \frac{\partial^2 \Phi}{\partial \theta^2} - \frac{\partial \Phi}{\partial r}\Gamma_{\theta}{}^{r}{}_{\theta} = \frac{\partial^2 \Phi}{\partial \theta^2} + r\frac{\partial \Phi}{\partial r},$$
$$\Phi_{,\phi\phi} = \frac{\partial^2 \Phi}{\partial \phi^2} - \frac{\partial \Phi}{\partial r}\Gamma_{\phi}{}^{r}{}_{\phi} - \frac{\partial \Phi}{\partial \theta}\Gamma_{\phi}{}^{\theta}{}_{\phi} = \frac{\partial^2 \Phi}{\partial \phi^2} + r\sin^2\theta\frac{\partial \Phi}{\partial r} + \sin\theta\cos\theta\frac{\partial \Phi}{\partial \theta}.$$

We have $\Phi_{,rr} = \Phi_{,\langle rr\rangle}$, $\Phi_{,\theta\theta} = r^2\Phi_{,\langle\theta\theta\rangle}$, $\Phi_{,\phi\phi} = r^2\sin^2\theta\,\Phi_{,\langle\phi\phi\rangle}$ in terms of physical components. That is,

$$\Phi_{,\langle rr\rangle} = \frac{\partial^2 \Phi}{\partial r^2},$$
$$\Phi_{,\langle\theta\theta\rangle} = \frac{1}{r^2}\frac{\partial^2 \Phi}{\partial \theta^2} + \frac{1}{r}\frac{\partial \Phi}{\partial r},$$
$$\Phi_{,\langle\phi\phi\rangle} = \frac{1}{r^2\sin^2\theta}\frac{\partial^2 \Phi}{\partial \phi^2} + \frac{1}{r}\frac{\partial \Phi}{\partial r} + \frac{\cot\theta}{r^2}\frac{\partial \Phi}{\partial \theta}.$$

Therefore, the Laplacian $\nabla^2\Phi$, which is the sum $\Phi_{,\langle rr\rangle} + \Phi_{,\langle\theta\theta\rangle} + \Phi_{,\langle\phi\phi\rangle}$ in physical components, becomes

$$\nabla^2\Phi = \frac{\partial^2 \Phi}{\partial r^2} + \frac{2}{r}\frac{\partial \Phi}{\partial r} + \frac{1}{r^2}\frac{\partial^2 \Phi}{\partial \theta^2} + \frac{1}{r^2\sin^2\theta}\frac{\partial^2 \Phi}{\partial \phi^2} + \frac{\cot\theta}{r^2}\frac{\partial \Phi}{\partial \theta}.$$

□

**Example A.2.10** We give the physical components of the divergence of a symmetric tensor field $T$ in the following coordinate systems:

a) Cartesian coordinate system $(x, y, z)$:

$$\begin{aligned}
(\operatorname{div} T)_{\langle x\rangle} &= \frac{\partial T_{\langle xx\rangle}}{\partial x} + \frac{\partial T_{\langle xy\rangle}}{\partial y} + \frac{\partial T_{\langle xz\rangle}}{\partial z},\\
(\operatorname{div} T)_{\langle y\rangle} &= \frac{\partial T_{\langle xy\rangle}}{\partial x} + \frac{\partial T_{\langle yy\rangle}}{\partial y} + \frac{\partial T_{\langle yz\rangle}}{\partial z},\\
(\operatorname{div} T)_{\langle z\rangle} &= \frac{\partial T_{\langle xz\rangle}}{\partial x} + \frac{\partial T_{\langle yz\rangle}}{\partial y} + \frac{\partial T_{\langle zz\rangle}}{\partial z}.
\end{aligned} \tag{A.81}$$

b) Cylindrical coordinate system $(r, \theta, z)$:

$$\begin{aligned}
(\operatorname{div} T)_{\langle r\rangle} &= \frac{\partial T_{\langle rr\rangle}}{\partial r} + \frac{1}{r}\frac{\partial T_{\langle r\theta\rangle}}{\partial \theta} + \frac{\partial T_{\langle rz\rangle}}{\partial z} + \frac{T_{\langle rr\rangle} - T_{\langle\theta\theta\rangle}}{r},\\
(\operatorname{div} T)_{\langle\theta\rangle} &= \frac{\partial T_{\langle r\theta\rangle}}{\partial r} + \frac{1}{r}\frac{\partial T_{\langle\theta\theta\rangle}}{\partial \theta} + \frac{\partial T_{\langle\theta z\rangle}}{\partial z} + \frac{2}{r}T_{\langle r\theta\rangle},\\
(\operatorname{div} T)_{\langle z\rangle} &= \frac{\partial T_{\langle rz\rangle}}{\partial r} + \frac{1}{r}\frac{\partial T_{\langle\theta z\rangle}}{\partial \theta} + \frac{\partial T_{\langle zz\rangle}}{\partial z} + \frac{1}{r}T_{\langle rz\rangle}.
\end{aligned} \tag{A.82}$$

c) Spherical coordinate system $(r, \theta, \phi)$:

$$\begin{aligned}
(\operatorname{div} T)_{\langle r\rangle} &= \frac{\partial T_{\langle rr\rangle}}{\partial r} + \frac{1}{r}\frac{\partial T_{\langle r\theta\rangle}}{\partial \theta} + \frac{1}{r\,\sin\theta}\frac{\partial T_{\langle r\phi\rangle}}{\partial \phi} \\
&\quad + \frac{1}{r}\Big(2\,T_{\langle rr\rangle} - T_{\langle \theta\theta\rangle} - T_{\langle \phi\phi\rangle} + \cot\theta\, T_{\langle r\theta\rangle}\Big), \\
(\operatorname{div} T)_{\langle \theta\rangle} &= \frac{\partial T_{\langle r\theta\rangle}}{\partial r} + \frac{1}{r}\frac{\partial T_{\langle \theta\theta\rangle}}{\partial \theta} + \frac{1}{r\,\sin\theta}\frac{\partial T_{\langle \theta\phi\rangle}}{\partial \phi} \\
&\quad + \frac{1}{r}\Big(3\,T_{\langle r\theta\rangle} + \cot\theta\,(T_{\langle \theta\theta\rangle} - T_{\langle \phi\phi\rangle})\Big), \\
(\operatorname{div} T)_{\langle \phi\rangle} &= \frac{\partial T_{\langle r\phi\rangle}}{\partial r} + \frac{1}{r}\frac{\partial T_{\langle \theta\phi\rangle}}{\partial \theta} + \frac{1}{r\,\sin\theta}\frac{\partial T_{\langle \phi\phi\rangle}}{\partial \phi} \\
&\quad + \frac{1}{r}\Big(3\,T_{\langle r\phi\rangle} + 2\,\cot\theta\, T_{\langle \theta\phi\rangle}\Big).
\end{aligned} \tag{A.83}$$

□

**Exercise A.2.8** Let $\boldsymbol{u}$ be a vector field. Show that

1) in the cylindrical coordinate system,

$$\operatorname{div} \boldsymbol{u} = \frac{\partial u_{\langle r\rangle}}{\partial r} + \frac{1}{r}\frac{\partial u_{\langle \theta\rangle}}{\partial \theta} + \frac{\partial u_{\langle z\rangle}}{\partial z} + \frac{1}{r}u_{\langle r\rangle};$$

2) in the spherical coordinate system,

$$\operatorname{div} \boldsymbol{u} = \frac{\partial u_{\langle r\rangle}}{\partial r} + \frac{1}{r}\frac{\partial u_{\langle \theta\rangle}}{\partial \theta} + \frac{1}{r\sin\theta}\frac{\partial u_{\langle \phi\rangle}}{\partial \phi} + \frac{2}{r}u_{\langle r\rangle} + \frac{\cot\theta}{r}u_{\langle \theta\rangle}.$$

**Exercise A.2.9** Let $\boldsymbol{u}$ be a vector field and $E = \frac{1}{2}(\nabla\boldsymbol{u} + \nabla\boldsymbol{u}^T)$. Express $E$ in cylindrical and spherical coordinate systems,

1) relative to the natural basis,
2) in terms of physical components.

**Exercise A.2.10** Let $T$ be a symmetric tensor field. Compute $\operatorname{div} T$, in cylindrical and spherical coordinate systems,

1) relative to the natural basis,
2) in terms of physical components. (Verify (A.82) and (A.83)).

**Exercise A.2.11** Let $\Phi : I\!R \to \mathcal{E}$ be a curve. Suppose that $\{\boldsymbol{e}_i(x)\}$ is the natural basis and $\phi^i(t)$ is the coordinate of $\Phi(t)$ in the coordinate system $(x^i)$. Show that

1) $\dot{\Phi}(t) = \dot{\phi}^i(t)\,\boldsymbol{e}_i(\phi(t))$,

2) $\ddot{\Phi}(t) = \Big(\ddot{\phi}^i(t) + \dot{\phi}^j(t)\dot{\phi}^k(t)\Gamma_j{}^i{}_k(\phi(t)\Big)\,\boldsymbol{e}_i(\phi(t))$.

# References


1. Antman, S. S.: *Nonlinear Problems of Elasticity*, Springer-Verlag, New York (1995).
2. Barbera, E.: On the principle of minimal entropy production for Navier-Stokes-Fourier fluids, *Continuum Mech. Thermodyn.* 11, 327–330 (1999)
3. Batra, R. C.: A thermodynamic theory of rigid heat conductors, *Arch. Rational Mech. Anal.* 53, 359–365 (1974).
4. Beatty, M. F.: A class of universal relations for constrained isotropic elastic materials, *Acta Mech.* 80, 299–312 (1989).
5. Berberian, S. K.: *Introduction to Hilbert Space*, Oxford University Press, New York (1961).
6. Bhatnagar, P. L.; Gross, E. P.; Krook, M.: A model for collision processes in gases. I. Small amplitude processes in charged and neutral one-component systems. *Phys. Rev.* 94. 511–525 (1954).
7. Boelher, J. P.: On irreducible representations for isotropic scalar functions, *ZAMM* 57, 323–327 (1977).
8. Capriz, G.; Podio-Guidugli, P.: Internal constraints, Appendix 3A, *Rational Thermodynamics*, 2nd edition, edited by C. Truesdell, Springer, New York-Berlin (1984).
9. Coleman, B. D.; Mizel, V. J.: On the general theory of fading memory, *Arch. Rational Mech. Anal.* 29, 18–31 (1968).
10. Coleman, B. D.; Noll, W.: Foundations of linear viscoelasticity, *Rev. Modern Phys.* 33, 239–249 (1961).
11. Coleman, B. D.; Noll, W.: The thermodynamics of elastic materials with heat conduction and viscosity, *Arch. Rational Mech. Anal.* 13, 167–178 (1963).
12. Courant, R.; Hilbert, D.: *Method of Mathematical Physics*, Vol. II. Interscience Publishers, New York (1962).
13. Dafermos, C.: *Hyperbolic Conservation Laws in Continuum Physics*, Springer, Berlin-Heidelberg (2000).
14. de Groot, S. R.; Mazur, P.: *Non-Equilibrium Thermodynamics*, North-Holland, Amsterdam (1963).
15. Dreyer, W.: Maximization of the entropy in non-equilibrium, *J. Phys. A: Math. Gen.* 20, 6505–6517 (1987).
16. Ericksen, J. L.: Deformations possible in every compressible, isotropic, perfectly elastic material. *J. Math. Phys.* 34, 126–128 (1955).
17. Ericksen, J. L.: Inversion of a perfectly elastic spherical shell, *Z. Angew. Math. Mech.* 35, 382–385 (1955).
18. Ericksen, J. L.: *Introduction to Thermodynamics of Solids*, Chapman & Hall, London (1991).
19. Eringen, A. C.: *Continuum Physics*, Vol. 1, Academic Press, New York (1976).
20. Eshelby, J. D.: The elastic energy-momentum tensor. *J. Elasticity*, 5, 321–335 (1975).

21. Fisher, A. E.; Marsden, J. E.: The Einstein evolution equation as a first order quasilinear symmetric hyperbolic system, *Comm. Math. Phys.* 28, 1–38 (1972).
22. Friedrichs, K. D.; Lax, P. D.: System of conservation equations with a convex extension, *Proc. Nat. Acad. Sci.* 68, 1686–1688 (1971).
23. Gibbs, J. W.: On the equilibrium of heterogeneous substances. *Scientific Papers of J. Willard Gibbs.* Vol. 1, Dover, New York (1961).
24. Grad, H.: Principles of the kinetic theory of gases, *Handbuch der Physik*, Vol. XII, edited by S. Flügge, Springer, Berlin-Heidelberg-New York (1958).
25. Gurtin, M. E.; Martins, L. C.: Cauchy's theorem in classical physics, *Arch. Rational Mech. Anal.* 60, 305–324 (1976).
26. Gurtin, M. E.; Mizel, V. J.; Williams, W. O.: A note on Cauchy's stress theorem, *J. Math. Anal. Appl.* 22, 398–401 (1968).
27. Gurtin, M. E.; Williams, W. O.: On the inclusion of the complete symmetry group in the unimodular group, *Arch. Rational Mech. Anal.* 23, 163–172 (1966).
28. Gurtin, M. E.; Williams, W. O.; Suliciu, I.: On rate-type constitutive equations and the energy of viscoelastic and viscoplastic materials, *Int. J. Solids Structures* 16, 607–617 (1980).
29. Hildebrand, F. B.: *Advanced Calculus for Applications*, 2nd edition, Prentice-Hall, Englewood Cliffs, New Jersey (1976).
30. Hutter, K.: On thermodynamics and thermostatics of viscous thermoelastic solids in electromagnetic fields, *Arch. Rational Mech. Anal.* 58, 339–386 (1975).
31. Jou, D.; Casas-Vázquez, J.; Lebon, G.: *Extended Irreversible Thermodynamics*, Springer, Berlin (1993).
32. Kearsley, E. A.: Asymmetric stretching of a symmetrically loaded elastic sheet, *Int. J. Solids Structures*, 22, 111–119 (1986).
33. Knops, R. J.; Wilkes, E. W.: *Elastic Stability*, Handbuch der Physik Vol. VIa/3, edited by S. Flügge, Springer, Berlin (1973).
34. Kogan, M. N.: *Rarefied Gas Dynamics*, Plenum Press, New York (1969)
35. Kremer, G. M.: Extended thermodynamics of ideal gases with 14 fields, *Ann. Inst. Henri Poincaré*, 45, 419–440 (1986)
36. Liu, I-Shih: *On irreversible thermodynamics*, Ph.D Dissertation, Department of Mechanics, The Johns Hopkins University, Baltimore, Maryland, USA (1972).
37. Liu, I-Shih: Method of Lagrange multipliers for exploitation of the entropy principle, *Arch. Rational Mech. Anal.*, 46, 131–148 (1972).
38. Liu, I-Shih: On representations of anisotropic invariants, *Int. J. Eng. Sci.* 20, 1099–1109 (1982).
39. Liu, I-Shih: On the structure of balance equations and extended field theories of mechanics, *Il Nouvo Cimento* 92B, 121–141 (1986).
40. Liu, I-Shih: Extended thermodynamics of viscoelastic materials, *Continuum Mech. Thermodyn.* 1, 143–164 (1989).
41. Liu, I-Shih: On interface equilibrium and inclusion problems, *Continuum Mech. Thermodyn.* 4, 177–186 (1992).
42. Liu, I-Shih: Stability of thick spherical shells, *Continuum Mech. Thermody.* 7, 249–258 (1995).
43. Liu, I-Shih: On entropy flux–heat flux relation in thermodynamics with Lagrange multipliers, *Continuum Mech. Thermody.* 8, 247–256 (1996).
44. Liu, I-Shih: Constitutive equations of extended thermodynamics from a hybrid pair of generator functions, *Continuum Mech. Thermodyn.* 13, 25–39 (2001)
45. Liu, I-Shih; Müller, I.: On the thermodynamics and thermostatics of fluids in electromagnetic fields, *Arch. Rational Mech. Anal.* 46, 149–176 (1972).
46. Liu, I-Shih; Müller, I.: Extended thermodynamics of classical and degenerate gases, *Arch. Rational Mech. Anal.* 83, 285–332 (1983).

47. Merritt, D. R.; Weinhaus, F.: The pressure curve of a rubber balloon, *Am. J. Phys.* 46, 976–977 (1978).
48. Moon, H.; Truesdell, C.: Interpretation of adscititious inequalities through the effects pure shear produces upon an isotropic elastic solid, *Arch. Rational Mech. Anal.* 55, 1–17 (1974).
49. Müller, I.: Zum Paradoxen der Wärmeleitungstheorie. *Zeitschrift für Physik*, 198, 329–344 (1967).
50. Müller, I.: On the entropy inequality, *Arch. Rational Mech. Anal.* 26, 118–141 (1967).
51. Müller, I.: A thermodynamic theory of mixtures of fluids, *Arch. Rational Mech. Anal.* 28, 1–39 (1968).
52. Müller, I.: Die Kältefunktion eine universelle Funktion in der Thermodynamik viskoser wärmeleitender Flüssigkeiten, *Arch. Rational Mech. Anal.*, 40, 1–36 (1971).
53. Müller, I.: The coldness, a universal function in thermoelastic bodies, *Arch. Rational Mech. Anal.*, 41, 319–332 (1971).
54. Müller, I.: A new approach to thermodynamics of simple mixtures, *Zeitschrift für Naturforschung*, 28, 1801–1813 (1973).
55. Müller, I.: *Thermodynamics*, Pitman Publishing, London (1985).
56. Müller, I.: Two instructive instabilities in nonlinear elasticity: biaxially loaded membrane and rubber balloons, *Meccanica*, 31, 387–395 (1996).
57. Müller, I., Ruggeri, T.: *Rational Extended Thermodynamics.* 2nd edition, Tracts in Natural Philosophy 37, Springer, New York (1998).
58. Noll, W.: A mathematical theory of the mechanical behavior of continuous media, *Arch. Rational Mech. Anal.* 2, 197–226 (1958).
59. Noll, W.: Proof of the maximality of the orthogonal group in the unimodular group, *Arch. Rational Mech. Anal.* 18, 100–102 (1965).
60. Prigogine, I.: *Introduction to Thermodynamics of Irreversible Processes*, Interscience Publishers, New York (1967)
61. Rajagopal, K. R.; Wineman, A. S.: New universal relations for nonlinear isotropic elastic materials, *J. Elasticity* 17, 75–83 (1987).
62. Rivlin, R. S.: Further remarks on the stress-deformation relations for isotropic materials, *J. Rational Mech. Anal.*, 4, 681–702 (1955).
63. Ruggeri, T.: Symmetric-hyperbolic system of conservative equations for a viscous heat conducting fluid, *Acta Mechanica* 47, 167–183 (1983).
64. Ruggeri, T.: Galilean invariance and entropy principle for systems of balance laws, *Continuum Mech. Thermodyn.* 1, 3–20 (1989).
65. Ruggeri, T.; Strumia, A.: Main field and convex covariant density for quasilinear hyperbolic systems. Relativistic fluid dynamics. *Ann. Inst. Henri Poincaré*, 34 A. 65–84 (1981).
66. Smith, G. F.: On isotropic functions of symmetric tensors, skew-symmetric tensors and vectors, *Int. J. Eng. Sci.* 9, 899–916 (1971).
67. Struchtrup, H.; Weiss, W.: Maximum of the local entropy production becomes minimal in stationary processes, *Phys. Rev. Lett.* 80, 5048–5051 (1998)
68. Treloar, L. R. G.: *The Physics of Rubber Elasticity*, Clarendon Press, Oxford (1975).
69. Truesdell, C.: *Rational Thermodynamics*, 2nd edition, Springer-Verlag, New York-Berlin (1984).
70. Truesdell, C.; Muncaster, R. G.: *Fundamentals of Maxwell's Kinetic Theory of a Simple Monatomic Gas*, Academic Press, New York (1980).
71. Truesdell, C.; Noll, W.: *The Non-Linear Field Theories of Mechanics*, Handbuch der Physik, edited by S. Flügger, Vol III/3, Springer, Berlin-Heidelberg-New York (1965).

72. Truesdell, C.; Toupin, R. A.: *The Classical Field Theories*, Handbuch der Physik, edited by S. Flügger, Vol III/1, Springer, Berlin-Heidelberg-New York (1960).
73. Wang, C.-C.: Stress relaxation and the principle of fading memory, *Arch. Rational Mech. Anal.* 18, 117–126 (1965).
74. Wang, C.-C.: A new representation theorem for isotropic functions, Part I and II, *Arch. Rational Mech. Anal.* 36, 166–197, 198–223 (1970); Corrigendum, *ibid.* 43, 392–395 (1971).
75. Wang, C.-C.; Liu, I-Shih: A note on material symmetry, *Arch. Rational Mech. Anal.* 74, 277–296 (1980).
76. Wang, C.-C.; Truesdell, C.: *Introduction to Rational Elasticity*, Noordhoff International Publishing, Leyden (1973).
77. Wilmański, K.: *Thermomechanics of Contínua*, Springer, Berlin-Heidelberg (1998)
78. Wilmański, K.: Toward an extended thermodynamics of porous and granular materials, *Trends in Applications of Mathematics to Mechanics*, edited by G. Iooss, Longman, New York (1999).
79. Woods, L. C.: The bogus axioms of continuum mechanics, *Bull. of Mathematics and its Applications*, 17, 98–102 (1981).

# Index